CHAPMAN & HALL/CRC
Monographs and Surveys in
Pure and Applied Mathematics **108**

CONSTRUCTIVE METHODS FOR LINEAR AND NONLINEAR BOUNDARY VALUE PROBLEMS FOR ANALYTIC FUNCTIONS

Theory and Applications

CHAPMAN & HALL/CRC
Research Notes in Mathematics Series

Main Editors

H. Brezis, *Université de Paris*
R.G. Douglas, *Texas A&M University*
A. Jeffrey, *University of Newcastle upon Tyne (Founding Editor)*

Editorial Board

H. Amann, *University of Zürich*
R. Aris, *University of Minnesota*
G.I. Barenblatt, *University of Cambridge*
H. Begehr, *Freie Universität Berlin*
P. Bullen, *University of British Columbia*
R.J. Elliott, *University of Alberta*
R.P. Gilbert, *University of Delaware*
R. Glowinski, *University of Houston*
D. Jerison, *Massachusetts Institute of Technology*
K. Kirchgässner, *Universität Stuttgart*

B. Lawson, *State University of New York at Stony Brook*
B. Moodie, *University of Alberta*
S. Mori, *Kyoto University*
L.E. Payne, *Cornell University*
D.B. Pearson, *University of Hull*
I. Raeburn, *University of Newcastle*
G.F. Roach, *University of Strathclyde*
I. Stakgold, *University of Delaware*
W.A. Strauss, *Brown University*
J. van der Hoek, *University of Adelaide*

Submission of proposals for consideration

Suggestions for publication, in the form of outlines and representative samples, are invited by the Editorial Board for assessment. Intending authors should approach one of the main editors or another member of the Editorial Board, citing the relevant AMS subject classifications. Alternatively, outlines may be sent directly to the publisher's offices. Refereeing is by members of the board and other mathematical authorities in the topic concerned, throughout the world.

Preparation of accepted manuscripts

On acceptance of a proposal, the publisher will supply full instructions for the preparation of manuscripts in a form suitable for direct photo-lithographic reproduction. Specially printed grid sheets can be provided. Word processor output, subject to the publisher's approval, is also acceptable.

Illustrations should be prepared by the authors, ready for direct reproduction without further improvement. The use of hand-drawn symbols should be avoided wherever possible, in order to obtain maximum clarity of the text.

The publisher will be pleased to give guidance necessary during the preparation of a typescript and will be happy to answer any queries.

Important note

In order to avoid later retyping, intending authors are strongly urged not to begin final preparation of a typescript before receiving the publisher's guidelines. In this way we hope to preserve the uniform appearance of the series.

CRC Press UK

Chapman & Hall/CRC Statistics and Mathematics
Pocock House
235 Southwark Bridge Road
London SE1 6LY
Tel: 0171 407 7335

CHAPMAN & HALL/CRC
Monographs and Surveys in
Pure and Applied Mathematics **108**

CONSTRUCTIVE METHODS FOR LINEAR AND NONLINEAR BOUNDARY VALUE PROBLEMS FOR ANALYTIC FUNCTIONS

Theory and Applications

VLADIMIR V. MITYUSHEV
SERGEI V. ROGOSIN

CHAPMAN & HALL/CRC

Boca Raton   London   New York   Washington, D.C.

Library of Congress Cataloging-in-Publication Data

Mityushev, Vladimir V.
    Constructive methods for linear and nonlinear boundary value problems for analytic functions : theory and applications / Vladimir V. Mityushev, Sergei V. Rogosin.
       p. cm. -- (Monographs and surveys in pure and applied mathematics ; v 108)
    Includes bibliographical references and index.
    ISBN 1-584-88057-0 (alk. paper)
    1. Boundary value problems. 2. Analytic functions. I. Title. II. Rogosin, Sergei V.
QA331.M675 1999
515'.35 21--dc21

                                                                                                                        99-044959

This book contains information obtained from authentic and highly regarded sources. Reprinted material is quoted with permission, and sources are indicated. A wide variety of references are listed. Reasonable efforts have been made to publish reliable data and information, but the author and the publisher cannot assume responsibility for the validity of all materials or for the consequences of their use.

Neither this book nor any part may be reproduced or transmitted in any form or by any means, electronic or mechanical, including photocopying, microfilming, and recording, or by any information storage or retrieval system, without prior permission in writing from the publisher.

The consent of CRC Press LLC does not extend to copying for general distribution, for promotion, for creating new works, or for resale. Specific permission must be obtained in writing from CRC Press LLC for such copying.

Direct all inquiries to CRC Press LLC, 2000 N.W. Corporate Blvd., Boca Raton, Florida 33431.

Trademark Notice: Product or corporate names may be trademarks or registered trademarks, and are used only for identification and explanation, without intent to infringe.

© 2000 by Chapman & Hall/CRC

No claim to original U.S. Government works
International Standard Book Number 1-584-88057-0
Library of Congress Card Number 99-044959
Printed in the United States of America   1 2 3 4 5 6 7 8 9 0
Printed on acid-free paper

Preface

The book is devoted to the description of a number of constructive methods developed in the framework of analytic functions. These methods extend the possibility of using mathematical constructions within different branches of mathematics, as well as mechanics, physics, chemistry, biology, economics etc. This field is too wide to be presented in one book. Thus we have limited ourselves by boundary value problems for analytic functions, and some related problems. The constructive ideas are always in our mind, so, the basic goal of this book is to be useful for experts in analytic function theory and for nonspecialists in it, and even for non-mathematicians who apply these methods in their research. The results presented here are prepared for direct applications.

The modern theory of boundary value problems goes back to the results of the giants of mathematics — B. Riemann, D. Hilbert, H. Poincaré. From the beginning two directions emerged from this subject — constructive and qualitative. One cannot divide, of course, the results too strongly. They could be mainly constructive or qualitative ones. Many results of the Soviet mathematical school were presented in the form of so-called *closed form solution*. It was a strategy of the books by F. D. Gakhov, and N. I. Muskhelishvili, which are now classical. This tradition was supported by the mathematical children and grandchildren of these scientists.

This tradition was kept mainly because of the connection of boundary value problems with different questions of elasticity, hydrodynamics, composite materials, diffraction, electrodynamics, queueing theory etc. Most of the results obtained in these directions by using analytic function methods deal with linear boundary value problems. There are several reasons for it. First, corresponding mechanical and physical models were stated in linear variants. Second, the linear theory of boundary value problems is closed to completeness. Last, it is highly suitable to have some formulae for solution of boundary value problems both for construction of solution of initial mechanical and physical problems, and for qualitative and numerical analysis of the latter problems.

There exists a great interest in nonlinear mechanical and physical problems now. They are more complicated than the linear ones. Therefore, it is not always possible to use the same methods for getting similar results. Some attempts in this direction were made the during last decades. We have collected here a number of results of mainly a constructive nature obtained in the framework of analytic functions methods. In the course of our description we try to answer the following principal questions:

1) how far can one go in solution of problems of nonlinear mechanics and physics using analytic (or analytic-like) functions' ideas?

2) what is the difference between linear and nonlinear cases from the qualitative point of view?

3) what kind of additional technique should be used in the investigation of nonlinear problems?

As it was already said we suppose to keep the tradition of having closed form solution. However, different interpretations can be given to the words *solution in closed form* or *analytic solution* . For example, analytic solution can be understood as the value of an integral operator whose density satisfies certain Fredholm integral equation. Another variant is the expansion of solution in the Taylor series with coefficients satisfying an infinite set of linear algebraic equations, i.e., the same Fredholm equation but in discrete form. Or simply, one can say that equation $Ax = y$ has a solution $x = A^{-1}y$, where A is an operator.

What is the meaning of the expression *closed form* for us? To get the closed form solution one obtains the formula which contains a finite set of elementary and special functions, arithmetic operations, compositions, integrals, derivatives and even series. Besides, all the objects in this formula have to have precise meaning (for instance, the type of convergence for integrals and series should be described). Last, domains of parameters, as well as of all functions, integrals etc., have to be explicitly determined. It should be shown also that they (or their intersections, if necessary) are non-empty.

This approach is slightly nontraditional. For instance, in the classical books a form in series was not supposed as a closed one. However, these books allow to have special functions in such formulae. But it is well known that not all special functions have integral representation; they can be defined in the form of series or even in an implicit form.

We suppose that closed form should give a way to work efficiently with mathematical objects involved in an analytic formula. In applications such formulae are very useful because they allow us not only to calculate unknown characteristics in special cases, but also investigate their dependence on parameters. For instance, it is important to know the distribution of the stresses in building constructions, but it is more important to choose such a construction in which the stresses do not exceed the critical values (so called an optimal design problem). Moreover, analytical solutions could be useful in advanced numerical computation and their qualitative analysis.

The subject of our book is the analytic type methods for linear and nonlinear boundary value problems and close questions of mechanics and physics. There are some interesting generalizations of this approach for more general than Cauchy–Riemann differential equations. We deal with these generalizations only briefly. It will be the subject of further investigations. As for the main subject of the book it should be noted that many results concerning it are spread in the Soviet scientific literature which are not always accessible to the mathematical community. Thus, one more special goal was behind us, namely, to gather most of the constructive results for boundary value problems, and present them in the unified modern form. One of the most interesting results described in the book is the complete solution of the Riemann–Hilbert problem for multiply connected domains. It is the first presentation of such a solution in the monographic literature.

The above described goals of the book directly influence its structure. We begin with a historical survey on nonlinear boundary value problems for analytic functions (Chapter 1). These described the most statements of such problems for which constructive results were obtained. We use a number of different (mostly technical) facts and constructions in the main part of the book. In order to make the text self-contained and useful for readers we present these components in a separate chapter (Chapter 2). A brief description of now classic linear boundary value problems for analytic functions is given here too. We suppose that it helps to understand the ideas applied in the next chapters. Analytic function methods for the basic nonlinear boundary value problems are presented in Chapter 3. The following problems are discussed: general nonlinear conjugation problem of power type, nonlinear conjugation problem of multiplication type (for closed, open, and compound contour), the general Riemann–Hilbert problem of power type, the modulus problem and its generalizations, linear fractional conjugation problem, and Cherepanov's mixed problem. In Chapter 4 we deal with the method of functional equations. On the basis of this method we obtain some results for different linear and nonlinear boundary value problems. First we discuss a model domain, namely, a doubly connected circular domain. Then we present a method of functional equations for general circular multiply connected domains. In particular the already mentioned presentation of the complete solution of the Riemann–Hilbert boundary value problem for multiply connected domains is given. Relations between the method of functional equations and the generalized method of Schwarz are exposed. A convergent modification of the method of Schwarz is proposed. Selected applications of the above described methods are discussed in Chapter 5. We choose among all possible applications the most attractive for us.

Mainly they connect with the problems for composite materials with a finite number of inclusions, the branch of mechanics which is highly developing nowadays.

The authors are grateful to Professor Yuri V. Obnosov for permission to use some descriptions of his ideas in the book. We are also indebted to Dr. V. V. Kashevskii for the stimulating discussion of the results. Our sincere thanks to the colleagues and friends from the Mechanical-Mathematical Department of Belarusian State University for their support of our work. We are pleased to express our deep gratitude to Professor Heinrich Begehr initiating of this book and for the help during its preparation. V.M. greatly appreciates Professor Pierre Adler for discussing applications included in Chapter 5. We appreciates Dr. A. E. Malevich for the prominent consultations concerning computer systems.

S. R. is thankful to the Belarusian Fund for Fundamental Scientific Research for the partial support of his work.

Contents

Preface		**v**
1	**Historical Survey**	**1**
2	**Notations and Auxiliary Results**	**9**
	2.1 Geometry of complex plane.	9
	2.2 Functional spaces.	12
	2.3 Operator equations in functional spaces.	18
	2.4 Properties of analytic and harmonic functions.	22
	2.5 Boundary properties.	23
	2.6 Cauchy type integral and singular integrals.	26
	2.7 Schwarz operator.	30
	2.7.1 Schwarz operator and complex Green function for simply connected domain.	30
	2.7.2 Complex Green function and Schwarz operator for multiply connected domain.	32
	2.8 $\mathbb{C}$-linear conjugation problem.	37
	2.8.1 Simply connected domain.	37
	2.8.2 Multiply connected domain.	40
	2.8.3 $\mathbb{C}$-linear conjugation problem on open arcs.	41
	2.9 Riemann–Hilbert boundary value problem.	42
	2.10 Entire functions.	45
	2.11 Conformal mappings.	50
	2.12 $\mathbb{R}$-linear problem and its applications	52
	2.13 Notes and Comments.	55
3	**Nonlinear boundary value problems**	**57**
	3.1 Conjugation problem of power type.	57

		3.1.1	Homogeneous problem 58

       3.1.2 Solution to nonlinear conjugation problem of power type with admissible zeros on the contour. 64

       3.1.3 Inhomogeneous problem. 70

  3.2 Problem of multiplication type. 73

  3.3 Entire functions methods. 75

       3.3.1 Multiplication type problem for an open arc. 75

       3.3.2 Multiplication type problem for a compound contour. 85

  3.4 General Riemann–Hilbert problem of power type. 100

       3.4.1 Homogeneous problem with positive integer exponent. 101

       3.4.2 Homogeneous problem with real exponent. 104

       3.4.3 Homogeneous problem with complex exponent. 106

  3.5 The modulus problem and its generalization. 108

       3.5.1 Simple problem on modulus. 108

       3.5.2 Generalized problem on modulus of an analytic function. 111

  3.6 Linear fractional problem. 113

  3.7 Cherepanov's mixed problem. 118

       3.7.1 Statement of the problem. Simple cases of solvability. 118

       3.7.2 Meromorphic solutions. 119

       3.7.3 Analytic solutions for $n \leq 2$. 124

  3.8 Notes and Comments. 127

4 Method of Functional Equations   **129**

  4.1 Dirichlet problem for a doubly connected domain. 129

  4.2 A nonlinear boundary value problem. 134

       4.2.1 General approach. 134

       4.2.2 A problem for a doubly connected domain. 137

       4.2.3 A nonlinear boundary value problem and functional equation. . 138

  4.3 Linear functional equations. 145

  4.4 Harmonic measures and Schwarz operator. 149

       4.4.1 Harmonic measures. 149

       4.4.2 Schwarz operator. 154

  4.5 Linear Riemann–Hilbert problem. 159

  4.6 Poincaré series. 164

  4.7 Mixed problem for multiply connected domains. 170

  4.8 Circular polygons with zero angles. 173

CONTENTS

4.9 Generalized method of Schwarz and other methods. 181
    4.9.1 Generalized method of Schwarz for a doubly connected domain. 181
    4.9.2 Generalized method of Schwarz for multiply connected domains and other methods. 184
    4.9.3 The modified method of Schwarz. 189
4.10 Notes and Comments. 194

5 Nonlinear Problems of Mechanics   199
5.1 Steady heat conduction. Nonlinear composites. 199
5.2 Linearized problem. 206
    5.2.1 Linearized problem for circular domains. 207
    5.2.2 A mixed problem for a circular domain. 209
    5.2.3 A mixed boundary value problem. 213
5.3 Constructive solution to integral equations. 219
5.4 Composite materials with reactive inclusions. 222
5.5 Steady heat conduction on configurations. 225
    5.5.1 A special surface. 226
    5.5.2 Configurations. 228
5.6 An elastic problem for composite materials. 235
5.7 Plane Stokes flow. 242
5.8 Notes and Comments. 255

Bibliography   **257**

Index   **281**

Chapter 1

Nonlinear Boundary Value Problems. Historical survey.

Boundary value problems for analytic functions seem to be the classic subject now. They were studied quite intensively for a long period beginning from the second decade of this century. The most developed is the linear theory. It is described in the background monographs by F. D. Gakhov [91], I. I. Daniljuk [63], E. Meister [155], I. N. Vekua [268], N. P. Vekua [269], and others. The modern state of this theory was discussed at the Conferences and Seminars on Boundary Value Problems. A couple of them were held in Minsk (for instance February 18–20, 1981, February 16–20, 1996, and 1999 September 14–18). Proceedings of these Conferences (see [225], [94]) give the idea on the recent results in the area. The reader should also turn his attention to the well known monographs on the theory of singular integral equations by I. Gohberg and N. Krupnik [98], S. G. Mikhlin and S. Prössdorf [159], N. I. Muskhelishvili [205], S. Prössdorf [232] etc., in which the questions of boundary value problems are of great interest.

As for the theory of nonlinear boundary value problems for analytic or harmonic functions, this branch appeared to be not unified and not so completed. There are many approaches in this area. First of all we have to point out a number of results of a "local" type. The pioneering article here was the paper by A. I. Guseinov [105]. He considered the question of conformal representation of near-circular domains. A number of functional-analytic methods were applied then by many authors to the investigation of nonlinear singular integral equations and some corresponding boundary value problems. Most of them were presented in the monograph [106]. Among the methods were those which used compactness arguments, monotone operators principle, implicit function theorem, and some numerical approaches. The nonlinear boundary value problems as well as the nonlinear singular integral equations were studied in [106] mainly in Hölder or generalized Hölder spaces.

Close to these results but quite different in nature are those which used certain geometrical methods. This direction had begun from the background article by A. I. Schnirel'man [249]. It was continued by different authors (see, e.g., [33] [34], [44], [71] – [74], [76], [237], [274], [278]). The brilliant description of geometrical approaches for nonlinear boundary value problems was given in the monograph [276]. One can find there a wide bibliography on the subject.

As for "constructive" (or "global") results, the achievements in this area highly depend on the type of problems under consideration. Most of these problems are generalizations $\mathbb{C}$-*linear conjugation problem* (known also as the *Riemann problem*), or the *Riemann–Hilbert problem* (known also as the *Hilbert problem*). The first of these linear problems consists of searching for two functions $\Phi^+(z)$, $\Phi^-(z)$, which are analytic in the interior ($int L = D^+$) and exterior ($ext L = D^-$) of certain Jordan curve L respectively, under the following $\mathbb{C}$-linear condition

$$\Phi^+(t) = G(t)\Phi^-(t) + g(t), \quad t \in L, \tag{1.1}$$

where G, g are given complex-valued functions. The second one consists of searching for a function $\Phi(z) = u(z) + iv(z)$, which is analytic in $int L = D^+$ (or in $ext L = D^-$), under condition

$$a(t)u(t) + b(t)v(t) = c(t), \quad t \in L, \tag{1.2}$$

where a, b, c are given real-valued functions. This condition can be also written in the form

$$\Re \overline{\lambda(t)\Phi(t)} = c(t), \quad t \in L, \tag{1.3}$$

where $\lambda(t) = a(t) + ib(t)$.

The $\mathbb{C}$-linear conjugation problem (1.1) and the Riemann–Hilbert problem (1.3) appeared and then were intensively investigated due to their numerous applications ([21], [56], [91], [92], [137], [155], [206]). The interest in these problems is partly also connected with the now classic approach allowed to get the closed form solution of (1.1) and (1.3) for simply connected domains in many cases. This approach was proposed and realized by F. D. Gakhov in 1936 ([91]). It was based on the method of analytic continuation. The *Cauchy index* of the coefficient G (known also as its *winding number*) played a great role.

In this book we also pay attention to the linear problem (1.3) for multiply connected domains. If $\lambda(t) \equiv 1$ in (1.3), then we have the *Schwarz problem* . G. M. Goluzin [99] solved the Schwarz problem for a multiply connected circular domain $\mathbb{D}$ under certain restrictions on the arrangement of the discs $\mathbb{D}_k$ completing $\mathbb{D}$ to the whole plane. He was first to apply functional equations and obtain the *Schwarz operator* in the form of the functional series which converges absolutely and uniformly in $cl\mathbb{D}$. Similar equations are systematically exploited in Chapters 4 and 5 of this book. V. A. Zmorovich [293], L. E. Dunduchenko [64], I. A. Aleksandrov & A. S. Sorokin [20] had completed Goluzin's idea. For instance in [20] Goluzin's series was transformed to an absolutely convergent one, but the closed form of the Schwarz operator was missed since the construction was complemented by a system of linear algebraic equations that appeared at the application of the method of truncation. Closed to Goluzin's result was obtained in [10] by using Poincaré series. Ultimately the Schwarz operator for an arbitrary multiply connected domain was constructed in [188] in closed form (see Section 4.4).

Various methods were applied to solve the problem (1.3) with the general coefficient $\lambda(t)$. They are presented in [30], [91], [294], [268], [280]. General qualitative theory is presented in [35], [282], including the nonlinear case. E. I. Zverovich [294] studied (1.3) using the principal functionals of the double of D^+. These functionals are implicitly determined as the solutions of certain integral equations. Therefore they do not give the closed form solution of (1.3). The problem was also studied in so called special case, $0 \le \chi \le n - 1$, where n is the connectivity of D^+. It was known that the solvability of (1.3) in this case depends on the solvability of a finite linear algebraic system (see Bojarski's addition to [268]). G. C. Wen & H. Begehr [35], [282] modified the problem (1.3) adding appropriate undertermined constants to expose qualitative dependence of the number of linear independent solutions on χ and n. The complete closed form solution to the Riemann–Hilbert problem (1.3) for multiply connected domains was given in [193], [201], where, in particular, the Bojarski's system was written in the explicit form.

Some problems of the mechanics of composite materials lead to the following boundary conditions for the Laplace equation

$$u^+(t) = u^-(t), \quad \frac{\partial u^+}{\partial n}(t) = \lambda \frac{\partial u^-}{\partial n}(t), \ t \in L, \tag{1.4}$$

where the unknown functions $u^+(z)$ and $u^-(z)$ are harmonic in D^+ and D^- respectively, and continuously differentiable in the closure of domains considered; $\partial/\partial n$ is the normal derivative, λ is a positive constant. Applying the Cauchy–Riemann equations one can reduce (1.4) to the condition

$$\Phi^+(t) = a(t)\Phi^-(t) + b(t)\overline{\Phi^-(t)} + c(t), \ t \in L, \tag{1.5}$$

where
$$a = 1, \ b = \frac{\lambda - 1}{\lambda + 1}, \ c = 0. \tag{1.6}$$

The condition (1.5) with arbitrarily given Hölder-continuous functions $a(t), b(t), c(t)$ is called the $\mathbb{R}$-*linear condition*. If $b(t) \equiv 0$ then we arrive at the $\mathbb{C}$-linear condition (1.1). It is hard to say who first noted that (1.4) and (1.5) are equivalent, and studied (1.5). Problem (1.5) with coefficients (1.6) was discussed by I. N. Vekua and S. Rukhadze [266] and by G. M. Goluzin [99]. These are the earliest works about problem (1.5)-(1.6) known to us. In 1946 A. I. Markushevich [151] proposed to study (1.5) with general coefficients, and investigated problem (1.5) with $a(t) \equiv c(t) \equiv 0$ and rational $b(t)$. Therefore problem (1.5) is sometimes called the *Markushevich problem*. The general theory of existence and normal solvability of (1.5) were discussed by B. Bojarski [43] and L. G. Mikhailov [156] in the Lebesgue spaces. Problem (1.5) is closely related to two-dimensional singular integral equations (see, e.g., [67], [68], [125]).

The first attempt to use the analytic continuation type technique for certain nonlinear generalizations of the boundary value problem (1.1) was made in 1941 by P. V. Solov'ev [253]. He had found partial solutions of the *nonlinear boundary value problem of power type*

$$a(t)(\Phi^+(t))^2 + b(t) = \Phi^-(t), \quad t \in L, \tag{1.7}$$

in the case $\text{wind}\, a(t) = -2\chi \leq 0$. Further steps ahead in the investigation of the nonlinear boundary value problems of power type were done at the beginning of the 1960s by G. V. Arzhanov [23], [24]. He had constructed the closed form solutions of the following problem

$$[\Phi^+(t)]^n = G(t)\Phi^-(t) + g(t), \quad t \in L, \ n \geq 2, n \in \mathbb{Z}, \tag{1.8}$$

where L is a simple smooth closed curve. E. K. Timofeev [258] in the framework of Arzhanov's approach studied the closed to (1.8) nonlinear problem of power type with a shift and conjugation:

$$[\Phi^+(\alpha(t))]^n = G(t)\overline{\Phi^-(t)} + g(t), \quad t \in L, \ n \geq 2, n \in \mathbb{Z}, \tag{1.9}$$

where $\alpha : L \to L$ is a certain Carleman type automorphism of a curve L (i.e., $\alpha(\alpha(t)) \equiv t$). Then many mathematicians paid attention to the following nonlinear boundary value problem

$$[\Phi^+(t)]^{\alpha^+(t)} = G(t)[\Phi^-(t)]^{\alpha^-(t)} + g(t), \quad t \in L, \tag{1.10}$$

where $\alpha^+(t), \alpha^-(t)$ are boundary values of functions $\alpha^+ : D^+ \to \mathbb{C}, \alpha^- : D^- \to \mathbb{C}$, analytic in the corresponding domains. Problem (1.10) is often called the *general nonlinear conjugation problem of power type*. It was studied by F. D. Gakhov [89], [90] in the homogeneous case ($g(t) \equiv 0$) with $\alpha^+(t) \equiv \alpha^+, \alpha^-(t) \equiv \alpha^- \in \mathbb{Z} \wedge \mathbb{Q}_+$, L a simple, smooth, closed curve. I. I. Komjak [124] considered the same problem with arbitrary positive indices ($\alpha^+, \alpha^- \in \mathbb{R}_+$). The homogeneous problem (1.10) with non-constant indices $\alpha^+(t), \alpha^-(t)$ was studied by A. Susea [256] and by G.V.Arzhanov [25]. M. E. Tolochko [259] solved some variants of problem (1.10) for multiply connected domains as well as the homogeneous problem in the exceptional case ([260]) (see also [239] for the solutions with addmissible zeros on the contour). In the articles by N. A .Rysjuk [241]– [243] was presented the solution of (1.10) in certain classes of automorphic functions. N. K. Kuznetzov studied the homogeneous problem (1.10) on an open arc ([133]) and on a system of two open arcs ([134]).

Another generalization of the linear conjugation problem (1.1) was proposed by A. Sh. Gabib-Zade. He considered the so-called *linear-fractional boundary value problem*

$$\Phi^+(t) - A(t)\Phi^-(t) - B(t)\Phi^+(t)\Phi^-(t) = G(t), t \in L, \qquad (1.11)$$

where L is a compound smooth curve. He reduced this problem to some complete linear singular integral equation and gave certain qualitative results for the latter (see [85]). Some further attempts to study the problems of type (1.11) were made (see, e.g., [112], [113], [121]–[123], [130], [223], [261]). But for the moment only some of its precise cases are completely investigated. It should be noted that this problem is connected with another interesting question of analysis. For instance, V. N. Gavdzinskii and I. M. Spitkovskii [96] reduced to (1.11) a variant of the problem of matrix-function factorization.

A series of results were obtained on the *boundary value problem of multiplication type*

$$\Phi^+(t)\Phi^-(t) = G(t), \quad t \in L. \qquad (1.12)$$

For the first time this problem was studied by G. P. Cherepanov in connection with certain elasticity problem (in the case of a simple smooth closed curve L) (see [49], [50]). He also constructed the solution of (1.12) in the case of an open arc L in the special class of so-called logarithmically-bounded functions, i.e., functions satisfying the following condition

$$0 < m_1 \leq |\Phi(z)| \leq m_2 < \infty, \quad z \in \mathbb{C} \backslash L. \qquad (1.13)$$

As for nonlinear analogs of the Riemann–Hilbert problem (1.3) there are several approaches to their study. Anyway it should be noted that different variants of nonlinear problem of Riemann–Hilbert type need to use quite different techniques. Besides not all of them are completely solved for general domains (in fact, only for a special type of them).

The *nonlinear Riemann–Hilbert problem of power type*

$$\Re\{\overline{A(t)}w^{\alpha}(t)\} = c(t), \qquad t \in \mathbb{T}, \tag{1.14}$$

was stated for the first time by V. K. Natalevich in [207]. Great attention to it was also paid by Yu. V. Obnosov (see [210], [211], [213], [214]). Articles [210], [213] were devoted to the solution of the homogeneous problem (1.14) (i.e., $c(t) \equiv 0$), but [211], [214] to the solution of the inhomogeneous one.

The predecessor of (1.14) is known as the *polynomial Riemann-Hilbert nonlinear problem*

$$\Re\{A_0(t)w^n(t) + \cdots + A_{n-1}(t)w(t)\} = c(t), \quad t \in \mathbb{T}, \tag{1.15}$$

was stated and solved under special conditions on the coefficients by V. K. Natalevich [207]–[209]. In [51] the outer problem of type (1.15) (i.e., the problem stated in the simply connected domain $\infty \in D$) was investigated in the case of analytically continued coefficients. It obtained a functional equation in the domain D instead of a boundary condition. This functional equation was solved in special cases using the method of indeterminate coefficients. Method of analytic continuation was applied to (1.15) in articles [12]–[14], [16], [17].

Geometric nature (cf., e.g., [102]) has a so-called *modulus problem*, i.e., the problem of searching for a function w analytic in a domain D knowing the boundary values of its modulus

$$|w(t)| = c(t), \qquad t \in \partial D, \tag{1.16}$$

where c is a given function on ∂D. It was formulated for the first time in [207]. One of the variants of its solution was presented in [217] (on the base of analytic continuation method). Problem (1.16) for harmonic functions was studied by R. Magnanini ([150]).

G. S. Litvinchuk [142],[143] has investigated problem (1.5) by reducing the boundary condition to the condition of *vector-matrix* $\mathbb{C}$-*linear problem*

$$\begin{pmatrix} \Phi_1^+(t) \\ \Phi_2^+(t) \end{pmatrix} = \frac{1}{\overline{a(t)}} \begin{pmatrix} |a(t)|^2 - |b(t)|^2 & tb(t) \\ -\overline{tb(t)} & 1 \end{pmatrix} \begin{pmatrix} \Phi_1^-(t) \\ \Phi_2^-(t) \end{pmatrix} + \frac{1}{\overline{a(t)}} \begin{pmatrix} \overline{a(t)c(t)} \\ -\overline{tc(t)} \end{pmatrix}, \tag{1.17}$$

where

$$\Phi_1^+(z) := \Phi^+(z), \Phi_2^+(z) := \frac{1}{z}\overline{\Phi^-(\frac{1}{\overline{z}})}, \Phi_1^-(z) := \Phi^-(z), \Phi_2^-(z) := \frac{1}{z}\overline{\Phi^+(\frac{1}{\overline{z}})}.$$

The detailed discussion of these results are presented in ([144])(cf. also [146]).

The $\mathbb{R}$-linear problem (1.5) is closely related to the *generalized modulus problem*

$$\mid w(t) - a(t) \mid = c(t), \qquad t \in \partial D. \tag{1.18}$$

This problem is of high importance because of its numerous applications in elasticity theory, electrodynamics etc. (see for instance [52], [78].)

In addition to the already mentioned reduction to vector-matrix $\mathbb{C}$-linear conjugation problem were proposed different methods to study it. One of them was developed within the framework of approximation theory. It goes back to the brilliant papers by Adamjan-Arov-Krejn [4], [5] and continued for instance in [233], [234] (cf. also [276]). Last, the method of linear boundary value problems combined with the integral equation approach was presented by Yu. V. Obnosov (see, e.g., [217]). He found explicit form solution of (1.18) in some special cases.

Applied problems lead also to the formulation of certain *mixed problems*. The most known among them is the so-called *mixed Cherepanov's problem* stated in 1962 [49], [50]. It consists of searching for analytic function in the upper half-plane Π under the following boundary conditions:

$$\begin{cases} \Re\{\overline{A(t)}w^\alpha(t)\} = 0, & t \in L = \cup_{k=1}^n (t_{2k-1}, t_{2k}), \\ \mid w(t) \mid = c(t), & t \in M = \mathbb{R} \setminus L, \end{cases} \tag{1.19}$$

where t_j are given real numbers, and $A(t), c(t)$ given functions on L and M respectively. In [49], [50] this problem was studied in connection with some questions of static elasto-plasticity. It was found that its solution highly depends on the prescribed behavior of an unknown function w near the branch points t_j and the boundary values of w on the line L as well as of the quantity of zeroes of this function in the half-plane Π. The complete solution of this problem for $n \leq 2$ was given in [212], [216] and for arbitrary n in the class of meromorphic functions in [215].

There are traditional applications of complex analysis to linear and nonlinear problems of mechanics. Many of them are well known in English literature (see, e.g., [1], [59], [206]). We pay our attention to elastic-plastic problems which are not sufficiently known to western mathematicians. The essence of the simplest elastic-plastic problems is in consideration of domains divided into elastic and plastic parts. This phenomena occurs when the stresses in some regions extend the critical quantities. Thus linear elastic state is transformed to the plastic flow. The stresses and displacement in the elastic region D_e can be derived by the complex potential of Kolosov–Muskhelishvili. In the plastic region D_p we have certain nonlinear equations. Moreover the curve separating D_e and D_p has to be determined. That is why such problems are closed to the inverse problems of hydrodynamics. It is worth while to note that thereare a lot of difficulties even at the stage of rigorous formulation of mathematical problems reflecting the effects of hysteresis in elasto-plasticity. We discuss briefly the current development in the study of correctly stated mathematical problems of elasto-plasticity (see [22], [93]). The most investigated are the problems of pure shear and torsion, when the displacement and stresses in the simply connected domain D_e are determined by an analytic function $\phi(z)$:

$$w(x,y) = \Re\phi(z), \quad \tau_{xz} + i\tau_{yz} = G\overline{\phi'(z)}.$$

Here G is an elastic constant, the vector $(0, 0, w(x,y))$ is the displacement; τ_{xz}, τ_{yz} are the components of the tensor of stresses. The remaining components of this tensor are supposed to be equal to zero. In the plastic region D_p the corresponding equations have the form

$$\frac{\partial \tau_{xz}}{\partial x} + \frac{\partial \tau_{yz}}{\partial y} = 0, \quad \tau_{yz}\frac{\partial w}{\partial x} - \tau_{xz}\frac{\partial w}{\partial y} = 0, \quad \tau_{xz}^2 + \tau_{yz}^2 = k^2, \qquad (1.20)$$

where the given constant k corresponds to the limit case, when the plastic flow occurs. On the curve L separating D_e and D_p the displacement and the normal stresses are continuous. The domain $D = D_e \cup L \cup D_p$ and the conditions on ∂D determine the analytic function $\phi(z)$ on the unknown curve L. For instance, one of the variants of the considered problem yields the condition (see [22])

$$t\phi'(t) = \frac{ik}{2}|t|, \quad t \in L.$$

Some questions of elasto-plasticity lead to the boundary value problems when on the known part the Riemann–Hilbert condition holds but on the unknown one the values of an analytic function are given. After transformation of the domain (the method of godograph) the inverse problem becomes a mixed boundary value problem (on a part of the boundary we have the modulus problem but on the rest the Riemann–Hilbert condition is stated). Galin [93] has considered similar problems of the torsion when the section of the cylinder is a polygon. Applying conformal mapping he gets a vector-matrix two-dimensional Riemann–Hilbert problem for the half-plane with piece-wise constant matrix.

Some other problems with unknown curves were solved (see, e.g., [22], [93]) in the case of small parameters by method of integral equations. We consider the model mathematical problems of analogous type in Section 3.7, 4.10. Such nonlinear problems are studied by the analytic functions methods.

Chapter 2

Notations and Auxiliary Results

A collection of necessary facts is gathered in this chapter. It is done in order to make text readable as well as to present main ideas more straightforward. The most of the results have an auxiliary character and can be omitted at the first reading. Less common material connected with general results on boundary value problems (especially on linear ones) is discussed here too. The reader can find it also in the classic monograph on the subject (for instance in books by F. D. Gakhov [91], N. I. Muskhelishvili [205].) Anyway, we suppose that this material presented in unified form can help to understand methods and approaches of the boundary value problems. In particular one can see the difference between linear and nonlinear boundary value problems and to compare the corresponding results.

2.1 Geometry of complex plane.

The *complex plane* (or the set of all complex numbers) is denoted by $\mathbb{C}$, but the *Riemann sphere* (extended complex plane) by $\widehat{\mathbb{C}} = \mathbb{C} \cup \{\infty\}$. The branch of an argument of a complex number is chosen usually in standard way: $0 \leq \arg z < 2\pi$ (if it is not supposed something else). Mostly used sets on the complex plane are *discs* (or balls), *circles* (or circumferences), *half-planes* and *lines*. Let

$$\mathbb{D}(a;r) := \{z \in C : |z - a| < r\}$$

denote an open disc with center $a \in \mathbb{C}$ and of radius $r \in \mathbb{R}_+$ (in particular $\mathbb{U} = \mathbb{D}(0;1)$ is a standard unit disc);

$$\mathbb{T}(a;r) := \{z \in \mathbb{C} : |z - a| = r\}$$

denotes a circle with center $a \in \mathbb{C}$ and of radius $r \in \mathbb{R}_+$ (in particular $\mathbb{T} = \mathbb{T}(0;1)$ is the standard unit circle). It is customary for us to omit the center and radius in the notation of a disc or a circle in some cases (if it does not lead to contradiction).

$$\Pi(a;\mathbf{n}) := \{z \in \mathbb{C} : (z-a)\mathbf{n} > 0\}$$

denotes an open half-plane with a point a on its boundary and with internal normal vector $\mathbf{n}$ to the boundary (in particular $\Pi = \Pi((0,0);(0,1))$ is the standard upper half-plane);

$$\mathrm{l}(a;\mathbf{n}) := \{z \in \mathbb{C} : (z-a)\mathbf{n} = 0\}$$

is a straight line containing a point a with normal vector $\mathbf{n}$ (in particular $\mathbb{R} = \mathrm{l}((0,0);(0,1))$ is the standard real line). An operation of *complex conjugation* is denoted by $\overline{(.)} : \mathbb{C} \to \mathbb{C}$ (i.e., $\overline{z} = \overline{(x+iy)} = x - iy$). We also use the notation $\mathbf{C}$ for an operator of complex conjugation. An operation of *symmetry with respect to* $\mathbb{T}(a,r)$ (or *symmetry with respect to* $\mathrm{l}(a;\mathbf{n})$) is denoted by $(.)^*_{\mathbb{T}(a;r)} : \mathbb{C} \to \mathbb{C}$ (or $(.)^*_{\mathrm{l}(a;\mathbf{n})} : \mathbb{C} \to \mathbb{C}$). The latter is defined by the formulae

$$z^*_{\mathbb{T}(a;r)} = (a + \rho e^{i\phi})^*_{\mathbb{T}(a;r)} = a + \frac{r^2}{\rho}e^{i\phi}, \quad 0 \le \rho \le \infty, 0 \le \phi < 2\pi,$$

or

$$z^*_{\mathrm{l}(a;\mathbf{n})} = (a + \xi + i\eta)^*_{\mathrm{l}(a;\mathbf{n})} = (a + \xi - i\eta); \quad \xi, \eta \in \mathbb{R}.$$

The *Schwarz's Reflection Principle* is useful for the applications: let a function $\varphi(z)$ be analytic in a domain D whose boundary contains an arc γ which is either a line segment or an arc of circle, and φ is continuous up to γ. If the image $\varphi(\gamma)$ belongs to an arc which is also either a line segment or an arc of circle, then φ can be analytically continued through γ into a domain symmetric for D with respect to γ possessing symmetric values in the symmetric points. We also apply the following result: let $\varphi(z)$ be analytic in a disc $|z-a| < r$, then the function $\overline{\varphi(z^*_{\mathbb{T}(a,r)})}$ is analytic in $|z-a| > r$ (these functions are not necessarily an analytic continuation of each other). A set $D \subset \mathbb{C}$ (or $\subset \widehat{\mathbb{C}}$) is called *connected* if there do not exist two disjoint sets $D_1, D_2 : D_1 \cup D_2 = D$. In some books on Complex Analysis the term *domain* is used to denote any open connected set D in $\mathbb{C}$ or in $\widehat{\mathbb{C}}$. The reader should be careful because we consider non-connected domains in some parts of the book as well. The type of the domain under consideration is described (if necessary) at the corresponding places. The set of all boundary points of D is called its *boundary* and denoted ∂D. Last, $\mathrm{cl}D := D \cup \partial D$ is the *closure* of D. A continuous mapping $p : [0,1] \to \widehat{\mathbb{C}}$ ($l : \mathbb{T} \to \widehat{\mathbb{C}}$) is called a *path* on $\widehat{\mathbb{C}}$ (respectively a *loop* on $\widehat{\mathbb{C}}$). Two

2.1. GEOMETRY OF COMPLEX PLANE.

paths p_1, p_2 (two loops l_1, l_2) are called *equivalent* if there exists a homeomorphism $\pi : [0,1] \to [0,1]$ (respectively $\lambda : \mathbb{T} \to \mathbb{T}$) such that $p_2(t) = p_1(\pi(t)), t \in [0,1]$ (resp. $l_2(t) = l_1(\lambda(t)), t \in \mathbb{T}$). A class of equivalent paths (loops) is then called a *continuous arc* γ (*continuous curve* Γ). Any of equivalent paths (or loops) is a *parametrization* of the arc (or the curve). Sometimes we will identify an arc (or a curve) with its *locus* $|\gamma|$ ($|\Gamma|$), i.e., with the image of one of the equivalent paths (or loops). If a curve Γ has no point of self-intersection, i.e., all the loops are injective, then it is called *simple*. Any curve (or arc) L is called *positively oriented*, if one of two possible orientations is chosen on it. It is customary to say that the chosen orientation is "positive" denoting the oriented curve L^+ or simply L. A curve (an arc) with opposite orientation is then denoted L^-. It should be noted that even in the case of a closed curve the positive orientation (in the above sense) does not necessarily coincide with the standard one (in counterclockwise sense). Sometimes it is customary for us to use a term curve with respect to a certain collection of curves and arcs (possibly having no common points, i.e., disconnected). To be more precise we use in some cases the term *"a closed curve"* for a curve as defined above, *"closed arc"* for an arc as defined above. *"Open arc"* is a class of equivalent continuous mappings of the type: $p : (0,1) \to \widehat{\mathbb{C}}$. Let D be a connected domain on $\widehat{\mathbb{C}}$. Two paths p_1, p_2 (two loops l_1, l_2) are called *homotopic* ($p_1 \cong p_2, l_1 \cong l_2$) with respect to domain D if there exists a homeomorphism $\pi : [0,1] \times [0,1] \to D$ (respectively $\lambda : [0,1] \times \mathbb{T} \to D$) such that $\pi(0,t) \equiv p_1(t), \pi(1,t) \equiv p_2(t), t \in [0,1]$; $\pi(\tau,0) = p_1(0) = p_2(0), \pi(\tau,1) = p_1(1) = p_2(1), \tau \in [0,1]$ (resp. $\lambda(0,t) \equiv l_1(t), \lambda(1,t) \equiv l_2(t), t \in \mathbb{T}$). In particular, a loop l is *"homotopic to zero"* with respect to D, if there exists $l_0 \cong l$ such that $l_0(t) \equiv z_0 \in D$. A domain D is called *simply connected* if any curve Γ, $|\Gamma| \subset D$, is homotopic to zero with respect to D. If it is not so for at least one curve Γ, $|\Gamma| \subset D$, then D is *multiply connected (finitely or infinitely)* . The boundary of finitely connected (more precisely, n-connected) domain consists of n simple continuous curves without common points. Due to conformal mapping arguments one can imagine an n-times connected domain as extended complex plane $\widehat{\mathbb{C}}$ with n "holes" (each encircled by a simple closed curve). A multiply connected domain can be bounded or unbounded. In the first case $D = D_0 \backslash (\cup cl D_k)$, where D_0 is a bounded domain enclosed by a simple closed curve L_0, and D_k are subsets D_0 such that $cl D_j \cap cl D_l = \emptyset, j \neq l$, enclosed by simple closed curves L_k. In the second case $D = \widehat{\mathbb{C}} \backslash (\cup cl D_k)$ (see Figure 2.1–2.1). Different kind of orientations are possible for multiply connected domains. In the first case (see Figure 2.1) we usually suppose that L_0 is oriented in the counterclockwise sense, but all L_k are oriented in the clockwise sense. In the second case all L_k are oriented in the clockwise sense (see Figure 2.1), or in the counterclockwise sense (see Figure 2.1). A curve $L \subset \widehat{\mathbb{C}}$ is called *a curve of class $C^{m,\alpha}$* if all the equivalent loops

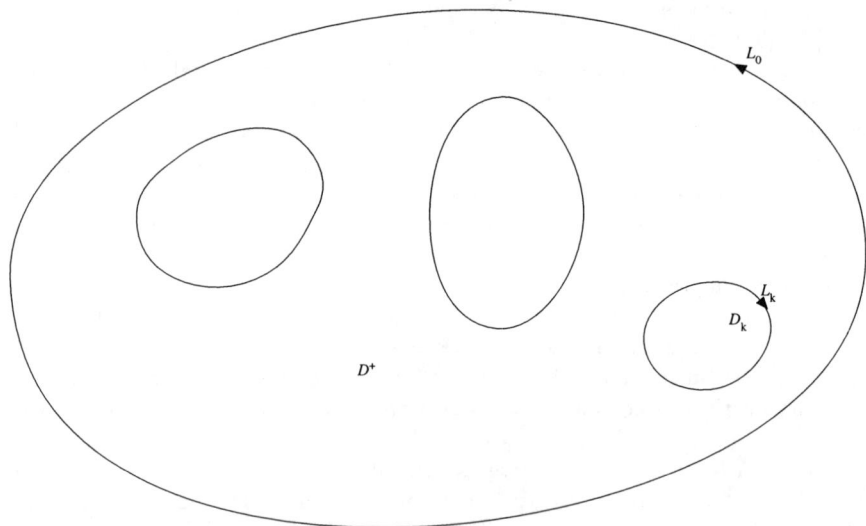

Figure 2.1: A bounded multiply connected domain

are of the class $C^{m,\alpha}$ (i.e., $|\, l^{(m)}(t_1) - l^{(m)}(t_2)\, | < C\, |\, t_1 - t_2\, |^\alpha$, for all $t_1, t_2 \in [0,1]$ and certain constants $C > 0$, $0 \leq \alpha \leq 1$.) If $m \geq 1$, then L is called *smooth*. A curve $L \subset \widehat{\mathbb{C}}$ of the class $C^{1,\alpha}$ is called the *Lyapunov curve*.

2.2 Functional spaces.

Let X be a connected subset of the extended complex plane $\widehat{\mathbb{C}}$ (in particular X is an open or closed arc, curve, simply or multiply connected domain, compact set, etc.) Then the collection

$$\mathcal{C}(X) := \{f : X \to \mathbb{C} \colon f \text{ is continuous on } X\}$$

forms the linear space of *continuous functions*. For X being either a closed curve, or a compact set, $\mathcal{C}(X)$ becomes a Banach space with the norm:

$$\|\, f\, \|_\mathcal{C} := \max_{x \in X} |\, f(x)\, |\, .$$

We also use the following known subspaces of the space of continuous functions:

$$\mathcal{H}^\alpha(X) := \{f \in \mathcal{C}(X) : \exists C > 0, |f(x_1) - f(x_2)| < C|x_1 - x_2|^\alpha, \forall x_1, x_2 \in X, 0 < \alpha \leq 1\}.$$

2.2. FUNCTIONAL SPACES.

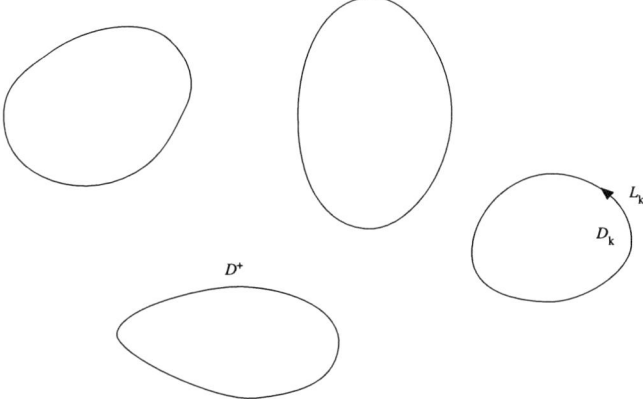

Figure 2.2: A bounded multiply connected domain D^+ (counterclockwise orientation on L_k)

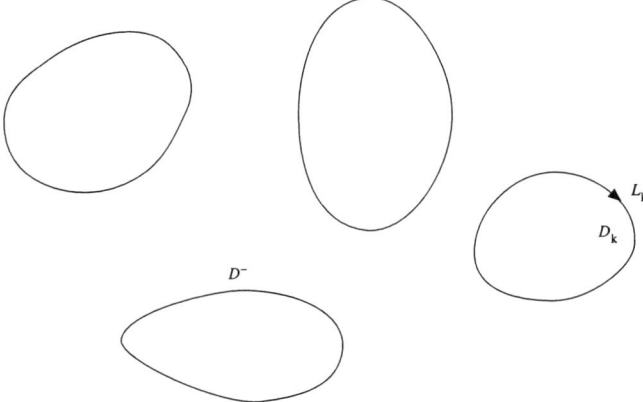

Figure 2.3: A bounded multiply connected domain D^- (clockwise orientation on L_k)

It is the linear space *of Hölder-continuous functions* (or Lipschitz-alpha functions) (the notations $Lip_\alpha(X), \mathcal{C}^{0,\alpha}(X)$ are also commonly used in this situation); in particular

$$Lip(X) := Lip_1(X) := \mathcal{C}^{0,1}(X)$$

is the linear space of *Lipschitz-continuous functions*. Again for X being either a closed curve, or a compact set, $\mathcal{H}^\alpha(X)$ becomes a Banach space with the norm:

$$\| f \|_\alpha := \| f \|_C + \sup_{x_1,x_2 \in X, x_1 \neq x_2} \frac{|f(x_1) - f(x_2)|}{|x_1 - x_2|^\alpha} = \| f \|_C + h(f,\alpha).$$

For X being an arc (open or closed), so-called *small Hölder spaces* are introduced:

$$\mathcal{H}_0^\alpha([a,b]) := \{f \in \mathcal{H}^\alpha([a,b]) : f(a) = f(b) = 0\},$$

$$\mathcal{H}_0^\alpha((a,b)) := \left\{f \in \mathcal{H}^\alpha((a,b)) : \lim_{x \to a} f(x) = \lim_{x \to b} f(x) = 0\right\}.$$

For X being a curve or an arc one can introduce *weighted Hölder spaces*:

$$\mathcal{H}^\alpha(X;\rho) := \{f : \exists f_0,\ f(x) = f_0(x)\rho(x),\ f_0 \in \mathcal{H}^\alpha(X)\}$$

with a given weight function ρ (for instance $\rho(t) := \prod_{k=1}^n |t - t_k|^{\beta_k}$, where $\beta_k \in \mathbb{R}$). The notation $\mathcal{H}(X)$ is also used in the case when the Hölder degree α is not too important. The collection of *Schauder spaces* is defined in the following way:

$$\mathcal{C}^{m,\alpha}(X) := \{f \in \mathcal{C}(X) : \exists C > 0, |f^{(m)}(x_1) - f^{(m)}(x_2)| < C |x_1 - x_2|^\alpha,$$
$$\forall x_1, x_2 \in X; m \in \mathbb{N},\ 0 \leq \alpha \leq 1\}.$$

Finally, the collection of all *infinitely differentiable functions*, i.e., those having derivatives of any positive order on X, is denoted by $\mathcal{C}^\infty(X)$:

$$\mathcal{C}^\infty(X) := \{f \in \mathcal{C}(X) : \forall m > 0, \forall x \in X, \exists f^{(m)}(x)\}.$$

The set of *analytic functions* in a domain D is denoted by $\mathcal{A}(D)$ (we do not use the term *holomorphic functions*; therefore do not apply another commonly used notation for this set, namely $\mathcal{H}(D)$.) For the space of continuous functions (as well for all its subspaces mentioned above) one can define the subspaces of the following type: let L be a smooth curve (consisting of not more than finite collection of connected component) dividing the complex plane into two domains D^+ (D^-) (which can be connected or not (cf., Figure 2.1-2.3) in accordance to the orientation on L). The set of all continuous (Hölder-continuous, Lipschitz-continuous, etc.) contains two subsets of functions analytically extended into D^+ (D^-). These subsets will be denoted $\mathcal{C}_+(L)$ $(\mathcal{C}_-(L))$ (respectively $\mathcal{H}_+^\alpha(L)$ $(\mathcal{H}_-^\alpha(L))$, $Lip_+(L)$ $Lip_-(L)$), etc.) It is known (cf.,, e.g.,,[255]) that they form Banach spaces with $\mathcal{C}$-norm (Hölder-norm, Lipschitz-norm, etc.), for instance

$$\mathcal{C}_+(L) := \{f \in \mathcal{C}(L) : \exists F \in \mathcal{A}(D^+) \cap \mathcal{C}(cl D^+),\ F(t) = f(t),\ t \in L\}.$$

2.2. FUNCTIONAL SPACES.

Maximum Modulus Principle for analytic functions gives us immediately that for any $f \in \mathcal{C}_+(L)$ the following equality takes place $\|f\|_{\mathcal{C}(L)} = \|F\|_{\mathcal{C}(cl(D^+))}$ in the above notations, i.e., $\mathcal{C}_+(L)$ is isometrically isomorphic to the space

$$\mathcal{C}_\mathcal{A}(D) := \{F \in \mathcal{A}(D^+) : \exists f \in \mathcal{C}(L), \lim_{z \to t} F(z) = f(t), \forall t \in L\}.$$

The same is true for another kind of orientation on L, as well as for another kind of domain D (simply-, or multiply-connected, or even non-connected (see Figure2.1-2.3)). Further equivalent functions f and F in the above sense are identified (if it does not lead to confusion). Such considerations can be applied also to the pairs of spaces $\mathcal{H}_+^\alpha(L)$ (or $\mathcal{H}_-^\alpha(L)$) and $\mathcal{H}_\mathcal{A}^\alpha(D)$, $Lip_+(L)$ (or $Lip_-(L)$) and $Lip_\mathcal{A}(D)$, etc. It is customary to define some classes of analytic functions with the prescribed number of zeros in a domain too. Let D be a simply- (or multiply-) connected domain with the smooth boundary L. Then

$$\mathcal{A}_m(D) := \{\phi \in \mathcal{C}_\mathcal{A}(D) : m \text{ is a divisor of multiplicity of any zero of } \phi\}.$$

In particular, $\mathcal{A}_m(D) = \mathcal{C}_\mathcal{A}(D)$. For functions of $\mathcal{A}_m(D)$ with the prescribed number of zeros we will use the following notation

$$\mathcal{A}_m^k(D) := \{\phi \in \mathcal{A}_m(D) : n_D(\phi) = m \cdot k\},$$

where $n_D(\phi)$ is the quantity (together with their multiplicities) of zeros of ϕ in D. In short, $\mathcal{A}^k(D) := \mathcal{A}_1^k(D)$, and $\mathcal{A}^0(D)$ is a class of nowhere vanishing functions of $\mathcal{C}_\mathcal{A}(D)$. Last, if L divides the complex plane $\mathbb{C}$ into two domains D^+, D^- in accordance with a chosen orientation on L, then the corresponding classes are denoted as follows:

$$\mathcal{A}_{m,n} := \{(\phi^+, \phi^-) : \phi^+ \in \mathcal{A}_m(D^+), \phi^- \in \mathcal{A}_n(D^-)\};$$

$$\mathcal{A}_{m,n}^{k,l} := \{(\phi^+, \phi^-) : \phi^+ \in \mathcal{A}_m^k(D^+), \phi^- \in \mathcal{A}_n^l(D^-)\}.$$

In particular,

$$\mathcal{A}^{k,l} := \mathcal{A}_{1,1}^{k,l}, \quad \mathcal{A}^{0,0} := \{(\phi^+, \phi^-) : \phi^+ \in \mathcal{A}^0(D^+), \phi^- \in \mathcal{A}^0(D^-)\}.$$

Let X be a connected set on $\widehat{\mathbb{C}}$. A triple (X, Ω, μ) is called a *measure space*, where Ω is a σ-algebra of sets on X and μ a (complex) measure on X. The following spaces are defined for X being either an arc, or a curve (simple or compound) on $\widehat{\mathbb{C}}$. Let us denote by $\mathcal{M}(X, \mu)$ a collection of all measurable (with respect to measure μ)

functions on X taking their values in $\widehat{\mathbb{C}}$. One can consider a set of all *p-summable functions* on X:

$$L_p(X,\mu) := \left\{ f \in \mathcal{M}(X,\mu) : \int_X |f(z)|^p\, d\mu\, |< +\infty \right\},$$

and introduce an equivalence relation on this set:

$$f \cong g \iff \int_X |f(z)|^p\, d\mu\, |= \int_X |g(z)|^p\, d\mu\, |\,.$$

A collection of all equivalence classes of p-summable functions is called the *Lebesgue space* $\mathcal{L}_p(X,\mu)$:

$$\mathcal{L}_p(X,\mu) := L_p(X,\mu)/\cong\,.$$

If μ is given (for instance μ is the standard Lebesgue measure) then one can use notations $L_p(X)$ and $\mathcal{L}_p(X)$ or simply L_p and $\mathcal{L}_p$. For any σ-finite with respect to a measure μ set X and for $p, 1 \leq p < \infty$, the Lebesgue spaces $\mathcal{L}_p(X,\mu)$ become the Banach spaces with the norm:

$$\|f\|_p := \left(\int_X |f(z)|^p\, d\mu\, | \right)^{1/p},\ 1 \leq p < \infty.$$

For $0 < p < 1$ the spaces $\mathcal{L}_p(X,\mu)$ are not more than metric spaces with the metric $\rho(\cdot,\cdot)$ defined by the formula

$$\rho(f,g) := \int_X |f(z) - g(z)|^p\, d\mu\, |\,.$$

The *Lebesgue space* $\mathcal{L}_\infty(X,\mu)$ is introduced in the same manner but on the basis of the following collection of functions:

$$L_\infty(X,\mu) := \{ f \in \mathcal{M}(X,\mu) : \text{ess sup}_{z \in X} |f(z)| < \infty \},$$

i.e.,

$$\mathcal{L}_\infty(X,\mu) := L_\infty(X,\mu)/\cong\,.$$

The equivalence $\cong$ means here that the essential supremum for corresponding functions takes the same value. $\mathcal{L}_\infty(X,\mu)$ for X being σ-finite with respect to a measure μ is also a Banach space with the norm:

$$\|f\|_\infty := \text{ess sup}_{z \in X} |f(z)|\,.$$

2.2. FUNCTIONAL SPACES.

The analogous Lebesgue-like spaces can be defined in the case $X = \bigcup_{k=1}^{n} X_k$, where X_k, $1 \leq k \leq n$, are disjoint connected subsets of $\widehat{\mathbb{C}}$:

$$\mathcal{L}_p(X, \mu) := \{\mathcal{L}_p(X_1, \mu), \cdots, \mathcal{L}_p(X_n, \mu)\}, \ 0 < p \leq \infty.$$

If all of sets X_k are σ-finite and $1 \leq p \leq \infty$, then $\mathcal{L}_p(X, \mu)$ becomes a Banach space with the norm:

$$\| f \|_p := \sup_k \| f_k \|_p,$$

where $f_k = f|_{X_k}$ are restrictions of the function f on the sets X_k. The latter definitions are used, for instance, for X being a boundary of a multiply connected domain in $\widehat{\mathbb{C}}$. The *Hardy space* $\mathcal{H}_p(\mathbb{U})$ is defined for any p, $0 < p < \infty$, as the space of all analytic functions on $\mathbb{U}$ satisfying the following condition:

$$\sup_{0<r<1} \int_0^{2\pi} | f(re^{i\theta}) |^p \, d\theta < \infty.$$

$\mathcal{H}_p(\mathbb{U})$ are Banach spaces for any p, $1 \leq p < \infty$. $\mathcal{H}_\infty(\mathbb{U})$ is the space of all bounded analytic functions on $\mathbb{U}$. Analogous spaces can also be defined for harmonic setting. For instance, the *harmonic Hardy spaces* $h_p(\mathbb{U})$, $0 < p < \infty$, are spaces of all harmonic functions u, satisfying the following condition:

$$\sup_{0<r<1} \int_0^{2\pi} | u(re^{i\theta}) |^p \, d\theta < \infty.$$

$h_\infty(\mathbb{U})$ is a space of all bounded harmonic functions on $\mathbb{U}$. In the most cases the spaces h_p and $\mathcal{H}_p$ are closely related to each other (see [63], [95], [102], [127].) Sometimes it is important to use notions of the following four spaces:

i) the space $BMO(\mathbb{T})$ of functions determined on the unit circle $\mathbb{T}$ with *bounded mean oscillation* which is defined as follows:

$$BMO(\mathbb{T}) := \left\{ f \in \mathcal{L}_1(\mathbb{T}) : \sup_I \left[\frac{1}{|I|} \int_I | f(t) - f_I | \, dt \right] < \infty \right\},$$

where $f_I = \frac{1}{|I|} \int_I f(t) dt$, $|I|$ is a length of any interval $I \in [0, 2\pi]$;

ii) the space $BMOA(\mathbb{T}) \subset BMO(\mathbb{T})$ of functions determined on the unit circle $\mathbb{T}$ with bounded mean oscillation which are analytically continued into the unit disc $\mathbb{U}$;

iii) the space $VMO(\mathbb{T}) \subset BMO(\mathbb{T})$ of functions with *vanishing mean oscillation* which is defined as follows:

$$VMO(\mathbb{T}) = \left\{ f \in \mathcal{L}_1(\mathbb{T}) : \lim_{|I| \to 0} \left[\frac{1}{|I|} \int_I | f(t) - f_I | \, dt \right] = 0 \right\},$$

where $f_I = \frac{1}{|I|} \int_I f(t)dt, |I|$ is a length of any interval $I \in [0, 2\pi]$;

iv) the space $VMOA(\mathbb{T}) \subset VMO(\mathbb{T})$ of functions determined on the unit circle $\mathbb{T}$ with vanishing mean oscillation which are analytically continued into the unit disc $\mathbb{U}$. All these spaces are Banach ones under the BMO-norm:

$$\| f \| := \sup_I \left[\frac{1}{|I|} \int_I | f(t) - f_I | \, dt \right].$$

2.3 Operator equations in functional spaces.

Standard results from operator theory are presented here. Let us begin with an operator equation of the type

$$x = \mathbf{A}x, \qquad (2.3.1)$$

where $\mathbf{A}$ is a linear operator acting on a metric (or normed) space $\mathcal{X}$.

Theorem 2.1 *([114, Banach theorem]) Let $\mathcal{X}$ be a Banach space and $\mathbf{A} \in \mathcal{L}(\mathcal{X}, \mathcal{X})$. If*

$$\|\mathbf{A}\| \leq q < 1, \qquad (2.3.2)$$

then the operator $\mathbf{I} - \mathbf{A}$ has a continuous inverse, and $\left\|(\mathbf{I} - \mathbf{A})^{-1}\right\| \leq \frac{1}{1-q}$.

Let us consider the equation of the type

$$x = \mathbf{A}x + y, \qquad (2.3.3)$$

where $\mathbf{A}$ is a continuous linear operator in a Banach space $\mathcal{X}$, y is given, and x is an unknown element of $\mathcal{X}$. A widely used method of the solution for equation (2.3.3) is the *method of successive approximation*. Taking an arbitrary element $x_0 \in \mathcal{X}$ (called an *initial point* of the approximation) one can determine the following approximative sequence

$$x_{n+1} = \mathbf{A}x_n + y, \ n = 0, 1, \ldots. \qquad (2.3.4)$$

The convergence of this sequence is connected with the convergence of the series

$$\mathbf{I} + \mathbf{A} + \mathbf{A}^2 + \ldots + \mathbf{A}^n + \ldots. \qquad (2.3.5)$$

2.3. OPERATOR EQUATIONS IN FUNCTIONAL SPACES.

Theorem 2.2 *([114, p. 209]) If the series (2.3.5) converges, then for any initial point $x_0 \in \mathcal{X}$ the successive approximation sequence (x_n) converges to the unique solution $x_* \in \mathcal{X}$ of the equation (2.3.3). Moreover the following estimate takes place:*

$$\|x_* - x_n\| \leq \|(\mathbf{I} - \mathbf{A})^{-1}\| \, \|\mathbf{A}^n\| \, \|x_1 - x_0\|, \; n = 1, 2, \ldots. \tag{2.3.6}$$

If the operator $\mathbf{A}$ does satisfy the inequality (2.3.2), then the estimate (2.3.6) can be rewritten in the form

$$\|x_* - x_n\| \leq \frac{q^n}{1-q} \|x_1 - x_0\|, n = 1, 2, \ldots.$$

For the equation depending on a parameter, namely

$$x = \lambda \mathbf{A} x + y, \tag{2.3.7}$$

the so-called spectral theory can be applied. Let us introduce some of its elements. In the linear operator theory it is proved (cf., e.g., [114, p.153]) that for any $\mathbf{A} \in \mathcal{L}(\mathcal{X}, \mathcal{X})$ ($\mathcal{X}$ is a normed space) there exists a finite number ρ_0:

$$\rho_0 = \lim_{n \to \infty} \sqrt[n]{\|\mathbf{A}^n\|}. \tag{2.3.8}$$

This number is called *the spectral radius* of the operator $\mathbf{A}$. It is evident that

$$\rho_0 \leq \|\mathbf{A}\|.$$

The spectral radius has the following property: if $|\lambda| < \rho_0$ then there exists a bounded inverse operator $(\mathbf{I} - \lambda \mathbf{A})^{-1}$, i.e., the following statement is valid:

Theorem 2.3 *([128], p. 75) If*

$$|\lambda| < \rho_0,$$

then the equation

$$x = \lambda \mathbf{A} x + y \tag{2.3.9}$$

has a unique solution

$$x_* = (\mathbf{I} - \lambda \mathbf{A})^{-1} y, \tag{2.3.10}$$

which is a limit of the successive approximations

$$x_{n+1} = \lambda \mathbf{A} x_n + y, n = 0, 1, \ldots. \tag{2.3.11}$$

Corollary 2.1 *If the equation (2.3.9) has a unique solution for all $y \in \mathcal{X}$, and all $\lambda \in \mathbb{C}, |\lambda| \le 1$, then this solution can be found by the method of successive approximation.*

Corollary 2.2 *Let $\mathbf{A}$ be a compact operator in $\mathcal{X}$. If the equation*

$$x = \lambda \mathbf{A} x$$

has only a trivial solution for all $\lambda \in \mathbb{C}, |\lambda| \le 1$, then equation (2.3.9) has a unique solution for all $y \in \mathcal{X}$, and all $|\lambda| \le 1$. This solution can be found by the method of successive approximation.

Theorem 2.4 *([128, p. 26]) Let spectral radius $\rho_0(\mathbf{A})$ of the operator $\mathbf{A}$ be such that $\rho_0(\mathbf{A}) < 1$. Then the successive approximations*

$$x_{n+1} = \mathbf{A} x_n + y, n = 0, 1, \ldots$$

are converging to the unique solution x_* of the equation

$$x = \mathbf{A} x + y,$$

and for any $\varepsilon, 0 < \varepsilon < 1 - \rho_0(\mathbf{A})$, the following estimate takes place

$$\|x_n - x_*\| \le c(\varepsilon)[\rho_0(\mathbf{A}) + \varepsilon]^n \|x_0 - \mathbf{A} x_0 - y\|. \qquad (2.3.12)$$

This estimate can be replaced by a more exact one:

$$\|x_n - x_*\| \le c\rho_0^n(\mathbf{A})\|x_0 - \mathbf{A} x_0 - y\|,$$

if the spectral radius $\rho_0(\mathbf{A})$ coincides with the norm of operator $\mathbf{A}$ for some equivalent norm in $\mathcal{X}$.

For $\mathbf{A}$ being a nonlinear operator one can use some other results on solvability of the above equations. Let $(\mathcal{X}, \rho)$ be a complete metric space, and $\Omega \subset \mathcal{X}$ be its closed subset. Let a (nonlinear) operator $\mathbf{A}$ be a mapping of the set Ω into itself. A point $x_* \in \Omega$ is called *a fixed point* of the operator $\mathbf{A}$, if

$$x_* = \mathbf{A}(x_*).$$

The operator $\mathbf{A}$ is called *contractive* (or *contraction operator*) if there exists a number $\alpha < 1$, such that

$$\rho(\mathbf{A} x, \mathbf{A} x') \le \alpha \rho(x, x'), \text{ for any } x, x' \in \Omega. \qquad (2.3.13)$$

2.3. OPERATOR EQUATIONS IN FUNCTIONAL SPACES.

Theorem 2.5 *[114, p. 609, (Banach–Cacciopolli)] Let $\mathbf{A}$ be a contractive operator on Ω. Then it has a unique fixed $x_* \in \Omega$, i.e., a unique solution x_* of the equation*

$$x = \mathbf{A}x \quad (2.3.14)$$

exists. This solution can be obtained as the limit of the successive approximation (x_n),

$$x_{n+1} = \mathbf{A}x_n, \; n = 0, 1, \ldots, \quad (2.3.15)$$

where x_0 is an arbitrary element of Ω.

For the differentiable operator the previous results lead to the following

Theorem 2.6 *([128, p. 28]) Let the operator $\mathbf{A}$ has the Frechet derivative $\mathbf{A}'$ at a point x_*, which is a solution of the equation*

$$x = \mathbf{A}x.$$

Let $\rho_0 < 1$, where ρ_0 is the spectral radius of the linear operator $\mathbf{A}'(x_*)$. Then the successive approximations

$$x_{n+1} = \mathbf{A}x_n, \; (n = 0, 1, \ldots)$$

are converging to x_* if the initial approximation is sufficiently closed to x_*. For any $\varepsilon > 0$ the following estimate takes place:

$$\|x_n - x_*\| \leq C(x_0; \varepsilon)(\rho_0 + \varepsilon)^n.$$

The "Pumping" Principle is a powerful tool for studying the equation of the type (2.3.14) and other closed problems in various spaces (see, e.g., [129]). We give a classic example here. Let the following be known for the operator $\mathbf{A}$: $\mathbf{A}$ is a contractive operator in the Lebesgue space $\mathcal{L}_p$, hence the equation (2.3.14) has a unique solution in $\mathcal{L}_p$; moreover, $\mathbf{A}$ acts from $\mathcal{L}_p$ into the space of continuous functions $\mathcal{C}$. Then the unique solution of (2.3.14) belongs to $\mathcal{C}$ in fact. The analogous consideration can be used for any other pair of spaces $\mathcal{X} \subset \mathcal{Y}$. A similar situation occurs frequently in the boundary value problems, when the given data are Hölder-continuous, and unknown functions are seeking to be continuous. Usually an inverse operator (if it exists) that solves the boundary value problem preserves the Hölder property. In this case the continuous solution belongs in reality to the smaller space of Hölder-continuous functions.

2.4 Properties of analytic and harmonic functions.

Some relations between analytic and harmonic functions in domains are presented here. Let D be a simply connected domain. The real part of a function analytic in D is a harmonic in D function. And conversely, each function $u(z)$ harmonic in D is the real part of a function $\varphi(z)$ analytic in D:

$$u(z) = \Re\varphi(z), \ z \in D.$$

The function $\varphi(z)$ is determined by $u(z)$ up to an arbitrarily purely imaginary additive constant due to formula:

$$\varphi(z) = u(z) + iv(z), \text{ where } v(z) = \int_w^z -\frac{\partial u}{\partial y}dx + \frac{\partial u}{\partial x}dy + C, \ C \in \mathbb{R}, \qquad (2.4.1)$$

w is an arbitrary point in D.

Now let D be a multiply connected domain. The real part of a function analytic in D is a harmonic function in D. But the converse is no longer true. This situation is described by the following two statements.

Theorem 2.7 (*Logarithmic Conjugation Theorem*) *[26, p.179] Let D be a finitely connected unbounded domain ($\infty \in D$). Let $K_0, K_1, \ldots, K_n$ be bounded components of $\widehat{\mathbb{C}}\setminus D$, and let $z_j \in K_j$ for $j = 0, 1, \ldots, n$. If u is real valued and harmonic on D, then there exists φ analytic in D and $A_1, A_2, \ldots, A_n \in \mathbb{R}$ such that*

$$u(z) = \Re\varphi(z) + \sum_{j=0}^{n} A_j \log |z - z_j| \qquad (2.4.2)$$

for all $z \in D$. Here $A_0 := -\sum_{j=1}^{n} A_j$.

The function $\varphi(z)$ that appeared in 2.7 is called the *complex potential*.

Theorem 2.8 (*Decomposition Theorem*) *[146], [26, p.173] Let D and K_j be sets defined in the previous theorem. If φ is analytic in D and $\varphi(\infty) = 0$, then φ has a unique decomposition*

$$\varphi(z) = \sum_{j=0}^{n} \varphi_j(z), \qquad (2.4.3)$$

where φ_j is analytic in $D \cup K_j$ and $\varphi_j(\infty) = 0$ for all $j = 0, 1, \ldots, n$.

2.5. BOUNDARY PROPERTIES.

The logarithmic Conjugation Theorem and Decomposition Theorem imply the following representation of a function u harmonic in D:

$$u(z) = \Re \sum_{j=0}^{n} [\varphi_j(z) + A_j \log(z - z_j)] + A, \qquad (2.4.4)$$

where φ_j is analytic in $D \cup K_j$, A_j and A are real constants, $\sum_{j=0}^{n} A_j = 0$. For L being a closed curve the integral $\int_L d\varphi(z)$ is called the *period* of $\varphi(z)$ along L. Calculating periods of the function $w(z) := \sum_{j=0}^{n} [\varphi_j(z) + A_j \log(z - z_j)]$ along each component L_m of the boundary ∂D one gets by virtue of logarithmic function properties that

$$\int_{L_m} dw(z) = 2\pi i A_m.$$

Therefore the constants $2\pi A_m$ which appear in (2.4.2), (2.4.4) are imaginary parts of the periods of the multi-valued analytic function $w(z)$ along the components of ∂D.

2.5 Boundary properties of analytic and harmonic functions.

Let us present the boundary properties of the analytic and harmonic function defined on the unit disc $\mathbb{U}$. Any function from $h_p(\mathbb{U}), 1 < p \leq \infty$, can be represented in the form of a *Poisson integral*, i.e., if $u \in h_p(\mathbb{U}), 1 < p \leq \infty$, then there exists a function $\phi \in L_p(0, 2\pi)$ such that

$$u(re^{i\theta}) = \frac{1}{2\pi} \int_0^{2\pi} \frac{1-r^2}{1+r^2 - 2r\cos(\theta-t)} \phi(t) dt = \frac{1}{2\pi} \int_0^{2\pi} P_r(\theta - t) \phi(t) dt, \qquad (2.5.1)$$

$0 \leq r < 1, \theta \in [-\pi, \pi]$, where $P_r(\tau) = \frac{1-r^2}{1+r^2 - 2r\cos\tau}$ is called the *Poisson kernel*. It should be noted that the Poisson kernel P_r is in fact the real part of the so-called *Schwarz kernel*:

$$P_r(\theta - t) = \frac{1-r^2}{1+r^2 - 2r\cos(\theta-t)} = \Re \frac{e^{it} + re^{i\theta}}{e^{it} - re^{i\theta}} = \Re \frac{e^{it} + z}{e^{it} - z}. \qquad (2.5.2)$$

It does not always hold for an arbitrary harmonic function in $\mathbb{U}$, but it is true for bounded harmonic functions in $\mathbb{U}$, hence for any function continuous up to the boundary. For a function from $h_p(\mathbb{U}), 1 < p \leq \infty$, we have to introduce the notion of a *non-tangential limit* to describe the behavior of the Poisson integral near the boundary of the unit disc. It will be said that a harmonic function $u(re^{i\theta})$ *converges non-tangentially* to a certain value u_0 at a point $e^{i\theta_0} \in \mathbb{T}$ if there exists the following limit:

$$\lim_{\Delta(\theta_0) \ni re^{i\theta} \to e^{i\theta_0}} u(re^{i\theta}) = u_0, \qquad (2.5.3)$$

where $\Delta(\theta_0)$ is a part of the unit disc $\mathbb{U}$ inside of any symmetric sector with a wedge at $e^{i\theta_0}$, and angle smaller than π. If $u(re^{i\theta}) \in h_p(\mathbb{U})$ for certain $p, 1 < p \leq \infty$, then for almost all $t \in [0, 2\pi]$ the function $u(re^{i\theta})$ (i.e., the Poisson integral (2.5.1)) has non-tangential limit denoted $u(e^{i\theta})$, besides

$$u(re^{i\theta}) = \frac{1}{2\pi} \int_0^{2\pi} P_r(\theta - t) u(e^{it}) dt. \qquad (2.5.4)$$

The function $u(e^{i\theta})$ is called a *non-tangential boundary function*. In particular, the Poisson integral for arbitrarily bounded harmonic functions has a non-tangetial limit in general only a.e. on $\mathbb{T}$. If a harmonic function u is continuous up to the boundary of the unit disc, then the non-tangential boundary function $u(e^{it})$ in (2.5.4) coincides with the usual boundary function (cf., e.g., [109, Ch.4])

$$u(e^{it}) = \lim_{re^{i\theta} \to e^{it}} u(re^{i\theta}).$$

Moreover, if $f(t) \in C(\mathbb{T})$ is an arbitrary continuous function on $\mathbb{T}$, then the Poisson integral

$$\frac{1}{2\pi} \int_0^{2\pi} P_r(\theta - t) f(t) dt$$

represents a harmonic function in $\mathbb{U}$ whose boundary function coincides with $f(t)$ everywhere on $\mathbb{T}$ (cf., e.g., [127, p.17]). All we have already said about harmonic functions from $h_p(\mathbb{U})$ is also true for analytic functions from $\mathcal{H}_p(\mathbb{U})$. The behavior of an analytic function in the unit disc can be partly described via its zeros. The notion of the *Blaschke product* is useful for it:

$$B(z) = \prod_n \frac{|z_n|}{z_n} \frac{z_n - z}{1 - \overline{z_n} z} \qquad (2.5.5)$$

where a sequence of points $(z_n), |z_n| < 1$, has accumulation points only on $\mathbb{T}$. A necessary and sufficient condition of the convergence of the Blaschke product in $\mathbb{U}$ is

2.5. BOUNDARY PROPERTIES.

the following:
$$\sum_n (1-|z_n|) < +\infty. \qquad (2.5.6)$$

In case (2.5.6) $B(z)$ is analytic in $\mathbb{U}$ and has only zeros at the points z_n. Moreover $|B(z)| < 1$ for all $|z| < 1$, and the non-tangential boundary function $B(e^{it})$ satisfies the condition
$$|B(e^{it})| = 1 \text{ a.e. on } \mathbb{T}. \qquad (2.5.7)$$

It should be noted that $B(z)$ is in fact analytic in $\mathbb{C} \setminus K$, where the compact set K consists of:

i) the points $z = \frac{1}{\bar{z}_n}$, which are the poles of $B(z)$;

ii) accumulation points of the sequence (z_n) (all laying on $\mathbb{T}$!), which are essential singularities of $B(z)$ (cf., e.g., [109, p.101]).

It appears that any function from analytic Hardy space has a convergent Blaschke product, more precisely: let $F(z) \in \mathcal{H}_p(\mathbb{U})$ for certain p, $0 < p \leq \infty$, (so, in particular, $F(t) \in C_+(\mathbb{T})$), then there exists the convergent Blaschke product $B(z)$ and nowhere vanishing function $G(z) \in \mathcal{H}_p(\mathbb{U})$ with the same p such that
$$F(z) = B(z)G(z).$$

As was already mentioned the analytic and harmonic Hardy-like spaces are closely related to each other. First, it is evident that $\mathcal{H}_p(\mathbb{U}) \subset h_p(\mathbb{U})$. The inverse is no longer true. For any h_p-function $u(z)$ one can consider the *harmonic conjugate* $v(z)$ which can be represented in the form:

$$v(re^{i\theta}) = \frac{1}{2\pi} \int_0^{2\pi} \frac{2r\sin(\theta-t)}{1+r^2-2r\cos(\theta-t)} \phi(t)dt = \frac{1}{2\pi} \int_0^{2\pi} Q_r(\theta-t)\phi(t)dt, \qquad (2.5.8)$$

where $Q_r(\tau) = \frac{2r\sin\tau}{1+r^2-2r\cos\tau}$ is a so-called *conjugate Poisson kernel*. It is easy to check that Q_r is an imaginary part of the Schwarz kernel:

$$Q_r(\theta-t) = \frac{2r\sin(\theta-t)}{1+r^2-2r\cos(\theta-t)} = \Im\frac{e^{it}+re^{i\theta}}{e^{it}-re^{i\theta}} = \Im\frac{e^{it}+z}{e^{it}-z}. \qquad (2.5.9)$$

The boundary values $v(e^{it})$ of the conjugate harmonic function $v(z)$ (if exist) give us the *conjugate boundary function* for $u(e^{it})$ (or simply *conjugate function*). The following notation is commonly used:
$$v(e^{it}) = \tilde{u}(e^{it}).$$

For any $p, 1 < p < \infty$, and $u(z) \in h_p(\mathbb{U})$ the harmonic conjugate $v(z)$ also belongs to $h_p(\mathbb{U})$ and conjugate function $\tilde{u}(e^{it})$ can be found using the *Hilbert transform* of $u(e^{it})$:

$$v(e^{it}) = \tilde{u}(e^{it}) := -v.p.\frac{1}{2\pi}\int_0^{2\pi} u(e^{i\tau})\cot\frac{\tau-t}{2}d\tau = -\mathbf{H}u. \qquad (2.5.10)$$

The spaces $\mathcal{H}_+^\alpha(\mathbb{T})$, $\mathcal{H}^\alpha(\mathbb{T})$, $0 < \alpha < 1$ are invariant under Hilbert transform. It is not true for the spaces $\mathcal{C}_+(\mathbb{T})$, and $\mathcal{C}(\mathbb{T})$, for which the following properties are valid:

$$\mathbf{H} : \mathcal{C}(\mathbb{T}) \to VMO(\mathbb{T}); \; \mathbf{H} : \mathcal{C}_+(\mathbb{T}) \to VMOA(\mathbb{T}).$$

The conjugate function (i.e., the image of the Hilbert transform) will be continuous only if the initial function (pre-image) has additional smoothness. One such condition is known as the *Dini condition*; namely, the following statement takes place:

If $\int_0^a \frac{\omega_u(t)}{t}dt < \infty$, where ω_u is the modulus of continuity of the function $u \in \mathcal{C}(\mathbb{T})$, then the conjugate function $\tilde{u} = \mathbf{H}u$ is determined at any point of $\mathbb{T}$, continuous on $\mathbb{T}$ and satisfies the following inequality:

$$\omega_{\tilde{u}}(\delta) \leq C \left(\int_0^\delta \frac{\omega_u(t)}{t}dt + \delta \int_\delta^\pi \frac{\omega_u(t)}{t^2}dt \right),$$

where C is an absolute constant. At last $\mathbf{H}(Lip(\mathbb{T})) = \cap_{0<\alpha<1}\mathcal{H}^\alpha(\mathbb{T})$; $\mathbf{H}(Lip_+(\mathbb{T})) = \cap_{0<\alpha<1}\mathcal{H}_+^\alpha(\mathbb{T})$.

2.6 Cauchy type integral and singular integrals.

The classic Cauchy integral formula (see, e.g., [240], [250]) can be formulated in the following way: let L be a simple, closed, piece-wise smooth curve on the complex plane $\mathbb{C}$. If a function $\Phi(z)$ is analytic in *int L* (respectively in *ext L*) and continuous up to the boundary, then it can be represented in the form of a *Cauchy integral*; namely, the next formulae are true:

$$\frac{1}{2\pi i}\int_L \frac{\Phi(t)}{t-z}dt = \begin{cases} \Phi(z), & z \in int\ L, \\ 0, & z \in ext\ L, \end{cases} \qquad (2.6.1)$$

respectively

$$\frac{1}{2\pi i}\int_L \frac{\Phi(t)}{t-z}dt = \begin{cases} \Phi(\infty), & z \in int\ L, \\ -\Phi(z)+\Phi(\infty), & z \in ext\ L. \end{cases} \qquad (2.6.2)$$

2.6. CAUCHY TYPE INTEGRAL AND SINGULAR INTEGRALS.

These formulae can be slightly generalized; namely, the simplicity of the curve can be avoided. To formulate the corresponding variant we introduce the definition of a winding number: let L be a closed Jordan curve on the complex plane $\mathbb{C}$ and $l(\mathbb{T})$ is one of its parametrizations. Let also $z \in \widehat{\mathbb{C}} \setminus L$; we choose then a branch of argument in order to have continuity of the following function: $\phi : [0, 2\pi] \longrightarrow \mathbb{R}, \tau \mapsto \arg(l(e^{i\tau} - z))$. A *winding number* of L about the point z is defined by the formula:

$$wind_z\, L := \frac{1}{2\pi}\left(\phi(2\pi) - \phi(0)\right).$$

If again L is a closed, piece-wise smooth curve on $\mathbb{C}$, a function $\Phi(z)$ is analytic in $int\, L$ and continuous up to the boundary then the first Cauchy formula has a form

$$\frac{1}{2\pi i}\int_L \frac{\Phi(t)}{t-z}dt = \Phi(z)\cdot wind_z\, L, \quad z \in \widehat{\mathbb{C}}\setminus L. \tag{2.6.3}$$

The corresponding result takes place in the case of function analytic outside of L. Another generalization is connected with differential properties of a boundary function $\Phi(t)$: if L is a closed, piece-wise smooth curve on $\mathbb{C}$, a function $\Phi(z)$ is analytic in $int\, L$ and its m-th derivative ($m \geq 1$) is continuous up to the boundary; then

$$\frac{1}{2\pi i}\int_L \frac{\Phi^{(m)}(t)}{t-z}dt = \Phi^{(m)}(z)\cdot wind_z\, L, \quad z \in \widehat{\mathbb{C}}\setminus L. \tag{2.6.4}$$

The condition on the curve and on the boundary behavior of a function as weakened by V. I. Smirnov. He proved (see, e.g., [102]) the following result: let L be a simple closed rectifiable Jordan curve on $\mathbb{C}$, and $\Phi(z)$ be an analytic in $int\, L$. The function Φ has a non-tangential limit $\Phi(t)$ almost everywhere on L for which the Cauchy formula

$$\frac{1}{2\pi i}\int_L \frac{\Phi(t)}{t-z}dt = \begin{cases} \Phi(z), & z \in int\, L, \\ 0, & z \in ext\, L, \end{cases}$$

holds iff $\Phi(z)$ belongs to the Smirnov space $\mathcal{E}_1\,(int\, L)$. Similar objects appeared in many problems of analysis as well as mechanical problems. Let L be a simple, rectifiable Jordan closed curve (or open arc) on $\mathbb{C}$ and $\phi \in \mathcal{L}_1\,(L)$. Then

$$\frac{1}{2\pi i}\int_L \frac{\phi(t)}{t-z}dt, \quad z \in \widehat{\mathbb{C}}\setminus L,$$

is called the *Cauchy type integral*. This integral defines two analytic functions $\Phi^+(z)$ and $\Phi^-(z)$ in the domains $int\, L$ and $ext\, L$ respectively if L is closed curve (or unique function $\Phi(z)$, analytic in $\widehat{\mathbb{C}} \setminus L$, if L is open arc). Thus the difference between the Cauchy integral (i.e., any integral in the Cauchy formulas) and the Cauchy type integral is the following: if the function ϕ can be analytically continued or in $int\, L$, or in $out\, L$ (in the case of closed curve L), then the Cauchy type integral becomes the Cauchy integral. Thus it is true for functions from the space $\mathcal{C}_+(L)$. It should be noted that the continuation can be understood here in the sense of a.e. coincidence of the boundary values of the corresponding analytic function with the values of the density $\phi(t)$. As for integrals along an open arc only the Cauchy type integral is defined in this case. Let again L be a simple, rectifiable Jordan closed curve (or open arc) on $\mathbb{C}$ and $\phi \in \mathcal{L}_1(L)$. The following limit

$$\lim_{\varepsilon \downarrow 0} \frac{1}{\pi i} \int_{L \setminus B(t,\varepsilon)} \frac{\phi(\tau)}{\tau - t} d\tau = v.p. \frac{1}{\pi i} \int_L \frac{\phi(\tau)}{\tau - t} d\tau := \mathbf{S}_L(\phi) \qquad (2.6.5)$$

is called the *Cauchy principal value of a singular integral* $\frac{1}{\pi i} \int_L \frac{\phi(\tau)}{\tau - t} d\tau$ or simply *singular integral* (with Cauchy kernel). It has some interesting properties (see e.g., [205], [91], [232]). As it was already said a singular integral with Cauchy kernel $\mathbf{S}_L(\phi)$ along a simple rectifiable closed curve L exists for almost all $t \in L$ if $\phi \in \mathcal{L}_1(L)$. As a matter of fact the acting properties of a singular integral operator $\mathbf{S}_L$ depend on two components: behavior of a density ϕ and properties of a curve L. Let us suppose for a moment that L is a smooth simple closed curve. Under a smooth curve we understand any curve of class $\mathcal{C}^1$ or equivalently a curve with a nowhere vanishing continuously changing tangent vector. Then one can give the meaning described above to a singular integral on many of known functional spaces (for instance, on any of Lebesgue spaces $\mathcal{L}_p(L)$, $1 \leq p \leq \infty$, on any of Hölder spaces $\mathcal{H}^\alpha(L)$, $0 < \alpha \leq 1$, on any of Schauder spaces $\mathcal{C}^{m,\alpha}(L)$, $m \in \mathbf{N}$, $0 \leq \alpha < 1$, on any of Sobolev spaces $W_p^k(L)$, $k \in \mathbf{N}$, $1 \leq p \leq \infty$, on the space $\mathcal{C}$ of continuous functions, etc.). Besides, an operator $\mathbf{S}_L$ is a bounded linear operator on every of the following spaces: $\mathcal{L}_p(L)$, $1 < p < \infty$; $\mathcal{H}^\alpha(L)$, $0 < \alpha < 1$; $\mathcal{C}^{m,\alpha}(L)$, $m \in \mathbf{N}$, $0 < \alpha < 1$; $W_p^k(L)$, $k \in \mathbf{N}$, $1 < p < \infty$. As for the remaining spaces the situation is quite close to the described one in Section 3.3, namely:

$$\mathbf{S}_L : Lip(L) \longrightarrow \bigcap_{0 < \alpha < 1} \mathcal{H}^\alpha(L), 0 < \alpha < 1, \qquad \mathbf{S}_L : \mathcal{C}(L) \longrightarrow VMO(L),$$

$$\mathbf{S}_L : \mathcal{C}^m(L) \longrightarrow \left\{ \phi : \phi^{(m)} \in VMO(L) \right\}, \qquad \mathbf{S}_L : \mathcal{C}^\infty(L) \longrightarrow \mathcal{C}^\infty(L).$$

2.6. CAUCHY TYPE INTEGRAL AND SINGULAR INTEGRALS.

In fact the properties presented above take place even for a weaker condition on the behavior of the curve L (see, e.g., [63], [95], [98], [232]). In the case of the unit circle there is a formula which connects a singular integral with the Cauchy kernel $\mathbf{S_T}$ and the Hilbert transform $\mathbf{H}$:

$$\mathbf{S_T}(\phi) = -i\mathbf{H}(\phi) + \frac{1}{2\pi}\int_0^{2\pi} \phi(e^{is})ds.$$

The norm of the operator $\mathbf{H}$ (hence the norm of a singular integral operator over the unit circle or real line) is calculated or at least estimated in the case of different functional spaces. The most known results are the following:

$$\|\mathbf{H}\|_{\mathcal{L}_p} = c_p = \begin{cases} \cot\frac{\pi}{2p}, & 1 < p \leq 2, \\ \tan\frac{\pi}{2p}, & 2 \leq p < \infty. \end{cases}$$

$$\|\mathbf{H}\|_{\mathcal{H}^\alpha} = C_\alpha = \frac{1}{\pi}B(\frac{\alpha}{2},\frac{1-\alpha}{2}),$$

where $B(\cdot,\cdot)$ is the Euler beta-function. If $\mathbf{S}_L$ is considered as an operator from one Hölder space into another, $\mathbf{S}_L : \mathcal{H}^\lambda \to \mathcal{H}^\mu, 0 < \mu < 2\lambda - 1 < 1$, then

$$\|\mathbf{S}_L\| \leq 1 + C_1 e^{\lambda-\mu} + C_2 e^\lambda,$$

where C_1, C_2 are absolute constants (see [45]). At last the boundary properties of the Cauchy type integral can be given in the more precise form. Let L be a simple, closed, smooth curve which divides the extended complex plane $\widehat{\mathbb{C}}$ onto two domains $D^+, D^- \ni \infty$. Then boundary functions $\Phi^+(t), \Phi^-(t)$ of the Cauchy type integral

$$\Phi^\pm(z) = \frac{1}{2\pi i}\int_L \frac{\phi(\tau)d\tau}{\tau - z}, \quad z \in D^\pm,$$

do satisfy the following identities (the *Sokhotsky-Plemelj formulae*):

$$\Phi^\pm(t) = \pm\frac{1}{2}\phi(t) + \frac{1}{2\pi i}\int_L \frac{\phi(\tau)d\tau}{\tau - t}. \tag{2.6.6}$$

These formulae take place for all $t \in L$ if (in particular) $\phi \in \mathcal{H}^\alpha(L), 0 < \alpha \leq 1$, or for almost all $t \in L$ if (in particular) $\phi \in \mathcal{L}_p(L), 1 \leq p \leq \infty$. Two operators are defined on the basis of these formulae:

$$\mathbf{P}_L := \frac{1}{2}(\mathbf{I} + \mathbf{S}_L), \quad \mathbf{Q}_L := \frac{1}{2}(\mathbf{I} - \mathbf{S}_L).$$

They have the following properties:

$$\mathbf{P}_L - \mathbf{Q}_L = \mathbf{S}_L;\quad \mathbf{P}_L^2 = \mathbf{P}_L;\quad \mathbf{Q}_L^2 = \mathbf{Q}_L;\quad \mathbf{P}_L\mathbf{Q}_L = \mathbf{Q}_L\mathbf{P}_L,$$

and thus they are called *standard complementary projectors*. They project, in particular the Banach spaces $\mathcal{L}_p(L), 1 < p < \infty$, and $\mathcal{H}^\alpha(L), 0 < \alpha < 1$, onto their closed subspaces, namely:

$$\mathbf{P}_L : \mathcal{L}_p(L) \to \mathcal{L}_p^+(L);\quad \mathbf{Q}_L : \mathcal{L}_p(L) \to \overset{\circ}{\mathcal{L}}_p^{\,-}(L);$$
$$\mathbf{P}_L : \mathcal{H}^\alpha(L) \to \mathcal{H}_+^\alpha(L);\quad \mathbf{Q}_L : \mathcal{H}^\alpha(L) \to \overset{\circ}{\mathcal{H}}_+^{\,\alpha}(L),$$

where $\mathcal{L}_p^+(L)$, $\mathcal{H}_+^\alpha(L)$ are the spaces of functions defined on L which are analytically continued into D^+ with boundary functions from $\mathcal{L}_p(L)$, $\mathcal{H}^\alpha(L)$ respectively; $\overset{\circ}{\mathcal{L}}_p^{\,-}(L)$, $\overset{\circ}{\mathcal{H}}_-^{\,\alpha}(L)$ are the spaces of functions defined on L which are analytically continued into D^- and vanishing at infinity with boundary functions from $\mathcal{L}_p(L)$, $\mathcal{H}^\alpha(L)$ respectively.

2.7 Schwarz operator.

2.7.1 Schwarz operator and complex Green function for simply connected domain.

The problem of the determination of an analytic function in a domain via boundary values of its real part is often called the *Schwarz problem* (see, e.g., [91]). In the case of a simply connected domain it can be simply reduced to the Dirichlet problem for harmonic functions. Anyway this problem as well as some special notations for it is used in the considerations connected with analytic functions (mostly due to the essential difference between Dirichlet and Schwarz problems in the case of multiply connected domains). Partly we have already seen this difference in Section 2.5. More precisely, for any continuous function f the Poisson integral with the density f solves the Dirichlet problem. An operator which stays in correspondence to a given on ∂D real-valued function u, an analytic in D function f, such that $\Re f_{|\partial D} = u$, is called the *Schwarz operator* $\mathbf{T}: u \mapsto f$. In the case of the unit disc $\mathbb{U}$ and the upper half-plane Π the Schwarz operator has an explicit form:

$$(\mathbf{T}u)(z) := \frac{1}{2\pi i} \int_\mathbb{T} u(\zeta) \frac{\zeta + z}{\zeta - z} \frac{d\zeta}{\zeta}; \qquad (2.7.1)$$

2.7. SCHWARZ OPERATOR.

$$(\mathbf{T}u)(z) := \frac{1}{\pi} \int_{-\infty}^{\infty} u(\xi) \frac{y}{(\xi - x)^2 + y^2} d\xi, \quad z = x + iy \in \Pi^+.$$

One can also use the following (equivalent) form for the Schwarz operator in the unit disc:

$$(\mathbf{T}u)(z) := \frac{1}{\pi i} \int_{\mathbb{T}} u(\zeta) \frac{d\zeta}{\zeta - z} - \frac{1}{2\pi i} \int_{\mathbb{T}} u(\zeta) \frac{d\zeta}{\zeta}, \quad z \in \mathbb{U}. \tag{2.7.2}$$

For general simply connected domains the Schwarz operator can be represented via a so-called *complex Green function* (see, e.g., [157]). Recall that the Green function of the Dirichlet problem for a domain $D \subset \hat{\mathbb{C}}$ is the function $G(z, \zeta)$ of two complex variables, satisfying the following properties: i) $G(z, \zeta) = \log \frac{1}{|z-\zeta|} + g(z, \zeta)$, where g is continuous in $D \times D$, harmonic in $z \in D$ for any fixed $\zeta \in D$, and in $\zeta \in D$ for any fixed $z \in D$; ii) $g(z, \zeta)$ is continuous in $\zeta \in \text{cl } D$ for any fixed $z \in D$, and besides $g(z, t) = -\log \frac{1}{|z-t|}, z \in D, t \in \partial D$.

Using the Green function one can solve the Dirichlet problem, i.e., find a harmonic function $u(z)$ in D, whose boundary values coincide with a given Hölder-continuous function $u(t)$, $t \in \partial D$:

$$u(z) = \frac{1}{2\pi} \int_{\partial D} u(\tau) \frac{\partial}{\partial n} G(z, \tau) |d\tau|, \quad z \in D.$$

It is known also that if $\omega(z)$ is an analytic function realizing the conformal mapping of the domain D onto the unit disc $\mathbb{U}$, then the Green function can be presented in the form:

$$G(z, \zeta) = \log |\omega_\zeta(z)|,$$

where $\omega_\zeta(z) = \frac{\omega(z) - \omega(\zeta)}{1 - \omega(z)\overline{\omega(\zeta)}}$.

Introduce a harmonic conjugate for $G(z, \zeta)$ in a variable z by a standard formula (cf.,(2.4.1)):

$$H(z, \zeta) := \int_w^z -\frac{\partial G}{\partial y} dx + \frac{\partial G}{\partial x} dy,$$

where w is an arbitrary point in D. Then the following formula gives the complex Green function for the domain D:

$$M(z, \zeta) := G(z, \zeta) + iH(z, \zeta).$$

Thus the Schwarz operator can be represented in the form:

$$(\mathbf{T}u)(z) := \frac{1}{2\pi} \int_{\partial D} u(\tau) \frac{\partial}{\partial n} M(z,\tau) |d\tau|, \ z \in D.$$

The above remark concerning conformal mapping can be applied to the complex Green function as well, namely,

$$M(z,\zeta) = \log \omega_\zeta(z).$$

2.7.2 Complex Green function and Schwarz operator for multiply connected domain.

Let D be a multiply connected domain bounded by the contours $L_1, \ldots, L_n$; every L_k, $k = 1, 2, \ldots, n$, is a simple closed smooth curve (see Figure 2.1). By definition the Green function of the domain D is a function $G : cl\, D \times cl(D) \to \mathbb{R}$ of two complex variables satisfying the following conditions:

$$G(z,\zeta) = g(z,\zeta) - \log|z - \zeta|, \qquad (2.7.3)$$

where g is harmonic in $\zeta \in D$ for any fixed $z \in D$, and

$$G(z,\tau) = 0, \ \text{if} \ \zeta = \tau \in L, z \in D. \qquad (2.7.4)$$

Evidently G is symmetric with respect to its variables, i.e.,

$$G(\zeta, z) = G(z, \zeta).$$

As in the case of a simply connected domain one can use the Green function to represent the solution of the Dirichlet problem for D, i.e., the harmonic in D function U via its given boundary function u:

$$U(z) = \frac{1}{2\pi} \int_L u(\tau) \frac{\partial G}{\partial \nu} d\sigma, \ d\sigma = |d\tau|, \qquad (2.7.5)$$

where ν is the inward normal vector to L at the point $\zeta = \tau$. By constructing the harmonic conjugate to G:

$$H(z,\zeta) := \int_w^z -\frac{\partial G}{\partial y} dx + \frac{\partial G}{\partial x} dy, \ z = x + iy, \qquad (2.7.6)$$

2.7. SCHWARZ OPERATOR.

where w is an arbitrary fixed point in clD, one can introduce the *complex Green function*

$$M(z,\zeta) := G(z,\zeta) + iH(z,\zeta), \qquad (2.7.7)$$

which is now multi-valued due to the logarithmic function in its definition. During the counterclockwise circuit of z around an internal curve L_k the function $H(z,\zeta)$ gets an increment $2\pi\alpha_k(\zeta)$:

$$\alpha_k(\zeta) := \frac{1}{2\pi} \int_{L_k} -\frac{\partial G}{\partial y} dx + \frac{\partial G}{\partial x} dy = \frac{1}{2\pi} \int_{L_k} \frac{\partial G}{\partial n} ds, \qquad (2.7.8)$$

where n is the inward normal vector to L_k at a point $z = t := x + iy$, and $\zeta \in D$. Therefore the function $M(z,\zeta)$ gets the increment $2\pi i\alpha_k(\zeta)$. It is equal to the increment of any branch of multi-valued function $\alpha_k(\zeta) \log(z - z_k)$, where z_k is an arbitrary fixed point inside L_k. Hence by the Logarithmic Conjugation Theorem (see Section 2.4) the complex Green function can be represented in the form:

$$M(z,\zeta) = M_0(z,\zeta) + \sum_{k=1}^{m} \alpha_k(\zeta) \log(z - z_k) - \log(\zeta - z) + A(\zeta), \qquad (2.7.9)$$

where $M_0(z,\zeta)$ is a single-valued analytic function of z in the domain D (for any $\zeta \in D$), as well as single-valued functions of $\zeta \in D$,

$$M_0(w,\zeta) = 0; \qquad (2.7.10)$$

$A(\zeta)$ is a real function on ζ. It is easy to see that the functions $\alpha_k(\zeta)$ are harmonic in ζ in the domain D. They can also be considered as the solution of the Dirichlet problems ($k = 1, 2, \ldots, n$):

$$\alpha_k(\tau) = \begin{cases} 1, & \text{if } \zeta = \tau \in L_k, \\ 0, & \text{if } \zeta = \tau \notin L_k. \end{cases}$$

It follows from the representation of the functions $\alpha_k(z)$ obtained by interchanging of variables z and ζ in (2.7.8):

$$\alpha_k(z) = \frac{1}{2\pi} \int_{L_k} \frac{\partial G(z,\tau)}{\partial \nu} d\sigma.$$

One can rewrite the last equality in the form:

$$\alpha_k(z) = \frac{1}{2\pi} \sum_{k=1}^{n} \int_{L_k} \alpha_k(\tau) \frac{\partial G(z,\tau)}{\partial \nu} d\sigma.$$

Comparing it with the definition of the Green function we get the necessary fact. The harmonic functions $\alpha_k(z)$, $k = 1, 2, \ldots, n$, are often called *harmonic measures of curves L_k with respect to a domain D* (see, e.g., [157]). We construct the harmonic measures for circular multiply connected domains in Section 4.4. Let z be a point of D, $\zeta = \tau \in L$. Setting

$$a_k(\zeta) := \frac{\partial \alpha_k(\zeta)}{\partial \nu}; \ T(z,\tau) := \frac{\partial M(z,\tau)}{\partial \nu}, \qquad (2.7.11)$$

we get the following equality:

$$T(z,\tau) := \frac{\partial M_0(z,\tau)}{\partial \nu} + \sum_{k=1}^{m} a_k(\tau) \log(z - z_k) - \frac{1}{\tau - z} \frac{\partial \tau}{\partial \nu} + \frac{\partial A}{\partial \nu}. \qquad (2.7.12)$$

The function $T(z,\tau)$ is called the *Schwarz kernel* for a multiply connected domain D. It is an analytic (but multi-valued) function on the variable z in D. Really, it has an increment $2\pi i a_k(\zeta)$ at the counterclockwise circuit along L_k. Besides $T(z,\tau)$ is single-valued, but non-analytic in τ. The real part of the Schwarz kernel is a normal derivative of the Green function. Finally, one of its branches satisfies the relation

$$\Im\{T(w,\tau)\} \equiv 0 \ \tau \in L := \partial D.$$

The Schwarz kernel (as in the case of a simply connected domain) determines the *Schwarz operator*:

$$(\mathbf{T}f)(z) := \frac{1}{2\pi} \int_L f(\tau) T(z,\tau) d\sigma, \qquad (2.7.13)$$

where $\tau = \tau(\sigma)$ is a complex parametrization of the curve ∂D. This operator stays in correspondence to any continuous real-valued function f, an analytic function with respect to z in D. But in the case of a multiply connected domain the only real part of (2.7.13) is single-valued in D. So, for the moment we have not enough information to get a (single-valued) solution of the Schwarz problem for multiply connected domains. Let us consider an analytic function $F(z) := u(z) + iv(z)$, $z \in D$. Suppose additionally that its real part $u(z)$ is single-valued, continuous up to the boundary L. Then the integral

$$\frac{1}{2\pi} \int_L u(\tau) T(z,\tau) d\sigma$$

gives a function which is analytic in z, whose real part coincides with $u(z)$. Therefore, it differs from $F(z)$ in an imaginary constant, namely,

$$\frac{1}{2\pi} \int_L u(\tau) T(z,\tau) d\sigma = F(z) - iv(w). \qquad (2.7.14)$$

2.7. SCHWARZ OPERATOR.

Even if the function $u(z)$ is single-valued, for its harmonic conjugate $v(z)$ we cannot say the same. The increment of the latter along the curve L_k is equal

$$\int_L u(\tau) a_k(\tau) d\sigma. \tag{2.7.15}$$

If the imaginary part $v(z)$ of $F(z)$ is single-valued, then a formula analogous to (2.7.14) takes place:

$$\frac{1}{2\pi} \int_L v(\tau) T(z,\tau) d\sigma = \frac{1}{i} F(z) - \frac{1}{i} u(w), \tag{2.7.16}$$

and the increment of $u(z)$ along the curve L_k is equal to

$$-\int_L v(\tau) a_k(\tau) d\sigma. \tag{2.7.17}$$

If $F(z)$ is single-valued then all integrals in (2.7.15), (2.7.17) should be equal to zero. Or more precisely the function $F(z)$, which is single-valued analytic in a multiply connected domain D and Hölder continuous up to the boundary, is determined by the Schwarz operator via formula (2.7.12) if the following formulae are valid:

$$\int_L F(\tau) a_k(\tau) d\sigma = 0, \ k = 1, 2, \ldots, m; \ \int_L \overline{F(\tau)} a_k(\tau) d\sigma = 0, \ k = 1, 2, \ldots, m. \tag{2.7.18}$$

Besides, for F being single-valued it follows from (2.7.14), (2.7.16) that

$$\begin{cases} \frac{1}{4\pi} \int_L F(\tau) T(z,\tau) d\sigma = F(z) - \frac{1}{2} F(w), \\ \frac{1}{4\pi} \int_L \overline{F(\tau)} T(z,\tau) d\sigma = \frac{1}{2} F(w). \end{cases} \tag{2.7.19}$$

In particular, it takes place for $F(z) \equiv 1$, which leads to the formula

$$\frac{1}{2\pi} \int_L T(z,\tau) d\sigma \equiv 1. \tag{2.7.20}$$

The following facts are of high importance for understanding the properties of the Schwarz operator.

- if a function $f(t), t \in L$, is continuous on L and m-times continuously differentiable on an arc $L' \subset L$, and the arc itself belongs to the same class of smoothness,

then the formula (2.7.12) determines the function which is $(m-1)$-times continuously differentiable on any subarc of L';
- the identity
$$g(\tau) \equiv h(\tau), \quad \tau \in L,$$
is valid for continuous functions g, h iff
$$\begin{cases} \frac{1}{4\pi} \int_L g(\tau) T(z,\tau) d\sigma = \frac{1}{4\pi} \int_L h(\tau) T(z,\tau) d\sigma, \\ \frac{1}{4\pi} \int_L \overline{g(\tau)} T(z,\tau) d\sigma = \frac{1}{4\pi} \int_L \overline{h(\tau)} T(z,\tau) d\sigma. \end{cases}$$

- if the contour L is of the class C^m, i.e., the mth derivative of an angle between the normal vector ν and real axis is continuous with respect to σ, then the following formula holds
$$\frac{1}{4\pi} T(z,\tau) d\sigma = \frac{1}{2\pi i} \frac{d\tau}{\tau - z} + P(z,\tau) d\sigma, \qquad (2.7.21)$$
where the function $P(z,\tau)$ (if its logarithmic singularities are removed) is continuous up to the contour and $(m-2)$-times differentiable on L with respect to $z = t$. As a consequence we obtain that this harmonic measure and the complex Green function belong to $C_z^\infty(\mathbb{D})$ for the circular domain $\mathbb{D}$. The complex Green function possesses an invariance property with respect to conformal maps too. Namely, let $w = \varepsilon(z)$ maps conformally the domain D on z-plane onto the domain D' on w-plane, and let $M(z,\zeta), M'(w,\omega)$ be the complex Green functions for these domains respectively. Then
$$M(z,\zeta) = M'(\varepsilon(z), \varepsilon(\zeta)).$$
Differentiating both sides of the last formula in ν we obtain the connection between the Schwarz kernels for different domains:
$$T(z,\tau) d\sigma = T'(\varepsilon(z), \varepsilon(\tau)) d\sigma', \, d\sigma' = |dw|. \qquad (2.7.22)$$
We construct the function $T(z,\tau)$ for an arbitrary circular multiply connected domain in Section 4.4. Therefore, if the conformal mapping $\varepsilon(z)$ of the domain D onto a circular domain is known, then using (2.7.22) one can construct the Schwarz operator for D. One of the constructive representations of the Schwarz operator is wellknown. This is the *Villat's formula* (see, e.g., [9, p.173]) for the circular annulus $\{z \in \mathbb{C} : r < |z| < 1\}$:
$$F(z) = \tfrac{i\omega}{\pi^2} \int_0^{2\pi} \Phi(s) \zeta(\tfrac{\omega}{\pi i} \log z - \tfrac{\omega}{\pi} s) ds$$
$$- \tfrac{i\omega}{\pi^2} \int_0^{2\pi} \phi(s) \left[\zeta(\tfrac{\omega}{\pi i} \log z - \tfrac{\omega}{\pi} s - \omega') + \eta' \right] ds + iC,$$

2.8. ℂ-LINEAR CONJUGATION PROBLEM.

where Φ, ϕ are given (Hölder-continuous) functions on $|z|=1, |z|=r$ respectively, C is an arbitrary real constant, ω is an arbitrary positive number, purely imaginary number ω' is found from the equality $e^{\frac{\pi i \omega'}{\omega}} = r$, and finally $\zeta(u) = \zeta(u|\omega,\omega')$ is the Weierstrass ζ-function. The problem of constructing the Schwarz operator for a multiply connected domain is connected with the so-called modified Dirichlet problem. According to [157] the *modified Dirichlet problem* for a multiply connected domain D is formulated as follows. To find a function $\varphi \in \mathcal{C}_A(D)$ with the boundary condition

$$\Re \varphi(t) = f(t) + c_k, \quad t \in L_k, \quad k=1,2,...,n, \tag{2.7.23}$$

where the function $f \in \mathcal{C}(L)$, is given, c_k are undetermined constants. If one of the constants c_k is fixed arbitrary, then the remaining constants c_k are determined uniquely. Mikhlin [157] reduced the problem (2.7.23) to an integral equation having a unique solution. The function φ is determined up to an additive pure imaginary constant.

2.8 ℂ-linear conjugation problem.

2.8.1 Simply connected domain.

Let L be a simple closed smooth curve on the complex plane $\mathbb{C}$ which divides $\mathbb{C}$ into two domains D^+ and $D^- \ni \infty$. The known *ℂ-linear conjugation problem for analytic functions* (or *linear Riemann boundary value problem*) consists of searching for two analytic functions $\Phi^+ \in \mathcal{C}_A(D^+)$, $\Phi^- \in \mathcal{C}_A(D^-)$, satisfying the boundary condition:

$$\Phi^+(t) = G(t)\Phi^-(t) + g(t), \quad t \in L, \tag{2.8.1}$$

where G, g are given complex valued functions. The statement of the problem depends on the behavior of the unknown functions near the boundary L of the domains D^+, D^-. The boundary condition can be considered at all points of L. Such continuous-like formulation becomes usually more precise, if we suppose that the given functions G, g are Hölder continuous (see, e.g., [91]). Then (analog of "Pumping" Principle, see Section 2.3) any solution $\mathcal{C}_A(D^+) \times \mathcal{C}_A(D^-)$ belongs to $\mathcal{H}_A^\alpha(D^+) \times \mathcal{H}_A^\alpha(D^-)$ by construction (cf., [91], [98], [205]). We have this remark in mind in our further considerations. The solvability of the problem (2.8.1) depends on the so-called *index of the problem*, coinciding with the *winding number* of the coefficient in the classic case (see, e.g., [91, Section 13-14]):

$$\chi := wind_L G(t). \tag{2.8.2}$$

2.8. C-LINEAR CONJUGATION PROBLEM.

where C is an arbitrary complex constant;

iii) if $\chi > 0$, then analytic solutions of (2.8.3) are given by the formula

$$\Phi^{\pm}(z) = P_{\chi}(z) X^{\pm}(z), \qquad (2.8.7)$$

where P_{χ} is an arbitrary polynomial of the order χ. The family of solutions depends on $\chi + 1$ arbitrary complex constants.

Representing the inhomogeneous boundary condition (2.8.1) in the form

$$\frac{\Phi^{+}(t)}{X_0^{+}(t)} = t^{\chi} \frac{\Phi^{-}(t)}{X_0^{-}(t)} + \frac{g(t)}{X_0^{+}(t)}, \quad t \in L,$$

one comes to the necessity of solving another factorization problem (or simply *jump problem*). Namely, we need to represent the given Hölder-continuous function as the difference of boundary values of functions analytic in D^{+}, D^{-} respectively:

$$\frac{g(t)}{X_0^{+}(t)} := \frac{g(t)}{X^{+}(t)} = \Psi^{+}(t) - \Psi^{-}(t), \ t \in L. \qquad (2.8.8)$$

This problem is solved by *the Cauchy type integral* with *the density* $\frac{g}{X^{+}}$:

$$\Psi^{\pm}(z) = \frac{1}{2\pi i} \int_L \frac{g(\tau)}{X^{+}(\tau)} \frac{d\tau}{\tau - z}, z \in D^{\pm}. \qquad (2.8.9)$$

The Analytic Continuation Principle gives us the following results for the inhomogeneous boundary value problem (2.8.1):

i) if $\chi < 0$, then problem (2.8.1) has an analytic solution iff *the solvability conditions* do satisfy, namely:

$$\int_L \frac{g(\tau)}{X^{+}(\tau)} \tau^{k-1} d\tau = 0, k = 1, \ldots, -\chi - 1. \qquad (2.8.10)$$

If so the unique solution of (2.8.1) has the form

$$\Phi^{\pm}(z) = X^{\pm}(z) \Psi^{\pm}(z), z \in D^{\pm}; \qquad (2.8.11)$$

ii) if $\chi \geq 0$, then all analytic solutions of the problem (2.8.1) are delivered by the formula

$$\Phi^{\pm}(z) = X^{\pm}(z) \left[\Psi^{\pm}(z) + P_{\chi}(z) \right], z \in D^{\pm}, \qquad (2.8.12)$$

where P_{χ} is an arbitrary polynomial of the order χ.

2.8.2 Multiply connected domain.

Let $L = \bigcup_{j=0}^{n} L_j$ be a collection of $n+1$ pair-wise disjoint simple closed smooth curves such that $\bigcup_{j=1}^{n} L_j \subset int L_0$, $ext\left(\bigcup_{j=1, j\neq k}^{n} L_j\right) \supset L_k$. We denote by D^+ $(n+1)$-connected domain inside L_0 and outside all $L_j, j = 1, \ldots, n$ (another types of orientation can be considered with small changes). We study the problem

$$\Phi^+(t) = G(t)\Phi^-(t) + g(t), \quad t \in L, \tag{2.8.13}$$

where G, g are given complex valued functions on L, Hölder-continuous on every curve $L_j, j = 0, \ldots, n$, $G(t) \neq 0, t \in L$. Denoting $\chi_j := wind_{L_j} G(t), j = 0, \ldots, n; \chi := \sum_{j=0}^{n} \chi_j$ (called the index of the problem under consideration) we have to study two factorization problems of the same type as in the case of the simply connected domain. Let $z = 0 \in D^+$ and z_j be arbitrary fixed points, $z_j \in int(L_j)$, then the canonical function, which solves the first factorization problem, can be represented in the form:

$$\begin{cases} X^+(z) = \prod_{j=1}^{n}(z - z_j)^{-\chi_j} \exp\{\Gamma^+(z)\}, z \in D^+, \\ X^-(z) = z^{-\chi} \exp\{\Gamma^-(z)\}, z \in D^-, \end{cases} \tag{2.8.14}$$

where

$$\Gamma^{\pm}(z) = \frac{1}{2\pi i} \int_L \frac{\log[\tau^{-\chi} \prod_{j=1}^{n}(\tau - z_j)^{-\chi_j} G(\tau)]}{\tau - z} d\tau, z \in D^{\pm}.$$

The solvability picture for homogeneous boundary value problem corresponding to (2.8.13) is analogous to that in the previous case:

i) if $\chi < 0$, then no non-trivial analytic solution exists;

ii) if $\chi = 0$, then a unique analytic solution is given by (2.8.6) with $X(z)$ presented by (2.8.14);

iii) if $\chi > 0$, then the general formula of analytic solutions is (2.8.7) with $X(z)$ from (2.8.14), and P_χ being an arbitrary polynomial of the degree not greater than χ.

The results for inhomogeneous problem (2.8.13) are also analogous to those in the case of simply connected domains. The only change that should be done is to replace the canonical function (2.8.5) by a new one delivered by the formula (2.8.14).

2.8. C-LINEAR CONJUGATION PROBLEM.

2.8.3 C-linear conjugation problem on open arcs.

Let us consider now the problem (2.8.1) on a system of open arcs L_j ($L := \bigcup_{j=1}^{n} L_j$, $L_i \cap L_j = \emptyset, i \neq j$), where $L_j := (a_j, b_j)$ are simple open Lyapunov arcs having one-sided tangent lines at the end-points of L_j (for simplicity the end-points are also-called *knots* of the curve L). Let G, g be Hölder-continuous on $cl(L)$, $G(t) \neq 0, t \in cl(L)$.

The solvability of such a problem highly depends on the classes of solutions to be found. Mainly the following possibilities are considered; namely, the solutions can be supposed to be bounded near some knots, or having integrable singularities near some other knots. *Almost bounded* functions, i.e., functions continuous on L outside of end-points, where logarithmic singularities are admitted, are used as a class of solutions.

First consider the homogeneous boundary value problem (2.8.3) for the above introduced type of contour L. The function

$$\Phi_0(z) := \exp\{\Gamma(z)\} := \exp \frac{1}{2\pi i} \int_L \frac{\log[G(\tau)]}{\tau - z} d\tau \qquad (2.8.15)$$

does satisfy the condition (2.8.3) except possibly the knots of L. One can describe the asymptotic behavior of $\Phi_0(z)$ near the knots by the formulae:

$$\begin{cases} \Phi_0(z) = (z - a_j)^{-\frac{\theta_j}{2\pi} + i\frac{\log \rho_j}{2\pi}} \exp\{\Gamma_j(z)\}, \\ \Phi_0(z) = (z - b_j)^{\frac{\theta_j + \Delta_j}{2\pi} - i\frac{\log \rho_j^*}{2\pi}} \exp\{\Gamma_j^*(z)\}, \end{cases} \qquad (2.8.16)$$

where $\rho_j e^{i\theta_j} =: G(a_j+0)$, $\rho_j^* e^{i(\theta_j + \Delta_j)} =: G(b_j-0)$; Γ_j, Γ_j^* are bounded near $z = a_j, z = b_j$ respectively. We choose the branch in an initial point a_j in one of the following ways: a) $-2\pi < \theta_j \leq 0$, if the solution of (2.8.3) is searching to be bounded at the point a_j; b) $0 < \theta_j < 2\pi$, if the solution of (2.8.3) is searching to be integrable at the point a_j. Besides we determine integer numbers $\chi_j := \left[\frac{\theta_j + \Delta_j}{2\pi}\right]$ in the case of a bounded solution at b_j, and $\chi_j := \left[\frac{\theta_j + \Delta_j}{2\pi}\right] + 1$ in the case of an integrable solution at b_j ($[\cdot]$ means the integral part of a real number). The number

$$\chi := \sum_{j=1}^{n} \chi_j$$

is called then the *index* of the problem (2.8.3) in the corresponding class of functions. The results for the homogeneous problem can be described in the following way:

i) if $\chi < 0$, then no non-trivial solution of (2.8.3) in the considered class exists;

ii) if $\chi = 0$, then the unique solution of (2.8.3) in the considered class can be given by the formula (2.8.6) with the function X of the type:

$$X(z) := \prod_{j=1}^{m}(z-b_j)^{-\chi_j}\exp\{\Gamma(z)\}. \qquad (2.8.17)$$

iii) if $\chi > 0$, then all solutions of (2.8.3) in the considered class are presented in (2.8.7) with $X(z)$ of the type (2.8.17).

The results for inhomogeneous problem (2.8.1) in this case are also analogous to those in the case of simply connected domains. The only changes should be done to replace the canonical function (2.8.5) by the new one delivered by formula (2.8.17), and to choose the index in accordance with the considered class.

Remark 2.1 *If the coefficient G has a finite collection of jump-discontinuities at certain finite number of points on L (of the first, second, or third type), then one can consider L as a finite collection of open subarcs (adding the jump-points to the set of knots). This situation is considered for instance in the same way as in the previous subsection (see [91]).*

2.9 Riemann–Hilbert boundary value problem.

Let L be a simple, closed, smooth curve, parametrized by the arc-length parameter $s \in [0, l]$ (i.e., s is the length of a part of L between the points $z(0)$ and $z(s)$ according to the orientation (for definiteness $D^+ = \text{int } L$, $\infty \notin D^+$)). Let a, b, c be given Hölder-continuous functions on L. The Riemann-Hilbert problem consists of finding an analytic function $F(z) := u(z) + iv(z) \in \mathcal{C}_A(D^+)$ in D^+, whose boundary values satisfy the following linear relation on L:

$$a(s)u(s) + b(s)v(s) = c(s), \ 0 \le s \le l. \qquad (2.9.1)$$

If the functions $a(s), b(s)$ have no common zeros on $[0, l]$, then without loss of generality we can suppose that

$$a^2(s) + b^2(s) = 1, \ 0 \le s \le l \qquad (2.9.2)$$

(if not we can divide (2.9.1) on $\sqrt{a^2(s) + b^2(s)}$). Under condition (2.9.2) the boundary relation (2.9.1) can be rewritten in the form

$$\Re\left[\overline{\lambda(s)}F(s)\right] = c(s), \ 0 \le s \le l, \qquad (2.9.3)$$

2.9. RIEMANN–HILBERT BOUNDARY VALUE PROBLEM.

where $\lambda(s) := a(s) + ib(s)$. The winding number of the function λ on L is called *the index of the problem* (2.9.1):
$$\chi := wind_L \, \lambda(t).$$
Let us begin with the study of the homogeneous problem:
$$\Re\left[\overline{\lambda(s)}F(s)\right] = 0, \ 0 \leq s \leq l, \tag{2.9.4}$$
corresponding to (2.9.3). First consider the following auxiliary problem: for a given non-vanishing Hölder-continuous function $\lambda : L \longrightarrow \mathbb{C}$ find a function $p : L \longrightarrow \mathbb{R}_+$ such that the expression $p(s)\lambda(s)$ represents the boundary values of a function $X^+(z)$ analytic and of zero order everywhere in D^+ except possibly the point $z = 0$, and $X^+(z) \sim z^\chi$, $z \to 0$ with χ being the winding number of λ on L. It follows from the statement of the problem that
$$X^+(z) = z^\chi e^{i\gamma(z)} = z^\chi e^{-\beta(z)} e^{i\alpha(z)}, \tag{2.9.5}$$
where $\gamma := \alpha + i\beta$. Boundary condition for X^+ gives immediately
$$\alpha(s) = \Re\gamma(s) = \arg\left\{[t(s)]^{-\chi}\lambda(s)\right\}, \ p(s) = |t(s)|^\chi e^{-\beta(s)}. \tag{2.9.6}$$
The first condition (2.9.6) shows that the function $\gamma(z)$ can be determined as the solution of the Schwarz problem for the domain D^+, i.e.,
$$\gamma(z) = \mathbf{T}\left[\arg \lambda - \chi \arg t(z)\right]. \tag{2.9.7}$$
Knowing $\gamma(z)$ one can find a function $p(s)$ using the second formula (2.9.6). The function $p(s)$ is called a *regularizing factor* of the function λ (in fact in real-valued form, because some other forms for this factor are possible (see, e.g., [91])). It should be noted that $p(s)$ in (2.9.6) is non-vanishing on L. Return now to the boundary condition (2.9.4). Dividing it by the found regularized factor $p(s)$, we can rewrite (2.9.4) (due to (2.9.2)) in the form:
$$\Re\left\{\frac{F(s)}{X^+(t(s))}\right\} = 0, \ 0 \leq s \leq l, \tag{2.9.8}$$

i) if $\chi < 0$, then the only analytic solution is identically zero function;

ii) if $\chi = 0$, then the function in braces in (2.9.8) can be analytically continued onto D^+ (Schwarz problem with identically zero data). The solution of such a problem is a purely imaginary constant. Therefore, the solution of (2.9.4) has the form:
$$F(z) = iCe^{i\gamma(z)}, \ C \in \mathbb{R}; \tag{2.9.9}$$

iii) if $\chi > 0$, then we have to search for a function analytic in D^+ except the pole of the order χ at the origin via the boundary condition (2.9.8). Simple calculations show us (see, e.g., [91, p.270]) that in the case of unit disc $\mathbb{U}$ the solution of such a problem is a rational function of the type

$$Q(z) = iC + \sum_{k=1}^{\chi} \left(c_k z^k - \overline{c_k} z^{-k}\right), \; c_k \in \mathbb{C}; \; C \in \mathbb{R}. \tag{2.9.10}$$

For arbitrary simply connected domain D^+ the solution of the above problem can be obtained from (2.9.10) by using the *Riemann map*

$$w := w(z) : D^+ \to \mathbb{U}, \; w(0) = 0, w'(0) > 0,$$

namely:

$$Q(z) = iC + \sum_{k=1}^{\chi} \left(c_k [w(z)]^k - \overline{c_k}[w(z)]^{-k}\right), \; c_k \in \mathbb{C}; C \in \mathbb{R}. \tag{2.9.11}$$

Then all analytic solutions of the Riemann–Hilbert problem can be delivered by the formula:

$$F(z) = X^+(z)Q(z), \tag{2.9.12}$$

where the function $Q(z)$ is an arbitrary rational function with respect to the Riemann map $f(z)$ of the type (2.9.11). In the case of an inhomogeneous problem we first represent the boundary condition (2.9.1) (or what is equal (2.9.3)) in the form:

$$\Re\left\{\frac{F(s)}{X^+(t(s))}\right\} = |t(s)|^{-\chi} e^{\beta(s)} c(s), \; 0 \le s \le l. \tag{2.9.13}$$

Again three different cases appear: i) if $\chi > 0$, then the solution can be found in the form:

$$F(z) = z^{\chi} e^{i\gamma(z)} \left[\mathbf{T}\left(|t(s)|^{-\chi} e^{\beta(s)} c(s)\right)(z) + iC\right], \; C > 0. \tag{2.9.14}$$

This solution is analytic iff the function in braces has zero of order χ at the origin. If we know the explicit formula for the Schwarz operator, then the analyticity conditions can be presented in the integral form, as in the case of the $\mathbb{C}$-linear conjugation problem);
ii) if $\chi = 0$, then the boundary condition (2.9.13) is the condition for the Schwarz problem. Its solution is given by the formula

$$F(z) = e^{i\gamma(z)} \left[\mathbf{T}\left(e^{\beta(s)} c(s)\right)(z) + iC\right], \; C \in \mathbb{R}; \tag{2.9.15}$$

2.10. ENTIRE FUNCTIONS.

iii) if $\chi > 0$, then the analogous considerations as in the case of the homogeneous problem lead us to the following general formula of solutions:

$$F(z) = z^{\chi} e^{i\gamma(z)} \left[\mathbf{T}\left(|t(s)|^{-\chi} e^{\beta(s)} c(s) \right)(z) + Q(f(z)) \right], \qquad (2.9.16)$$

where Q is an arbitrary rational function of the type (2.9.10), and $f(z)$ is the Riemann map for the domain D^+. There is a deep connection between the Riemann–Hilbert problem and the $\mathbb{C}$-linear conjugation problem (stated in continuous-like form, see remark at the beginning of Section 2.8). In particular, if the contour L is the real line $\mathbb{R}$, or the unit circle $\mathbb{T}$, then one can extend the unknown function $F(z) := F^+(z)$ into a symmetric domain due to the formula

$$F^-(z) := \overline{F^+(z^*)} \qquad (2.9.17)$$

$((\cdot)^*$ denotes the symmetry relation with respect to the contour L). We can rewrite condition (2.9.3) in the form

$$(a - ib)F^+(t) + (a + ib)\overline{F^+(t)} = 2c,$$

or in notation (2.9.17):

$$F^+(t) = -\frac{a(t) + ib(t)}{a(t) - ib(t)} F^-(t) + \frac{2c(t)}{a(t) - ib(t)}, \quad t \in L. \qquad (2.9.18)$$

The latter is the boundary condition of the $\mathbb{C}$-linear conjugation problem.

2.10 Entire functions.

Functions which are analytic in the whole complex plane $\mathbb{C}$ are called *entire functions*. They can be characterized in three different ways (asymptotic behavior, series representation and zeros distribution). *Asymptotic behavior* of an entire function $f = f(z)$ is described mainly in terms of its characteristics. The most important ones are *order, type, indicator function*. Entire functions, which is growing not faster than a power of r are polynomials; the rest are called *transcendental* entire functions. Transcendental entire functions are said to be *of finite order* if there exist positive numbers r_0 and α such that

$$|f(re^{i\theta})| \leq e^{r^\alpha} \text{ for all } r \geq r_0; \qquad (2.10.1)$$

lower bound ρ of such numbers α is called an *order* of the function f (in particular, if for a function f we have $\inf\{\alpha : (2.10.1) \text{ holds}\} = 0$ then f has zero-order). If

(2.10.1) does not hold for any $\alpha > 0$, then the function f has an infinite order. Let us denote
$$M(f,r) := \max_{|z|=r} |f(z)|. \qquad (2.10.2)$$
If $f \neq constant$ then $M(f,r)$ is an increasing function of the parameter r. In terms of $M(f,r)$ one can determine an order ρ of an entire function due to the formula
$$\rho := \limsup_{r \to \infty} \frac{\log \log M(f,r)}{\log r}. \qquad (2.10.3)$$
An entire function f of an order $0 \leq \rho < \infty$ is said to be *of finite type* if there exist positive numbers r_0 and A such that
$$|f(re^{i\theta})| \leq e^{Ar^\rho} \text{ for all } r \geq r_0; \qquad (2.10.4)$$
lower bound σ of such numbers A is called *a type* of the function f (in particular, if for a function f we have $\inf \{A : (2.10.4) \text{ holds}\} = 0$ then f has *minimal type* at the order ρ). If (2.10.4) does not hold for any $A > 0$, then the function f has a *maximal type* at the order ρ. In terms of the function $M(f,r)$ the type σ of an entire function f is determined by the formula
$$\sigma := \limsup_{r \to \infty} \frac{\log M(f,r)}{r^\rho}. \qquad (2.10.5)$$
Inequalities (2.10.1), (2.10.4) describe the asymptotics of an entire function f in the following way: the growth of f is not faster than $e^{(\sigma+\varepsilon)r^\rho}$, and, inversely, there exists a sequence of positive numbers (r_n) such that
$$M(f,r_n) \geq e^{(\sigma-\varepsilon)r^\rho}, \; n \geq n_0.$$
This description is often expressed in the following symbolic way:
$$e^{(\sigma-\varepsilon)r^\rho} \overset{n}{\lesssim} |f(re^{i\theta})| \overset{as}{\lesssim} e^{(\sigma+\varepsilon)r^\rho}. \qquad (2.10.6)$$
In order to describe behavior of an entire function f along rays $\{z \in \mathbb{C} : \arg z = \theta\}$ one can use *an indicator function*
$$h_f(\theta) = \limsup_{r \to \infty} \frac{\log |f(re^{i\theta})|}{r^\rho}, \qquad (2.10.7)$$
or in symbols:
$$e^{(h_f(\theta)-\varepsilon)r^\rho} \overset{n}{\lesssim} |f(re^{i\theta})| \overset{as}{\lesssim} e^{(h_f(\theta)+\varepsilon)r^\rho}.$$

2.10. ENTIRE FUNCTIONS.

The basic property of the indicator function is its ρ-trigonometric convexity (see, e.g., [139], [140]), i.e.,

$$\begin{vmatrix} \sin \rho\theta_1 & \sin \rho\theta_2 & \sin \rho\theta_3 \\ \cos \rho\theta_1 & \cos \rho\theta_2 & \cos \rho\theta_3 \\ h_f(\theta_1) & h_f(\theta_2) & h_f(\theta_3) \end{vmatrix} \geq 0.$$

As entire functions are analytic in the whole complex plane $\mathbb{C}$ their *power series*

$$f(z) = \sum_{k=1}^{\infty} a_k z^k \qquad (2.10.8)$$

are converging everywhere, i.e.,

$$\lim_{k \to \infty} \sqrt[k]{|a_k|} = 0. \qquad (2.10.9)$$

The asymptotic behavior at $k \to \infty$ of the coefficients of series representation (2.10.8) for an entire function f is connected with its asymptotics at $|z| \to \infty$; namely, the following formulae are valid for an order ρ and a type σ (cf., (2.10.3), (2.10.5))

$$\rho = \limsup_{k \to \infty} \frac{\log \frac{1}{|a_k|}}{k}, \qquad (2.10.10)$$

$$(\rho e \sigma)^{\frac{1}{\rho}} = \limsup_{k \to \infty} k \sqrt[k]{|a_k|}. \qquad (2.10.11)$$

An entire function represented in the form of power series (2.10.8) can be characterized also in terms of the following quantities:

$$\mu_f(r) := \max_k |a_k| r^k; \quad \nu_f(r) := \{n : \mu_f(r) = |a_n| r^n\}, \qquad (2.10.12)$$

which are called *maximal term* and *central index* respectively of the power series (2.10.8). For an entire function of finite order the following inequalities take place:

$$\mu_f(r) \leq M_f(r) \leq r_f^{\rho+\varepsilon} \mu_f(r), \quad \forall \varepsilon > 0, \, r > r_0(\varepsilon). \qquad (2.10.13)$$

Asymptotic behavior of entire functions is closely related to their *zeros distribution*. If (z_n) is a sequence of zeros of an entire function f of finite order ρ, then the latter can be represented in the following form:

$$f(z) = e^{h(z)} \prod_{n=1}^{\infty} G\left(\frac{z}{z_n}, [\rho]\right) e^{q_0 + q_1 z + \ldots + q_{[\rho]} z^{[\rho]}}, \qquad (2.10.14)$$

where
$$G(w,p) := \begin{cases} (1-w), & \text{if } p = 0, \\ (1-w)e^{w+\frac{w^2}{2}+\cdots+\frac{w^p}{p}}, & \text{if } p > 0, \end{cases}$$

is a prime Weierstrass factor, h is a polynomial of an order not greater than $[\rho]$, and $[\cdot]$ denotes an integer part of a number. The sequence of zeros (z_n) is finite or denumerable (with accumulating point at infinity in the last case). Its behavior is characterized first by a *counting function* $n_f(t)$, which is nothing more than the quantity of points z_n (with their multiplicity) in the disc $\{z \in \mathbb{C} : |z| \leq t\}$; namely it can be introduced as an *exponent of convergence* ρ_1 of a sequence (z_n):

$$\rho_1 := \limsup_{t \to \infty} \frac{\log_+ n_f(t)}{\log t}, \qquad (2.10.15)$$

$(\log_+ x = \begin{cases} \log x, & \text{if } x > 1, \\ 0, & \text{if } 0 \leq x \leq 1, \end{cases})$ which in fact coincides with the following quantity:

$$\rho_1 = \inf\left\{\lambda > 0 : \int_0^\infty \frac{n_f(t)dt}{t^{\lambda+1}} < +\infty\right\}.$$

It can be also introduced as a *density* Δ of the sequence (z_n) at the exponent ρ_1, which is analogous to the type of an entire function:

$$\Delta := \limsup_{t \to \infty} \frac{n_f(t)}{t^{\rho_1}}. \qquad (2.10.16)$$

The classic Hadamard's inequality $n_f(r) \leq \log M(f, er), r > 0$ gives immediately that $\rho_1 \leq \rho$, and if $\rho_1 = \rho$, then $\Delta \leq \sigma e^\rho$. For *canonical Weierstrass products*, i.e., for the functions represented in the form

$$f(z) = \prod_{n=1}^\infty G\left(\frac{z}{z_n}, p\right) e^{q_0 + q_1 z + \cdots + q_p z^p},$$

where p is such that

$$\sum_{n=1}^\infty |z_n|^{-p} = \infty, \sum_{n=1}^\infty |z_n|^{-p-1} < \infty,$$

the following *Borel–Valiron inequality* takes place which is in a certain sense the inverse to that by Hadamard:

$$\log M(f, r) \leq C_p \left\{ r^p \int_0^r \frac{n_f(t)dt}{t^{p+1}} + r^{p+1} \int_r^\infty \frac{n_f(t)dt}{t^{p+2}} \right\}.$$

We formulate here the results as for inhomogeneous problem (2.8.1), as for corresponding homogeneous problem

$$\Phi^+(t) = G(t)\Phi^-(t), \quad t \in L. \tag{2.8.3}$$

The basic constructive elements of the solution are two *factorizations* of a special type. First we have to represent the coefficient $G(t)$ in the form

$$G(t) = \frac{X^+(t)}{X^-(t)}, \tag{2.8.4}$$

where the functions $X^+(z), X^-(z)$ are analytic and of zero order in the domains $D^+, D^-\backslash\{\infty\}$ respectively. These functions, or the functions

$$X_0^+(z) := X^+(z), \quad X_0^+(z) := z^\chi X^-(z),$$

which solve the factorization problem:

$$G(t) = t^\chi \frac{X_0^+(t)}{X_0^-(t)},$$

and are analytic and of zero order everywhere in D^+, D^- respectively, are often called *canonical functions of the homogeneous problem*. The solution of the problem (2.8.4) has the form:

$$\begin{cases} X^+(z) = \exp\left\{\frac{1}{2\pi i} \int_L \frac{\log[\tau^{-\chi}G(\tau)]}{\tau - z} d\tau\right\} =: \exp\{\Gamma^+(z)\}, z \in D^+, \\ X^-(z) = z^{-\chi}\exp\left\{\frac{1}{2\pi i} \int_L \frac{\log[\tau^{-\chi}G(\tau)]}{\tau - z} d\tau\right\} =: z^{-\chi}\exp\{\Gamma^-(z)\}, z \in D^-. \end{cases} \tag{2.8.5}$$

Representing condition (2.8.3) in the form

$$\frac{\Phi^+(t)}{X_0^+(t)} = t^\chi \frac{\Phi^-(t)}{X_0^-(t)}, t \in L,$$

one can get the following results for the homogeneous problem on the basis of the Analytic Continuation Principle:

    i) the problem (2.8.3) has no non-trivial (i.e., $\equiv 0$) analytic solution if $\chi < 0$;
    ii) if $\chi = 0$, problem (2.8.3) has one-parametric family of analytic solutions

$$\Phi^\pm(z) = CX^\pm(z) = CX_0^\pm(z), \tag{2.8.6}$$

2.10. ENTIRE FUNCTIONS.

This inequality gives inverse results for comparison of ρ and ρ_1, as well as of σ and Δ. It should be noted that for entire functions of an integer order, an exponent of convergence of the sequence of zeros can be smaller than an order of entire function, and even if they coincide, a density can be of another type (e.g., the function can have $\Delta = 0$, but be of the maximal type ($\sigma = \infty$)). A more complicated distribution of zeros in their arguments is difficult to describe for general entire functions. B. Ya. Levin and A. Pfluger found the class of entire functions (called the class of functions of *completely regular growth*) in which the distribution of zeros in their arguments is closely related to the indicator function (see, e.g., [139], [140], [110] .) We do not present these results here because little from this theory will be used in our further considerations. Lastly we have to introduce the following notion: a function f will be called an *entire function with respect to a point* $a \in \mathbb{C}$ if $f(\frac{1}{z-a})$ is an entire function in the standard sense. The following collection of theorems is known as the Phragmen–Lindelöf type theorems. They play the role of maximum modulus principle for analytic functions in unbounded domains.

Theorem 2.9 *Let G be a simply connected domain, $z = 0, z = \infty \in \partial G$. Let S_ρ be a section of G by the circle $\mathbb{T}(0, \rho)$, and $s(\rho)$ is its length. If f is single-valued analytic in G, continuous up to $\partial G \setminus \{\infty\}$, and satisfies the conditions*

$$|f(z)| \leq 1, z \in \partial G; \quad \lim_{\rho \to \infty} \frac{\log M_f(\rho)}{\sigma(\rho)} = 0,$$

where $M_f(\rho) := \max_{z \in S_\rho} |f(z)|$, $\sigma(\rho) = \exp\left\{\pi \int_1^\rho \frac{du}{s(u)}\right\}$, then $|f(z)| \leq 1, z \in G$.

The next results are consequences of the above theorems (see, e.g., [82]).

Theorem 2.10 *Let a function f be single-valued analytic in a wedge domain G of an angle $\frac{\pi}{\alpha}$ and continuous up to its sides. If $|F(z)| \leq M$ on ∂G, then $|F(z)| \leq M$ on G, or*

$$\log \max_{|z|=\rho} |f(z)| > C^\alpha \rho^\alpha, \rho > \rho_0,$$

for certain $C > 0$.

Theorem 2.11 *Let a function f be single-valued analytic in the half-plane $\Re z > 0$, continuous in $\Re z \geq 0$, and satisfy the inequality*

$$\log |f(z)| < -\nu(|z|), \Re z \geq 0,$$

where $\nu(t)$ is a continuous positive function for $t \geq 0$. If

$$\int_1^\infty \frac{\nu(t)}{t^2} dt = +\infty,$$

then $f(z) \equiv 0$.

Theorem 2.12 *Let a function f be single-valued analytic in a strip $|\Im z| < \frac{\pi}{2}$, continuous in $|\Im z| \leq \frac{\pi}{2}$, and satisfy the inequality*

$$\log|f(x+iy)| < -\nu(x), \quad -\frac{\pi}{2} < y < \frac{\pi}{2},$$

where $\nu(x)$ is a continuous positive function. If $\int_0^\infty \nu(x) e^{-x} dx = +\infty$, then $f(z) \equiv 0$.

2.11 Conformal mappings.

Let $f : D \to \widehat{\mathbb{C}}$ be a continuous function of complex variable z defined in a domain $D \subset \widehat{\mathbb{C}}$. It will be said following [247] that it is to be determined a *conformal mapping* from D onto $f(D)$ if it is injective (i.e., $f^{-1}(w_1) \neq f^{-1}(w_2)$ for any two different points $w_1, w_2 \in f(D)$, and "preserves an angles" (i.e., if γ_1, γ_2 are two smooth paths beginning in a point $z_0 \in D$, then an angle between them is equal to an angle between their images $f(\gamma_1), f(\gamma_2)$). The conformal mappings are often identified with the mappings determined by univalent analytic (or meromorphic) functions. Let us present the main properties of conformal mappings. By definition conformal mapping gives 1-1 correspondence between f and $f(D)$; moreover it is a homeomorphism of these domains, i.e., $f(D)$ is open and both functions f, f^{-1} are continuous in $D, f(D)$, respectively. If the boundaries $\partial D, \partial f(D)$ of f and $f(D)$, respectively, are simple Jordan curves then by Caratheodory's theorem (see, e.g., [102]) it is continued up to the homeomorphism of the closed domains (i.e., $cl\, D$ onto $cl\, f(D)$). Besides, this continuation remains the smoothness class of boundaries. It means that if for instance $\partial D, \partial f(D)$ are of a class $C^{m,\alpha}, 0 < \alpha < 1, n \in \mathbb{N} \cup \{0\}$, then $f_{|\partial D}$ also is. The main result concerning conformal mapping is the well known *Riemann mapping theorem*, which states the following: Let D be an arbitrary simply connected domain on the complex plane C, whose boundary consists of more than one point. Then there exists an analytic function f conformally mapping this domain onto the unit disc. This function is unique if the following conditions are satisfied: $f(z_0) = 0$ for a certain

2.11. CONFORMAL MAPPINGS.

point $z_0 \in D$, and $f'(z_0) > 0$. If a domain D is encircled by an analytic Jordan curve then the conformal mapping $w = f(z)$ of this domain onto the unit disc is continued up to ∂D as the differentiable function and $f'(z) \neq 0, z \in \partial D$. The behavior of conformal mappings near a boundary is described by the theorems by Alhfors and Warshawski (see e.g., [86]). Analogs of the Riemann Mapping Theorem for multiply connected domain are much more complicated (see, e.g., [102], [117]). There are many types of canonical domains onto which a multiply connected domain can be mapped. In the present book we use the circular domain

$$\mathbb{D} := \left\{ z \in \widehat{C} : |z - a_k| > r_k \text{ for all } k = 1, 2, ..., n \right\}$$

as the canonical domain. Here the discs $|z - a_k| \leq r_k$ are mutually disjoint.

Theorem 2.13 *[102, p. 235] Let D be an arbitrary multiply connected domain. Then there exists a circular domain $\mathbb{D}$ and an analytic function f conformally mapping D onto $\mathbb{D}$. This function is unique if a fixed point $a \in D$ is mapped onto infinity, and the principal part of f at infinity has the form*

$$\frac{1}{z-a} + f_1(z-a) + \ldots \text{ if } a \in \mathbb{C}, \text{ and } z + \frac{f_1}{z} + \ldots \text{ if } a = \infty.$$

Theorem 2.14 *[102, p. 232] Each conformal mapping of a circular multiply connected domain onto another circular multiply connected domain is a Möbius transformation.*

As is well known [157], if for a certain simply connected domain D we know a solution of the Dirichlet problem for the Laplace equation, then it is possible to deduce the function conformally transforming this domain onto the unit disc. Let $w(z)$ be the unknown conformal mapping and $w(z_0) = 0$, where the point $z_0 = 0$ belongs to D. The function $w(z)$ has to satisfy the boundary condition $|w(z)| = 1$, $t \in \partial D$, where ∂D is the boundary of D. Let us introduce the auxiliary function $\varphi(z) = \log \frac{1}{z-z_0} w(z)$ which is analytic and univalent in D, and continuous in $\overline{D}$. The branch of the logarithm is fixed in an arbitrary way. The function $w(z)$ satisfies the boundary condition $\Re\varphi(t) = -\log|t - z_0|$, $t \in \partial D$. The latter problem has a unique solution up to an arbitrary additive pure imaginary constant $i\gamma$ [157]. Then the conformal mapping $w(z)$ is determined up to the multiplier $\exp i\gamma$, which corresponds to a rotation of the unit disc.

2.12 $\mathbb{R}$-linear problem and its applications

We discuss the conductivity of the plane composite material, when the complex plane $z = x + iy$ is divided into domains D^+ and D^- occupied by materials of unit and $\lambda_k > 0$ conductivity, respectively (Figure 2.1). The domains D_k are called inclusions. The potentials $u(x,y)$ and $u_k(x,y)$ are harmonic in D^+ and D_k [1], [236]. We use a complex variable instead of real ones in the harmonic functions. For instance, we identify the functions $u(x,y)$ and $u(z)$. The external field can be modeled by the classic Dirichlet or Neumann condition

$$u = g \text{ on } L_0; \text{ or } \frac{\partial u}{\partial n} = g \text{ on } L_0,$$

where $\partial/\partial n$ is the outward normal derivative. The perfect contact between materials is modeled by the boundary conditions

$$u = u_k, \quad \frac{\partial u}{\partial n} = \lambda_k \frac{\partial u_k}{\partial n} \text{ on } L_k. \quad (2.12.1)$$

Let us discuss (2.12.1) in the language of the steady heat conduction, when the potential means the temperature distribution and the normal derivative multiplied by conductivity determines the heat flux normal to a curve. In this case the relations (2.12.1) provide transversal to L_k continuity of the temperature and normal flux. Thus (2.12.1) means the perfect thermal contact between different materials. The same formalism is applicable to many other physical phenomena, such as electrical conductivity, magnetic permeability, flow in porous media [1], [6], [236]. We reduce (2.12.1) to a complex conjugate condition. Let us fix a curve L_k. The normal and tangent derivatives can be written in the form

$$\frac{\partial}{\partial n} = n_1 \frac{\partial}{\partial x} + n_2 \frac{\partial}{\partial y}; \quad \frac{\partial}{\partial s} = -n_2 \frac{\partial}{\partial x} + n_1 \frac{\partial}{\partial y},$$

where $\mathbf{n} = (n_1, n_2)$ is the normal vector to L_k. Applying the operator $\partial/\partial s$ to the first relation (2.12.1) we obtain

$$-n_2 \frac{\partial u}{\partial x} + n_1 \frac{\partial u}{\partial y} = -n_2 \frac{\partial u_k}{\partial x} + n_1 \frac{\partial u_k}{\partial y}. \quad (2.12.2)$$

The second relation (2.12.1) can be written in the form

$$n_1 \frac{\partial u}{\partial x} + n_2 \frac{\partial u}{\partial y} = \lambda_k n_1 \frac{\partial u_k}{\partial x} + \lambda_k n_2 \frac{\partial u_k}{\partial y}. \quad (2.12.3)$$

2.12. R-LINEAR PROBLEM AND ITS APPLICATIONS

Let us introduce the complex potentials

$$\psi^+(z) := \frac{\partial u}{\partial x} - i\frac{\partial u}{\partial y} \text{ and } \psi^-(z) := \frac{\lambda_k + 1}{2} \left(\frac{\partial u_k}{\partial x} - i\frac{\partial u_k}{\partial y} \right) \qquad (2.12.4)$$

analytic in D^+ and D^-, respectively, and continuous in the closures of the domains considered. Substituting

$$\frac{\partial u_k}{\partial x} = \frac{1}{\lambda_1 + 1}(\psi^- + \overline{\psi^-}), \quad \frac{\partial u_k}{\partial y} = \frac{i}{\lambda_1 + 1}(\psi^- - \overline{\psi^-}),$$

$$\frac{\partial u}{\partial x} = \frac{1}{2}(\psi^+ + \overline{\psi^+}), \quad \frac{\partial u}{\partial y} = \frac{i}{2}(\psi^+ - \overline{\psi^+})$$

into (2.12.2), (2.12.3) we obtain the following conjugation condition

$$\psi^+(t) = \psi^-(t) + \rho_k \overline{\mathbf{n}^2 \psi^-(t)}, \ t \in L_k, \qquad (2.12.5)$$

where we identify $\mathbf{n}$ with the complex number $n_1 + in_2$, $\rho_k := (\lambda_k - 1)(\lambda_k + 1)^{-1}$. We obtain a conjugation condition on the derivatives (2.12.4) of the complex potentials (complex velocity in [1]). Integrating (2.12.5) by s along L_k we obtain

$$\phi^+(t) = \phi^-(t) - \rho_k \overline{\phi^-(t)}, \ t \in L_k. \qquad (2.12.6)$$

We put the constant of integration to be zero, because any potential is determined up to an additive constant which can be included in $\phi^-(z)$. Note that the function $\phi^-(z)$ is single-valued in each simply connected domain D_k. Equality (2.12.6) immediately implies that $\phi^+(z)$ has zero periods along L_k; hence it is single-valued in D^+. Let us consider now the unbounded multiply connected domain displayed on Figure 2.1. Given Hölder continuous functions $a(t)$, $b(t)$ and $c(t)$ on $L = \cup_{k=1}^n L_k$, $a(t) \neq 0$, find functions $\phi^{\pm}(t) \in \mathcal{C}_A(D^+)$ with the boundary condition

$$\phi^+(t) = a(t)\phi^-(t) + b(t)\overline{\phi^-(t)} + c(t), \ t \in L. \qquad (2.12.7)$$

If $b(t) \equiv 0$ then (2.12.6) becomes the $\mathbb{C}$-linear conjugation problem (2.8.1). One can see that $\phi^{\pm}(t)$ are related by an $\mathbb{R}$-linear transformation $Z \to aZ + b\overline{Z} + c$. That is why we call (2.12.6) by $\mathbb{R}$-linear conjugation problem. The term $\mathbb{R}$-linear conjugation condition is also used in appropriate case. The conditions of the perfect contact (2.12.5) and (2.12.6) refer to the $\mathbb{R}$-linear conjugation condition (2.12.7).

Theorem 2.15 *[43], [156]. Let $\mathrm{wind}_L a(t) = 0$ and $\sup_L \left| \frac{a(t)}{b(t)} \right| < 1$ in the problem (2.12.7). If we additionally assume that $\phi^+(\infty) = 0$, then the problem (2.12.7) has a unique solution.*

Corollary 2.3 *Let $a(t) = a_k$, $c(t) = \gamma_k$ and $b(t) = -\rho_k$ be constants on each L_k, $|a_k| = 1$, $|\rho_k| < 1$. Then the problem (2.12.7) has only constant solutions.*

Let us note that without loss of generality we can take $a_k = 1$. Then the problem becomes
$$\phi^+(t) = \phi^-(t) - \rho_k \overline{\phi^-(t)} + \gamma_k, \ t \in L_k, \ k = 1, 2, ..., n. \quad (2.12.8)$$
Make the change $\omega^+(z) = \phi^+(z) - \phi^+(\infty)$ in (2.12.8). Then the problem becomes
$$\omega^+(t) = \phi^-(t) - \rho_k \overline{\phi^-(t)} + \delta_k, \ t \in L_k, \ k = 1, 2, ..., n, \ \omega^+(\infty) = 0, \quad (2.12.9)$$
where $\delta_k := \gamma_k - \phi^+(\infty)$. According to Theorem 2.15 the problem (2.12.9) has a unique solution. We can present this solution
$$\omega^+(z) = 0, \ \phi^-(z) = -\frac{\delta_k + \rho_k \overline{\delta_k}}{1 - \rho_k^2}, \ z \in D_k, \ k = 1, 2, ..., n.$$
Therefore, $\phi^+(z) = \phi^+(\infty)$ and $\phi^-(z)$ are constants. Along similar lines we can consider mixed problems for domains displayed on Figure 2.1, when the condition (2.12.7) is given on L_k, $k = 1, 2, ..., n$ and one of the conditions (2.8.1) or (2.9.3) is given on L_0. If the conductivity of the inclusions D_k tends to infinity, then (2.12.1) becomes the modified Dirichlet problem $u = u_k$ on L_k, where u_k is an undetermined constant. If the inclusion D_k is an isolator, then $\lambda_k = 0$ and we arrive at the Neumann problem $\frac{\partial u}{\partial n} = 0$ on L_k. These cases correspond to $\rho_k = 1$ and $\rho_k = -1$. For instance for $\rho_k = 1$, (2.12.5) becomes
$$\psi^+(t) = \psi^-(t) + \overline{n^2 \psi^-(t)}, \ t \in L_k. \quad (2.12.10)$$
Integrating (2.12.10) we obtain
$$\phi^+(t) = \phi^-(t) - \overline{\phi^-(t)} + c_k, \ t \in L_k, \quad (2.12.11)$$
where c_k are the constants of integration. Here we cannot take $c_k = 0$, since we can include only an imaginary part of c_k into $\phi^-(z)$. Hence, we assume that c_k are undetermined real constants. Calculating the real part of (2.12.11) we arrive at the modified Dirichlet condition (see Section 2.7.2)
$$\Re \phi^+(t) = c_k, \ t \in L_k. \quad (2.12.12)$$
Conversely, let ϕ^+ in (2.12.12) be given. Introduce the functions $\phi_k(z) \in \mathcal{H}_A(D_k)$ satisfying the problems
$$2\Im \phi_k(t) = \Im \phi^+(t), \ t \in L_k, \quad (2.12.13)$$
for each $k = 1, 2, ..., n$. Then the functions $\phi^+(z)$, $\phi^-(z) := \phi_k(t)$ in D_k ($k = 1, 2, ..., n$) satisfy (2.12.11). Therefore, we establish the equivalence of the conditions (2.12.11) and (2.12.12), (2.12.13).

2.13 Notes and Comments.

2.1.° Most of the definitions here are standard. Some non-standard notations (in particular special type of orientation for multiply connected domains) are given in order to simplify the presentation of certain results in the main text.

2.2.° The described collection of the functional spaces is more or less natural for the books connected with boundary value problems for analytic functions. In the description of the spaces $\mathcal{C}_+(L)$, $\mathcal{C}_+(D)$, etc. we follow the ideas presented in [95], [109]. The notations and results concerning some spaces of Lebesgue integrable functions and Hardy-like spaces seem to be far out of our main ideas. The special notations for the subspaces of analytic functions with the prescribed number of zeros in domains are due to Yu. V. Obnosov (see, e.g., [217]). Generalized Hölder spaces were studied by different authors (see, e.g., [38], [39], [106]).

2.3.° We follow here the standard description presented in [114], [128] (see also [292]).

2.4.° Both theorems are taken from the book by S. Axler et al. ([26]) (see also [146]).

2.5.° Different questions of harmonic analysis are described now from the position of Hardy-like spaces as one can see in the now classic books by K. Hoffman [109], P. Duren [65] As our main text is mostly connected with the smooth functions, therefore we try to collect also the results concerning such functions. Not all of them are clearly formulated in the literature. The role of the spaces VMO, $VMOA$, as well as acting properties of Hilbert transform is presented in the book by K. Zhu ([292]).

2.6.° The Cauchy integral, Cauchy type integral, and singular integral with Chauchy kernel are among the subjects in the books by F. D. Gakhov ([91]) and N .I. Muskhelishvili ([205]). The exact norm of Hilbert transform (hence of singular integral over the unit circle) was calculated by S. K. Pichorides ([224]) for the $\mathcal{L}_p$-spaces, and by A. B. Aleksandrov ([19]) for the $\mathcal{H}^\alpha$-spaces. Norm estimate for a singular integral operator acting between two Hölder-spaces was obtained by M. Kh. Brenerman and B. A. Kac ([45]). The results of Hardy–Littlewood type in generalized Hölder-spaces can be found in [106].

2.7.° In the description of the Schwarz operator for the simply connected domain we follow the standard approach (see, e.g., [1], [91]). The most straightforward and understandable presentation of the results for multiply connected domains presented in the book by S. G. Mikhlin ([157]) is taken as the basis of Subsection 2.7.2. We and Mikhlin consider different types of the domain D^+ : Figure 2.1 and Figure 2.1. The structure of the Schwarz operator is derived in [188]. The multi- and single-valued components of the Schwarz operator are written in explicit form (see Section 4.4).

Villat's formula of the Schwarz operator for circular annulus is given, e.g., in [9].

2.8.° ℂ-linear conjugation problem (or Riemann boundary value problem) is presented in the classic way as given in the book by F. D. Gakhov ([91]). The problem of determination of an analytic function in a domain via boundary values of its real part is often called the *Schwarz problem* (see, e.g., [91]).

2.9.° The described results are now classic and given for instance in [91]. Some generalizations of the Riemann–Hilbert problem are discussed also in the books by E. Meister [155] (background description and applications), by E. Wegert [276] (linear and nonlinear case on the basis of geometric approach), by I. N. Vekua [268] (Riemann–Hilbert problem for generalized analytic functions), etc.

2.10.° This subsection contains the main part of the entire functions theory. The results are taken from the classic books by B. Ja. Levin [139], by R. P. Boas [41] (see also [82], [140]).

2.11.° The book by G. M. Goluzin [102] is now a standard reference text for geometric function theory. The results concerning conformal mapping of multiply connected domains goes back to the article by M. V. Keldysh [117]

2.12.° The ℝ-linear conjugation problem was studied in two non-overlapping approaches: the general pure mathematical problem (2.12.7) and applied problem (2.12.1) which is equivalent to (2.12.6) (see Chapter 1). From some sources it is known also as the *general conjugation problem, generalized conjugation problem*, or *Markushevich's problem*. The history of the names of the problem (2.12.7) is as long as the proper history of the problem. V.M. and others used many names until E. I. Zverovich proposed the names "complex conjugation problem" for (2.8.1) and "real conjugation problem" for (2.12.7). Later V.M. modified these names as it is in the titles of Section 2.8 and 2.12. Anyway, the term ℝ-linear conjugation problem (ℝ -linear problem) is now widely used. Relations between the ℝ-linear problem (2.12.7) with $|a(t)| \equiv |b(t)|$ and the Riemann–Hilbert problem are studied by L. G. Mikhajlov [156]. Equivalence (2.12.11) and (2.12.12), (2.12.13) is established in [166]. The Nöther theory and the estimates of the defect numbers of the ℝ- linear conjugation problem were studied by B. Bojarski [43], by L. G. Mikhailov [156], and by G. S. Litvinchuk and I. M. Spitkovskii [145]. The substantial survey on the subject is presented in [144], [146]. The constructive approach related to applications is discussed in Chapter 5.

Chapter 3

Nonlinear Boundary Value Problems. Analytic function methods.

A number of constructive methods applied to the nonlinear boundary problems for analytic and harmonic functions are presented here. In order to describe the ideas "out of technical noise" we frequently restrict our study to the problems stated on the standard domains (in particular, on discs or half-planes). We have to note the following feature of some of the problems under consideration: we are looking for the solutions analytic in domains, continuous up to their boundaries, but taking the given data with additional smoothness (say Hölder-continuous). Therefore (cf. Pumping Principle, Section 2.3) any continuous solution becomes in most cases in fact Hölder-continuous.

3.1 General nonlinear conjugation problem of power type.

Let us consider the nonlinear boundary value problem which is a natural generalization of the $\mathbb{C}$-linear conjugation problem (cf. Section 2.8); namely, let L be a simple closed smooth curve on the complex plane $\mathbb{C}$, divided into two domains $D^+ \ni 0$ and $D^- \ni \infty$; the problem is in searching for two functions Φ^+ and Φ^-, $(\Phi^+, \Phi^-) \in \mathcal{C}_A(D^+) \times \mathcal{C}_A(D^-)$, satisfying the boundary condition:

$$\left(\Phi^+(t)\right)^\alpha = G(t)\left(\Phi^-(t)\right)^\beta + g(t), \ t \in L, \qquad (3.1.1)$$

where $G(t) \neq 0$, $g(t)$ are given Hölder-continuous functions, and $\alpha \neq 0$, $\beta \neq 0$ are given comlex numbers.

3.1.1 Homogeneous problem

Let us begin with a homogeneous problem corresponding to (3.1.1), i.e., $g(t) \equiv 0$:

$$\left(\Phi^+(t)\right)^\alpha = G(t)\left(\Phi^-(t)\right)^\beta, \quad t \in L. \tag{3.1.2}$$

First we consider problem (3.1.2) only in the subclass $\mathcal{C}_\mathcal{A}^0$ of functions from $\mathcal{C}_\mathcal{A}(D^+) \times \mathcal{C}_\mathcal{A}(D^-)$ (cf. Section 2.2) with nonvanishing boundary functions, i.e., $\Phi^+(t) \neq 0$, $\Phi^-(t) \neq 0$, for all $t \in L$. Note that the case of exponents α, β of different signes (i.e., $\alpha\beta < 0$) is closer to the problem of product type which is considered in the next section. If both exponents are negative then problem (3.1.2) is equivalent to the problem with positive exponents $-\alpha, -\beta$ but with coefficient G^{-1}. Therefore we can study here only the case of positive exponents. Let us distinguish some different cases for the values of exponent. The most simple one is

a) $\alpha = m, \beta = n$ are positive integer numbers. After changing of unknown functions

$$\phi^+(z) = \left(\Phi^+(z)\right)^m, \phi^-(z) = \left(\Phi^-(z)\right)^n$$

one get the equivalent homogeneous $\mathbb{C}$-linear conjugation problem

$$\phi^+(t) = G(t)\phi^-(t), t \in L, \tag{3.1.3}$$

$(\phi^+, \phi^-) \in \mathcal{A}_{m,n}$ (see Section 2.2). The solvability of the latter depends on its *index* (cf. Section 2.8):

$$\chi := wind_L\, G. \tag{3.1.4}$$

i) $\chi < 0$. In this case problem (3.1.3) has no analytic solution, hence no solution in $\mathcal{A}_{m,n}$. Therefore problem (3.1.2) with $\chi < 0$ has no solution in the $\mathcal{C}_\mathcal{A}^0$.

ii) $\chi = 0$. The unique solution of problem (3.1.3) has the following form:

$$\phi^\pm(z) = C\exp\{\frac{1}{2\pi i}\int_L \frac{\log G(\tau)}{\tau - z}d\tau\}, z \in D^\pm, \tag{3.1.5}$$

where $C \neq 0$ is an arbitrary complex constant. This solution has no zero in $D^\pm$; hence it is also the solution in $\mathcal{A}_{m,n}^{0,0}$. The corresponding solution of problem (3.1.2) can be

3.1. CONJUGATION PROBLEM OF POWER TYPE.

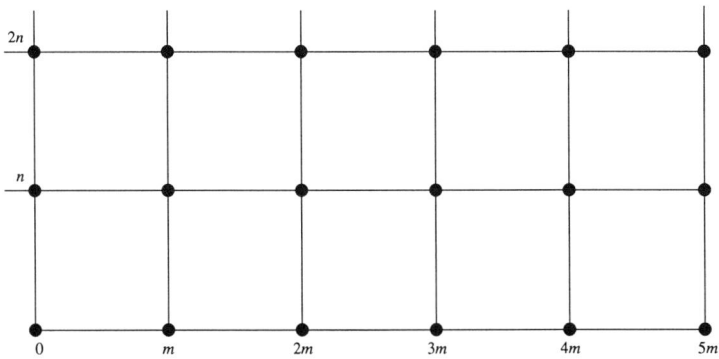

Figure 3.1: The cell of solvability

given by the formula:

$$\begin{cases} \Phi^+(z) = C^{1/m} \exp\{\frac{1}{2\pi i m} \int_L \frac{\log G(\tau)}{\tau - z} d\tau\}, & z \in D^+, \\ \Phi^-(z) = C^{1/n} \exp\{\frac{1}{2\pi i n} \int_L \frac{\log G(\tau)}{\tau - z} d\tau\}, & z \in D^-. \end{cases} \quad (3.1.6)$$

In the right-hand sides one can take any single-valued branch of corresponding (in general multi-valued) functions. Hence the general solution of problem (3.1.2) depends on at most mn complex constants. It can be smaller if the numbers m, n have a common divisor $\neq 1$.

iii) $\chi > 0$. In this case problem (3.1.3) has solutions in the class $\mathcal{A}_{m,n}$ iff the following equation is solvable with respect to non-negative integer numbers N^+, N^- (it is a direct consequence of an Argument Principle (see, e.g., [240])):

$$\chi = mN^+ + nN^-. \quad (3.1.7)$$

If the equation (3.1.7) has a solution, then the corresponding numbers N^+, N^- are the quantities of zeros of the solutions $\Phi^+(z)$, $\Phi^-(z)$ for problem (3.1.1) in the domains D^+, D^- respectively (including a zero at infinity if exists). To understand the meaning of solvability of (3.1.7) one can imagine a cell in the first quadrant with the vortices at the points with coordinates (km, ln), $k, l \in \mathbb{N}_0$. To each point correspond a "distance" $\rho = km + ln$. Hence, equation (3.1.7) is solvable if the value of χ coincides with the

CHAPTER 3. NONLINEAR BOUNDARY VALUE PROBLEMS

distance to at least one of the points of the cell. If so then all solutions of problem (3.1.3) can be represented in the following form:

$$\begin{cases} \phi^+(z) = \left(\prod_{j=1}^{k}(z-z_j^+)\right)^m \phi_1^+(z), & z \in D^+, \\ \phi^-(z) = z^{-l_0 n}\left(\prod_{j=1}^{l_1}(1-\tfrac{z_j^-}{z})\right)^n \phi_1^-(z), & z \in D^-, l = l_0 + l_1, \end{cases} \quad (3.1.8)$$

where $(\phi^+, \phi^-) \in \mathcal{A}^{0,0}$ (i.e., are analytic and nonvanishing in D^+, D^- respectively, continuous up to the boundary), with ϕ_1^+, ϕ_1^- satisfying the linear boundary condition:

$$\phi_1^+(t) = \frac{\left(t^{-l_0}\prod_{j=1}^{l_1}(1-\tfrac{z_j^-}{t})\right)^n}{\left(\prod_{j=1}^{k}(t-z_j^+)\right)^m} G(t)\phi_1^-(t), \quad t \in L, \quad (3.1.9)$$

or equivalently (due to (3.1.7)):

$$\phi_1^+(t) = \frac{\left(\prod_{j=1}^{l_1}(t-z_j^-)\right)^n}{\left(\prod_{j=1}^{k}(1-\tfrac{z_j^+}{t})\right)^m} (G(t)t^{-\chi})\phi_1^-(t), \quad t \in L. \quad (3.1.10)$$

Ultimately we have

$$\phi_2^+(t) = \left[G(t)t^{-\chi}\right]\phi_2^-(t), \quad t \in L, \quad (3.1.11)$$

where

$$\phi_2^+(z) := \frac{\phi_1^+(z)}{\left(\prod_{j=1}^{l_1}(z-z_j^-)\right)^n}, \quad \phi_2^-(z) := \frac{\phi_1^-(z)}{\left(\prod_{j=1}^{k}(1-\tfrac{z_j^+}{z})\right)^m}.$$

The index of the $\mathbb{C}$-linear conjugation problem (3.1.11) is equal to zero. Thus (3.1.11) is to be solved in the class of nowhere vanishing analytic functions. Hence, its unique solution can be given by the formula (3.1.5) with $G(t)$ changed by $G_1(t) := G(t)t^{-\chi}$.

3.1. CONJUGATION PROBLEM OF POWER TYPE.

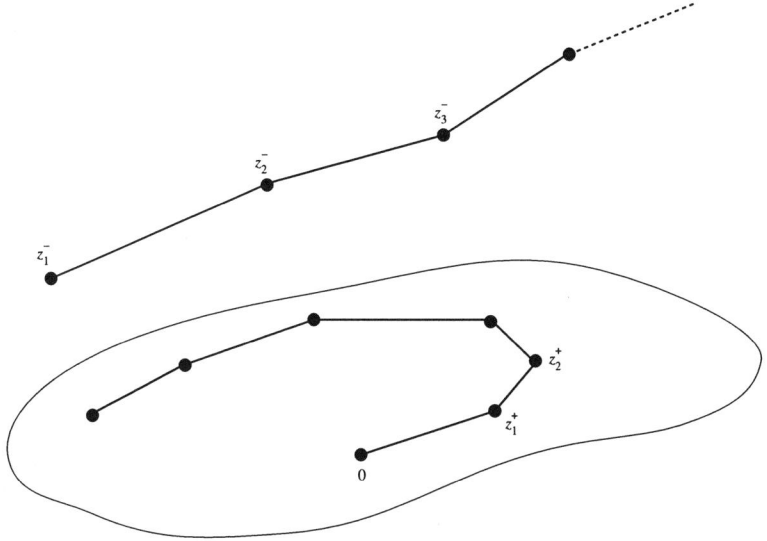

Figure 3.2: The cuts connected internal and external branch points

Consequently, the solutions of starting problem (3.1.2) are

$$\begin{cases} \Phi^+(z) = C^{\frac{1}{m}} \prod_{j=1}^{k}(z - z_j^+) \left(\prod_{j=1}^{l_1}(z - z_j^-) \right)^{\frac{n}{m}} \exp\{\frac{1}{2\pi i m} \int_L \frac{\log G_1(\tau)}{\tau - z} d\tau\}, \\ \Phi^-(z) = C^{\frac{1}{n}} z^{-l_0} \prod_{j=1}^{l_1}(1 - \frac{z_j^-}{z}) \left(\prod_{j=1}^{k}(1 - \frac{z_j^+}{z}) \right)^{\frac{m}{n}} \exp\{\frac{1}{2\pi i n} \int_L \frac{\log G_1(\tau)}{\tau - z} d\tau\}, \end{cases} \quad (3.1.12)$$

where $C \neq 0$ is an arbitrary constant.

The second products in both formulas are in general multi-valued functions. One can draw a cut L^- connecting the points $z_1^-, \ldots, z_{l_1}^-, \infty$ ($L^- \subset D^-$), and a cut L^+ connecting the points $0, z_1^+, \ldots, z_k^+$ ($L^+ \subset D^+$), and choose any branch of the function $\left(\prod_{j=1}^{l_1}(z - z_j^-) \right)^{\frac{n}{m}}$ in $\widehat{\mathbb{C}} \setminus L^-$, as well as any branch of the function $\left(\prod_{j=1}^{k}(1 - \frac{z_j^+}{z}) \right)^{\frac{m}{n}}$ in $\widehat{\mathbb{C}} \setminus L^+$ (for any finite collection of zeros $(z_j^+), (z_j^-)$ one can draw the lines L^+, L^-

being simple and having no common point with L (see Figure 3.2)). Arbitrary pairs of these branches correspond to different solutions of type (3.1.12) of problem (3.1.2) (cf. [242].)

b) $\alpha = \frac{p}{q}, \beta = \frac{r}{s}$ are positive rational numbers.

First we have to note that problem (3.1.2) with rational exponent could not be reduced equivalently to that with integer exponents. Really, the problem

$$(\Phi^+(t))^{ps} = (G(t))^{qs} (\Phi^-(t))^{rq}, \ t \in L, \qquad (3.1.13)$$

is equivalent not to unique problem (3.1.2), but to the collection of problems

$$(\Phi^+(t))^{\frac{p}{q}} = \varepsilon_{qs} G(t) (\Phi^-(t))^{\frac{r}{s}}, \ t \in L, \qquad (3.1.14)$$

where ε_{qs} are all roots of qs-th power of 1. Our starting problem belongs to the series of problems (3.1.14). Anyway one can start with problem (3.1.13), but take only those solutions of this problem, which correspond to $\varepsilon_{qs} = 1$ in (3.1.14). Again there are three different cases. Denoting as before (cf. (3.1.4)) $\chi := wind_L \, G(t)$ we get the following results.

i) $\chi < 0$. Then $wind_L \, (G(t))^{qs} < 0$, i.e., the index of problem (3.1.13) is negative, and this problem has no analytic solution. Thus none of problem (3.1.14) has a solution (in particular, the problem (3.1.2)).

ii) $\chi = 0$. Then $wind_L \, (G(t))^{qs} = 0$. Therefore, problem (3.1.13) has only one analytic (and nonvanishing) solution

$$\begin{cases} \Phi^+(z) = C^{\frac{1}{ps}} \left(\exp\{\frac{1}{2\pi i p} \int\limits_L \frac{\log G(\tau)}{\tau - z} d\tau\} \right)^{qs}, \ z \in D^+, \\ \Phi^-(z) = C^{\frac{1}{rq}} \left(\exp\{\frac{1}{2\pi i r} \int\limits_L \frac{\log G(\tau)}{\tau - z} d\tau\} \right)^{qs}, \ z \in D^-, \end{cases} \qquad (3.1.15)$$

where we consider all single-valued branches of $\left(\exp\{\frac{1}{2\pi i} \int\limits_L \frac{\log G(\tau)}{\tau - z} d\tau\} \right)^{\frac{1}{p}}$ in D^+, and all single-valued branches of $\left(\exp\{\frac{1}{2\pi i} \int\limits_L \frac{\log G(\tau)}{\tau - z} d\tau\} \right)^{\frac{1}{r}}$ in D^-. It follows from boundary condition (3.1.2) that the pair of functions (3.1.15) gives the solutions of the starting problem if for the constant $C = \rho e^{i\theta}, \theta \in [0, 2\pi)$ the values $C^{\frac{1}{ps}}, C^{\frac{1}{rq}}$ are equal respectively to $\rho^{\frac{1}{ps}} e^{\frac{i\theta}{ps}}, \rho^{\frac{1}{rq}} e^{\frac{i\theta}{rq}}$.

3.1. CONJUGATION PROBLEM OF POWER TYPE.

iii) $\chi > 0$. The solvability of problem (3.1.13) is equivalent (as in case a)) to solvability with respect to non-negative integer numbers N_+, N_- in the following equation:

$$\chi qs = N_+ ps + N_- rq. \tag{3.1.16}$$

If some $N_+, N_- \in \mathbb{N}_0 = \mathbb{N} \cup \{0\}$ satisfy (3.1.16), then the solution of problem (3.1.13) has N_+ zeros in D^+, and N_- zeros in D^- (including that at ∞). Applying the same approach as in case a) we get the solutions of problem (3.1.13) in the form

$$\begin{cases} \Phi^+(z) = C^{\frac{1}{ps}} \prod_{j=1}^{k}(z-z_j^+) \left(\prod_{j=1}^{l_1}(z-z_j^-)\right)^{\frac{rq}{ps}} \exp\{\frac{q}{p}\Gamma_1(z)\}, \\ \Phi^-(z) = C^{\frac{1}{rq}} z^{-l_0} \prod_{j=1}^{l_1}(1-\frac{z_j^-}{z}) \left(\prod_{j=1}^{k}(1-\frac{z_j^+}{z})\right)^{\frac{ps}{rq}} \exp\{\frac{s}{r}\Gamma_1(z)\}, \end{cases} \tag{3.1.17}$$

where $\Gamma_1(z) := \frac{1}{2\pi i}\int_L \frac{\log G_1(\tau)}{\tau-z}d\tau$. If we choose the branches of roots of products and exponents arbitrarily, but calculate the roots of a constant C in the same way as in ii), then we get in formula (3.1.17) all the solutions of the starting problem (3.1.2).

c) α, β are real positive, and at least one of them is irrational.

Under condition

$$\chi = \alpha N^+ + \beta N^-, \tag{3.1.18}$$

where χ is determined in (3.1.4), we will obtain the results which are qualitatively the same as in previous cases.

i) $\chi < 0$. Problem (3.1.2) has no nontrivial solution.

ii) $\chi = 0$. The only pair $N^+ = 0$, $N^- = 0$ satisfies the equation (3.1.18). Therefore the unique solution of problem (3.1.2) can be delivered by the formula

$$\begin{cases} \Phi^+(z) &= C^{\frac{1}{\alpha}} \exp\{\frac{1}{2\pi i \alpha} \int_L \frac{\log G(\tau)}{\tau-z}d\tau\} =: C^{\frac{1}{\alpha}} \exp\{\frac{1}{\alpha}\Gamma^+(z)\}, \quad z \in D^+, \\ \Phi^-(z) &= C^{\frac{1}{\beta}} \exp\{\frac{1}{2\pi i \beta} \int_L \frac{\log G(\tau)}{\tau-z}d\tau\} =: C^{\frac{1}{\beta}} \exp\{\frac{1}{\beta}\Gamma^-(z)\}, \quad z \in D^-, \end{cases} \tag{3.1.19}$$

$C \neq 0$ is an arbitrary constant, the values of $C^{\frac{1}{\alpha}}, C^{\frac{1}{\beta}}$ are chosen as before.

iii) $\chi > 0$. Let $k = N^+, l = N^-$ be the solutions of (3.1.18) in non-negative integer numbers, then the solution of problem (3.1.2) has the form

$$\begin{cases} \Phi^+(z) = C^{\frac{1}{\alpha}} \exp\left\{\frac{1}{\alpha}\Gamma^+(z)\right\} \prod_{j=1}^{k}(z-z_j^+) \left(\prod_{j=1}^{l_1}(z-z_j^-)\right)^{\frac{\beta}{\alpha}}, \\ \Phi^-(z) = C^{\frac{1}{\beta}} \exp\left\{\frac{1}{\beta}\Gamma^-(z)\right\} z^{-l_0} \prod_{j=1}^{l_1}\left(1-\frac{z_j^-}{z}\right) \left(\prod_{j=1}^{k}\left(1-\frac{z_j^+}{z}\right)\right)^{\frac{\alpha}{\beta}}. \end{cases} \quad (3.1.20)$$

d) α, β are complex numbers with non-negative real parts.

Again in the case of validity of (3.1.18) the general results are the same as in c).

3.1.2 Solution to nonlinear conjugation problem of power type with admissible zeros on the contour.

As it was already said the solvability of homogeneous problem (3.1.2)

$$\left(\Phi^+(t)\right)^\alpha = G(t)\left(\Phi^-(t)\right)^\beta, \quad t \in L, \quad (3.1.21)$$

in the class $\mathcal{C}_\mathcal{A}^0$ of functions $(\Phi^+, \Phi^-) \in \mathcal{C}_\mathcal{A}(D^+) \times \mathcal{C}_\mathcal{A}(D^-)$, whose boundary values $\Phi^+(t), \Phi^-(t)$ are nonvanishing on L (i.e., $\Phi^\pm(t) \neq 0$, $t \in L$), depends on solvability in the positive integer numbers N^+, N^- of the equation

$$\chi = \alpha N^+ + \beta N^-, \quad (3.1.22)$$

where $\chi = wind_L\, G(t)$. If (3.1.22) is not valid, then one can look for the solution of (3.1.2) in $\mathcal{C}_\mathcal{A}(D^+) \times \mathcal{C}_\mathcal{A}(D^-)$, but not in $\mathcal{C}_\mathcal{A}^0$. Consider first the model case:

a) $\alpha = m$, $\beta = n$ are positive integer numbers.

After substitution

$$\phi^+(z) = \left[\Phi^+(z)\right]^+, \; z \in D^+; \; \phi^-(z) = \left[\Phi^-(z)\right]^n, \; z \in D^-, \quad (3.1.23)$$

we obtain the $\mathbb{C}$-linear conjugation homogeneous problem

$$\phi^+(t) = G(t)\phi^-(t), \quad t \in L, \quad (3.1.24)$$

which should be studied in the class $\widetilde{\mathcal{A}}_{m,n} = \bigcup_{k,l \in \mathbb{N}_0} \widetilde{\mathcal{A}}_{m,n}^{k,l}$, where

$$\widetilde{\mathcal{A}}_{m,n}^{k,l} := \{\phi^\pm \in \mathcal{A}_{m,n} : N^+ = mk,\; N^- = nl\} \quad (3.1.25)$$

3.1. CONJUGATION PROBLEM OF POWER TYPE.

and "wave" means that $\phi^{\pm}(t)$ possibly have zeros (not necessarily of integer order) on the contour L. If k, l are already fixed then the following cases should be considered:

i) $\chi_0 := \chi - km - ln < 0$. Choosing arbitrary k points z_j^+ in D^+ and $l_1 (l_1 \le l, l = l_0 + l_1)$ points z_j^- in D^- we reduce problem (3.1.24) to the following one

$$\phi_1^+(t) = \frac{\left(\prod_{j=1}^{l_1}(t - z_j^-)\right)^n}{\left(\prod_{j=1}^{k}(1 - \frac{z_j^+}{t})\right)^m} G(t) t^{-km-nl} \phi_1^-(t), \quad t \in L, \qquad (3.1.26)$$

which should be studied with respect to

$$(\phi_1^+(z), \phi_1^-(z)) := \left(\frac{\phi^+(z)}{\left(\prod_{j=1}^{k}(z - z_j^+)\right)^m}, \frac{\phi^-(z)}{z^{l_0}\prod_{j=1}^{l_1}(1 - \frac{z_j^-}{z})^n}\right) \in \widetilde{\mathcal{A}}^{0,0},$$

i.e., in the class of analytic functions $\phi_1^+(z), \phi_1^-(z)$ nonvanishing in D^+, D^- respectively (but with admissible zeros on L). Index of this auxiliary problem is equal to χ_0:

$$\chi_0 = \chi - km - ln = wind_L \ G_1(t) := wind_L \ G(t) t^{-km-ln}.$$

Hence (3.1.26) has no analytic solution. Therefore no solution exists in $\widetilde{\mathcal{A}}^{0,0}$. Thus, the starting problem (3.1.2) has no nontrivial solution in $\widetilde{\mathcal{A}}^{k,l}_{m,n}$.

ii) $\chi_0 = 0$. It is exactly the case, when the numbers $k = N^+, l = N^-$ are solutions of problem (3.1.22). The solvability of problem (3.1.2) in this situation was already discussed. The solutions are delivered by formula (3.1.12). They have no zero on L.

iii) $\chi_0 = \chi - km - ln > 0$. Again we reduce problem (3.1.2) to (3.1.26). The latter has to be studied in $\widetilde{\mathcal{A}}^{0,0}$. But now the index of (3.1.26) is positive. Choosing arbitrary χ_0 points on L ($t_1, \ldots, t_{\chi_0} \in L$) we represent the boundary condition (3.1.26) in the form:

$$\phi_2^+(t) = t^{-\chi_0} G_1(t) \phi_2^-(t), t \in L, \qquad (3.1.27)$$

where $\phi_2^+(z) := \phi_1^+(z) \left(\prod_{s=1}^{\chi_0}(z - t_s)\right)^{-1}$, $\phi_2^+(z) := \phi_1^+(t) \left(\prod_{s=1}^{\chi_0}(1 - \frac{t_s}{z})\right)^{-1}$. Problem (3.1.27) is of zero index. Hence it has the unique solution of the form:

$$\phi_2^{\pm}(z) = C \exp\left\{\frac{1}{2\pi i} \int_L \frac{\log[\tau^{-\chi_0} G_1(\tau)]}{\tau - z} d\tau\right\} =: C \exp\{\Gamma_2^{\pm}(z)\}, \ z \in D^{\pm}, \qquad (3.1.28)$$

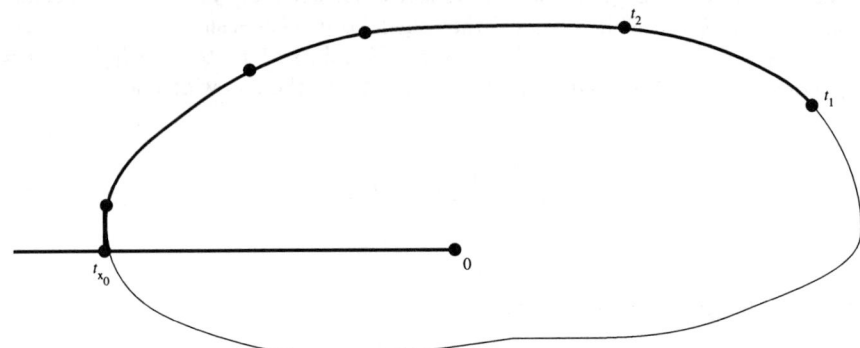

Figure 3.3: The cuts connected boundary points

which is nowhere vanishing. Therefore the solution of initial problem (3.1.2) having k zeros in D^+, and l zeros in D^- with admissible zeros on the contour L, is given by the formula:

$$\begin{cases} \Phi^+(z) = C^{\frac{1}{m}} \prod_{j=1}^{k}(z-z_j^+) \left(\prod_{j=1}^{l_1}(z-z_j^-)\right)^{\frac{n}{m}} \left(\prod_{s=1}^{\chi_0}(z-t_s)\right)^{\frac{1}{m}} \exp\{\Gamma_2^+(z)\} \\ \Phi^-(z) = C^{\frac{1}{n}} z^{-l_0} \prod_{j=1}^{l_1}(1-\frac{z_j^-}{z}) \left(\prod_{j=1}^{k}(1-\frac{z_j^+}{z})\right)^{\frac{m}{n}} \left(\prod_{s=1}^{\chi_0}(1-\frac{t_s}{z})\right)^{\frac{1}{n}} \exp\{\Gamma_2^-(z)\}. \end{cases}$$
(3.1.29)

The branches of the corresponding multi-valued functions in the right hand-sides can be fixed arbitrarily in $\mathbb{C}\setminus L^+$, $\mathbb{C}\setminus L^-$ respectively.

Here L^+ is a part of L from t_1 till t_{χ_0}, and the ray from t_{χ_0} till ∞; L^- is a part of L from t_1 till t_{χ_0}, and the ray from t_{χ_0} till 0 (without loss of generality one can consider L to be star-like with respect to $z = 0$ (cf., e.g., [102])). It is clear that the same considerations remain useful in the case of α, β being any real numbers for which

$$\chi_0 := \chi - k\alpha - l\beta$$

is an integer number for certain $k, l \in \mathbb{N}_0 = \mathbb{N} \cup \{0\}$. The last step coincides with the above proposed, but the preparative substitutions are the same as in the case $\chi = k\alpha + l\beta$ for α, β real, i.e., c) in Subsection 3.1.1.). Hence, we can discuss the solvability results only briefly. Let us consider the case

3.1. CONJUGATION PROBLEM OF POWER TYPE.

b) $\alpha = \frac{p}{q}, \beta = \frac{r}{s}$ are positive rational numbers.

As it was already said we have to study problem (3.1.13), i.e.,

$$\left(\Phi^+(t)\right)^{ps} = (G(t))^{qs} \left(\Phi^-(t)\right)^{rq}, \ t \in L,$$

which is in fact equivalent to a collection of problems (3.1.14), i.e.,

$$\left(\Phi^+(t)\right)^{\frac{p}{q}} = \varepsilon_{qs} G(t) \left(\Phi^-(t)\right)^{\frac{r}{s}}, \ t \in L,$$

where ε_{qs} are all roots of qs-power of 1. Our starting problem (3.1.2) is one of problems of collection (3.1.14). To get the solution of (3.1.2) one has to choose accurately the constant in the final formula for the solutions of (3.1.13). Further we always take into account this consideration. Let us study problem (3.1.13) in the classes $\widetilde{\mathcal{A}}_{ps,rq}^{k,l}$. Same substitutions as before reduce problem (3.1.13) to the following one (cf. [239]):

$$\phi_2^+(t) = (G(t))^{qs} \, t^{-kps-lrq} \phi_2^-(t), \ t \in L, \tag{3.1.30}$$

which has to be solved in the class $\widetilde{\mathcal{A}}^{0,0}$. The solvability of this problem depends on the "index" of (3.1.30):

$$\chi_0 := \chi qs - kps - lrq.$$

As χ_0 is an integer number, so three cases appear:

i) $\chi_0 < 0$. Problem (3.1.30) has no solution in $\widetilde{\mathcal{A}}^{0,0}$. Hence no solution of (3.1.2) in the considered class exists.

ii) $\chi_0 = 0$. Problem (3.1.30) has a unique solution. Corresponding solution of (3.1.2) is delivered by formula (3.1.12).

iii) $\chi_0 > 0$. The solution of problem (3.1.2) is given by formula (3.1.29) with $m := ps$, $n := rq$, $G(t) := (G(t))^{qs}$.

c) α, β are real positive numbers, and at least one of them is irrational.

If $\chi < 0$ then problem (3.1.2) has no analytic solution, hence no solution in the above chosen classes. If $\chi \geq 0$ then we have to study the possiblity to construct the solution of (3.1.21) with k zeros in D^+, l zeros in D^-, and with (possibly) vanishing boundary functions $\Phi^\pm(t)$, i.e., in the class $\widetilde{\mathcal{A}}^{k,l}$. Denoting as before the internal zeros by z_j^+, the external ones by z_j^- (arbitrary, not necessarily different points in D^+, D^- respectively) we introduce the new unknown functions $\phi_1^\pm(z)$ by the formulas

$$\begin{cases} \Phi^+(z) = \prod_{j=1}^{k}(z - z_j^+)\phi_1^+(z), \\ \Phi^-(z) = z^{-l_0} \prod_{j=1}^{l_1}(1 - \frac{z}{z_j^-})\phi_1^-(z). \end{cases} \tag{3.1.31}$$

The functions $\phi_1^+(z)$, $\phi_1^-(z)$ are nonvanishing in D^+, D^- respectively, and satisfy the following boundary condition:

$$\left(\phi_1^+(t)\right)^\alpha = \frac{\left(\prod_{j=1}^{l_1}(t - z_j^-)\right)^\beta}{\left(\prod_{j=1}^{k}(1 - \frac{z_j^+}{t})\right)^\alpha} G(t) t^{-k\alpha-l\beta} \left(\phi_1^-(t)\right)^\beta, \quad t \in L, \qquad (3.1.32)$$

where the branches of corresponding multi-valued functions are chosen as before; namely, $\left(\prod_{j=1}^{l_1}(z - z_j^-)\right)^\beta$ is any branch of multi-valued function in $\mathbb{C}\setminus L^-$, where $L^- \subset D^-$ is a simple smooth curve through the points $z_1^-, \ldots, z_{l_1}^-, \infty$; $\left(\prod_{j=1}^{k}(1 - \frac{z}{z_j^+})\right)^\alpha$ is any branch of multi-valued function in $\mathbb{C} \setminus L^+$, where $L^+ \subset D^+$ is a simple smooth curve through the points $0, z_1^+, \ldots, z_k^+$ (see Figure 3.2). The branch of $z^{-k\alpha-l\beta}$ will be chosen (if necessary) later. Making one more substitution

$$\begin{cases} \phi_2^+(z) = \left(\prod_{j=1}^{l_1}(z - z_j^-)\right)^{-\beta} \left(\phi_1^+(z)\right)^\alpha, \quad z \in D^+, \\ \phi_2^-(z) = \left(\prod_{j=1}^{k}(1 - \frac{z}{z_j^+})\right)^{-\alpha} \left(\phi_1^-(z)\right)^\beta, \quad z \in D^-, \end{cases} \qquad (3.1.33)$$

one gets the following $\mathbb{C}$-linear conjugation problem:

$$\phi_2^+(t) = G_1(t)\phi_2^-(t), \quad t \in L, \qquad (3.1.34)$$

where $G_1(t) = G(t)t^{-k\alpha-l\beta}$. The coefficient of this problem is in general discontinuous in at least one point of L.

For the moment we can fix this point arbitrarily, say, $t_0 \in L$. It means that $t^{-k\alpha-l\beta}$ is the boundary value of the single-valued branch of the function $z^{-k\alpha-l\beta}$ determined in $\mathbb{C}\setminus\{L^+ \cup L_0 \cup L^-\}$, where cuts L^+, L^- were already defined, and L_0 is any simple smooth curve connected z_k^+, t_0, z_1^-, so that $\widehat{z_k^+, t_0} \subset D^+$, $\widehat{t_0, z_1^-} \subset D^-$ (see Figure 3.4). The function $t^{-k\alpha-l\beta}$ is continuous if $k\alpha + l\beta$ is an integer number. In this case problem (3.1.21) is solvable in $\widetilde{\mathcal{A}}^{k,l}$ if and only if problem (3.1.34) is solvable in $\widetilde{\mathcal{A}}^{0,0}$. Such a situation is completely analogous to that described in Subsection 3.1.1. If $k\alpha + l\beta$ is not an integer number then we receive the solution of (3.1.21) in $\widetilde{\mathcal{A}}^{k,l}$

3.1. CONJUGATION PROBLEM OF POWER TYPE.

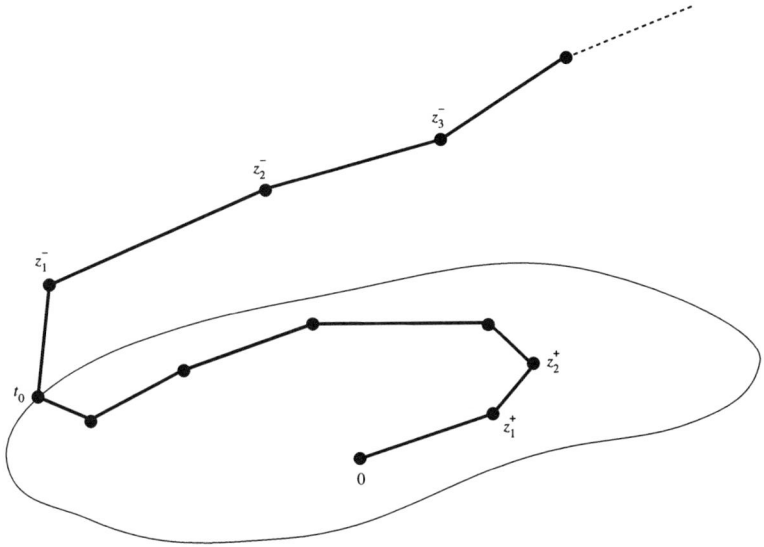

Figure 3.4: The cuts through z_j^+, t_0, z_j^-

if the corresponding solution of (3.1.34) in $\widetilde{\mathcal{A}}^{0,0}$ has t_0 as zero-point for $\phi_2^+(t)$, $\phi_2^-(t)$. Two different cases appear:

i) $\chi_0 = \chi - k\alpha - l\beta < 0$ ($\chi_0 \notin \mathbb{Z}$). As it follows from [91, p. 433-435] problem (3.1.34) has no analytic solution continuous up to the boundary. Hence, no solution of the above described type exists. Therefore, starting problem (3.1.2) has no solution in $\widetilde{\mathcal{A}}^{k,l}$.

ii) $\chi_0 = \chi - k\alpha - l\beta > 0$ ($\chi_0 \notin \mathbb{Z}$). Factorizing the function

$$G(t)t^{-\chi}$$

in the standard way (see, e.g., [91, p. 109]; cf. Section 2.8):

$$G(t)t^{-\chi} = \frac{X^+(t)}{X^-(t)},$$

where

$$X^\pm(z) := \exp\left\{\frac{1}{2\pi i}\int_L \frac{\log[G(\tau)\tau^{-\chi}]}{\tau - z}d\tau\right\} =: \exp\left\{\Gamma^\pm(z)\right\},$$

one gets the following model problem

$$\phi_3^+(t) = t^{\chi_0}\phi_3^-(t), \ t \in L, \tag{3.1.35}$$

which has to be solved in $\widetilde{A}^{0,0}$ (the branch of t^{χ_0} is chosen as before; $\phi_3^{\pm}(z) = \phi_2^{\pm}(z)/X^{\pm}(z)$). The only solution of this problem has the form:

$$\phi_3^+(z) = C(z-t_0)^{\chi_0}, \ \phi_3^-(z) = C(1-\frac{t_0}{z})^{\chi_0}, \tag{3.1.36}$$

where the branches of corresponding functions are chosen as follows: $(z-t_0)^{\chi_0}$ is any branch of multi-valued function in $\widehat{\mathbb{C}} \setminus \left\{\widehat{(t_0,z_1^-)} \cup L^-\right\}$; $(1-\frac{t_0}{z})^{\chi_0}$ is that branch of multi-valued function in $\widehat{\mathbb{C}} \setminus \left\{L^+ \cup \widehat{(z_k^+,t_0)}\right\}$ for which $(1-\frac{t_0}{z})^{\chi_0} = \frac{(z-t_0)^{\chi_0}}{z^{\chi_0}}$, where the branches for $(z-t_0)^{\chi_0}$, z^{χ_0} are already chosen. Returning to the starting problem we can represent the solution of (3.1.2) in $\widetilde{A}^{k,l}$ in the following form:

$$\begin{cases} \Phi^+(z) = C^{\frac{1}{\alpha}}(z-t_0)^{\frac{\chi_0}{\alpha}} \prod_{j=1}^{k}(z-z_j^+) \left(\prod_{j=1}^{l_1}(z-z_j^-)\right)^{\frac{\beta}{\alpha}} \exp\{\frac{1}{\alpha}\Gamma^+(z)\}, \\ \Phi^-(z) = C^{\frac{1}{\beta}}\left(1-\frac{t_0}{z}\right)^{\frac{\chi_0}{\beta}} z^{-l_0} \prod_{j=1}^{l_1}(1-\frac{z_j^-}{z}) \left(\prod_{j=1}^{k}(1-\frac{z_j^+}{z})\right)^{\frac{\alpha}{\beta}} \exp\{\frac{1}{\beta}\Gamma^-(z)\}. \end{cases} \tag{3.1.37}$$

Remark 3.1 *The solution of problem (3.1.2) in $\widetilde{A}^{k,l}$ can have a prescribed finite number of zero-points on the boundary in the case $\chi_0 = \chi - k\alpha - l\beta > 0$ (say $t_1, \ldots, t_d$). To construct such a solution one has to represent the function t^{χ_0} in the form $t^{\chi_1} \cdot \ldots \cdot t^{\chi_d}$, where $\chi_0 = \chi_1 + \ldots + \chi_d$, $\chi_j > 0$, choosing the cuts for corresponding multi-valued functions which intersect L at $t_1, \ldots, t_d$ respectively.*

3.1.3 Inhomogeneous problem.

We consider here the solvability of the inhomogeneous problem (3.1.1)

$$\left(\Phi^+(t)\right)^\alpha = G(t)\left(\Phi^-(t)\right)^\beta + g(t), \ t \in L,$$

in the class $\widetilde{A}^{k,l}$ in the most general case c) (α, β are real positive numbers, and at least one of them is irrational). The simpler cases can be studied with corresponding changes described in detail in the previous subsection. Introducing new unknown

3.1. CONJUGATION PROBLEM OF POWER TYPE.

functions $\phi_1^\pm$, $\phi_2^\pm$ in the same manner as before (i.e., by the formulae (3.1.31), (3.1.33)) one can reduce problem (3.1.1) to the following one

$$\phi_2^+(t) = G(t)t^{-k\alpha-l\beta}\phi_2^-(t) + \frac{g(t)}{\left(\prod_{j=1}^{k}(1-\frac{z_j^+}{t})\right)^\alpha \left(\prod_{j=1}^{l_1}(t-z_j^-)\right)^\beta}, \quad t \in L. \quad (3.1.38)$$

This problem has to be solved in the class $\tilde{\mathcal{A}}^{0,0}$. Let first $k\alpha + l\beta$ be a non-negative integer number. Then the general theory of the $\mathbb{C}$-linear conjugation problem (see, e.g., [91], Ch. 2; cf. Section 2.8) leads us to three different situations: i) $\chi_0 := \chi - k\alpha - l\beta < 0$; ii) $\chi_0 = 0$; iii) $\chi_0 > 0$.

In the first case the corresponding homogeneous problem has no nontrivial solution (hence no solution in $\tilde{\mathcal{A}}^{0,0}$ besides of identically zero). Anyway one can factorize the coefficient of problem (3.1.38) in such a way (see [91, p. 112]):

$$G(t)t^{-k\alpha-l\beta} = \frac{\exp\{\Gamma^+(t)\}}{t^{-\chi_0}\exp\{\Gamma^-(t)\}} =: \frac{X^+(t)}{X^-(t)},$$

where $\Gamma^\pm(z) := \frac{1}{2\pi i}\int_L \frac{\log[G(\tau)\tau^{-\chi}]}{\tau - z}d\tau$, $z \in D^\pm$. Therefore the boundary condition (3.1.38) can be rewritten in the form

$$\frac{\phi_2^+(t)}{X^+(t)} = \frac{\phi_2^-(t)}{X^-(t)} + \frac{g(t)}{\left(\prod_{j=1}^{k}(1-\frac{z_j^+}{t})\right)^\alpha \left(\prod_{j=1}^{l_1}(t-z_j^-)\right)^\beta X^+(t)}, \quad (3.1.39)$$

or

$$\frac{\phi_2^+(t)}{X^+(t)} - \Psi^+(t) = \frac{\phi_2^-(t)}{X^-(t)} - \Psi^-(t), \quad (3.1.40)$$

where

$$\Psi^\pm(z) := \frac{1}{2\pi i}\int_L \frac{g(\tau)}{\left(\prod_{j=1}^{k}(1-\frac{z_j^+}{\tau})\right)^\alpha \left(\prod_{j=1}^{l_1}(\tau - z_j^-)\right)^\beta X^+(\tau)}\frac{d\tau}{\tau - z}.$$

This leads us to an analytic solution of problem (3.1.38) if and only if the following solvability conditions do satisfy:

$$\int_L \frac{g(\tau)}{\left(\prod_{j=1}^{k}(1-\frac{z_j^+}{\tau})\right)^\alpha \left(\prod_{j=1}^{l_1}(\tau - z_j^-)\right)^\beta} \frac{\tau^{h-1}}{X^+(\tau)}d\tau = 0, \quad h = 1,\ldots,-\chi_0 - 1. \quad (3.1.41)$$

If so, the unique solution of problem (3.1.38) is given by the formula:
$$\phi_2^\pm(z) = X^\pm(z)\Psi^\pm(z). \qquad (3.1.42)$$

It should be noted that the fulfillment of conditions (3.1.41) depends in general on the choice of points z_j^+, z_j^-. The solution (3.1.42) is nonvanishing in $cl\,D^\pm$ iff the point $z = 0$ does not belong to the closure of the image of the domains D^+, D^- under the mappings Ψ^+, Ψ^- respectively, i.e.,

$$0 \notin cl\left(\Psi^+(D^+)\right) \cup cl\left(\Psi^-(D^-)\right). \qquad (3.1.43)$$

It can happen because $cl\left(\Psi^+(D^+)\right) \cup cl\left(\Psi^-(D^-)\right) \neq \mathbb{C}$ by the conditions on the coefficients of problem (3.1.1) (see, e.g., [255, v. 1]). Under conditions (3.1.41), (3.1.43) the unique solution of the starting problem (3.1.1) is delivered by the formula

$$\begin{cases} \Phi^+(z) = \prod_{j=1}^{k}(z-z_j^+)\left(\prod_{j=1}^{l_1}(z-z_j^-)\right)^{\frac{\beta}{\alpha}} \exp\{\frac{1}{\alpha}\Gamma^+(z)\}(\Psi^+(z))^{\frac{1}{\alpha}} \\ \Phi^-(z) = z^{-l_0}\prod_{j=1}^{l_1}(1-\frac{z_j^-}{z})\left(\prod_{j=1}^{k}(1-\frac{z_j^+}{z})\right)^{\frac{\alpha}{\beta}} \exp\{\frac{1}{\beta}\Gamma^-(z)\}(\Psi^-(z))^{\frac{1}{\beta}}. \end{cases} \qquad (3.1.44)$$

ii) $\chi_0 = 0$. The same considerations as before lead us to the following formula of the solution of (3.1.1)

$$\begin{cases} \Phi^+(z) = \prod_{j=1}^{k}(z-z_j^+)\left(\prod_{j=1}^{l_1}(z-z_j^-)\right)^{\frac{\beta}{\alpha}} \exp\{\frac{1}{\alpha}\Gamma^+(z)\}(\Psi^+(z)+C)^{\frac{1}{\alpha}}, \\ \Phi^-(z) = z^{-l_0}\prod_{j=1}^{l_1}(1-\frac{z_j^-}{z})\left(\prod_{j=1}^{k}(1-\frac{z_j^+}{z})\right)^{\frac{\alpha}{\beta}} \exp\{\frac{1}{\beta}\Gamma^-(z)\}(\Psi^-(z)+C)^{\frac{1}{\beta}}. \end{cases} \qquad (3.1.45)$$

In order to have the solution from $\widetilde{\mathcal{A}}^{k,l}$ one ought to choose the constant C in the following way:

$$C \notin cl\left(-\Psi^+(D^+)\right) \cup cl\left(-\Psi^-(D^-)\right). \qquad (3.1.46)$$

iii) $\chi_0 > 0$. Then one has to factorize the coefficient of the problem (3.1.38) as in the corresponding case for the homogeneous problem:

$$G(t)t^{-k\alpha-l\beta} = \frac{\prod_{s=1}^{m}(t-t_s)^{d_s}\exp\{\Gamma^+(t)\}}{\prod_{s=1}^{m}(1-\frac{t_s}{t})^{d_s}\exp\{\Gamma^-(t)\}} =: \frac{X^+(t)}{X^-(t)}. \qquad (3.1.47)$$

3.2. PROBLEM OF MULTIPLICATION TYPE.

The solution of starting problem (3.1.1) will be presented then in the form:

$$\begin{cases} \Phi^+(z) = \prod_{j=1}^{k}(z-z_j^+)\left(\prod_{j=1}^{l_1}(z-z_j^-)\right)^{\frac{\beta}{\alpha}} [X^+(z)]^{\frac{1}{\alpha}}\left(\Psi^+(z)+C\right)^{\frac{1}{\alpha}}, \\ \Phi^-(z) = z^{-l_0}\prod_{j=1}^{l_1}\left(1-\frac{z_j^-}{z}\right)\left(\prod_{j=1}^{k}\left(1-\frac{z_j^+}{z}\right)\right)^{\frac{\alpha}{\beta}} [X^-(z)]^{\frac{1}{\beta}}\left(\Psi^-(z)+C\right)^{\frac{1}{\beta}}. \end{cases} \quad (3.1.48)$$

The analyticity condition for this solution coincides with (3.1.46). Combining the above obtained results one can formulate the following

Theorem 3.1 *Nonlinear inhomogeneous problem (3.1.1) is solvable in the classes $\tilde{A}^{k,l}$ with $k\alpha + l\beta > \chi$ (in particular for $\chi < 0$ in every such class) if and only if the solvability conditions (3.1.41) do satisfy. The unique solution of (3.1.1) is presented by the formula (3.1.44) under additional analyticity condition (3.1.43). Nonlinear inhomogeneous problem (3.1.1) is unconditionally solvable in the classes $\tilde{A}^{k,l}$ with $k\alpha + l\beta \leq \chi$. The solution has the form (3.1.48), where the constant C should be chosen according to condition (3.1.46).*

3.2 Nonlinear conjugation problem of multiplication type.

Let L be a simple closed smooth curve dividing $\mathbb{C}$ into two domains $D^+ \ni 0$, and $D^- \ni \infty$. Suppose that $f \in \mathcal{H}(L), f(t) \neq 0$. We consider here problem of searching for two functions $(\Phi^+, \Phi^-) \in \mathcal{C}_A(D^+) \times \mathcal{C}_A(D^-)$ via the boundary condition:

$$\Phi^+(t) \cdot \Phi^-(t) = f(t), \ t \in L. \quad (3.2.1)$$

It follows from the Argument Principle that the winding number of the function f is equal to the difference between the quantity of zeros of Φ^+ into D^+, and Φ^- into D^-. Therefore it is customary to study problem (3.2.1) in the classes $\mathcal{A}^{k,l}$ of analytic functions Φ^+, and Φ^-, having k and l zeros in these domains. Then the pair k,l ought to be a solution in positive integer numbers of the equation

$$k - l = \chi := \text{wind}_L f(t). \quad (3.2.2)$$

If z_j^+, z_j^- are zeros (not necessarily different) of the functions Φ^+ into D^+, and Φ^- into D^- respectively, then one can introduce new analytic functions

$$\Psi^+(z) := \frac{\Phi^+(z)}{\prod_{j=1}^{k}(z-z_j^+)}, \quad \Psi^-(z) := \frac{\Phi^-(z)}{z^{-l_0}\prod_{j=1}^{l_1}\left(1-\frac{z_j^-}{z}\right)}, \quad (3.2.3)$$

where k, l are any non-negative numbers, satisfying (3.2.2), $l = l_0 + l_1$. The functions Ψ^+, Ψ^- have no zeros in D^+, D^-, and satisfy the boundary condition

$$\Psi^+(t) \cdot \Psi^-(t) = f_1(t) := \frac{f(t)t^l}{\prod_{j=1}^{k}(z-z_j^+)\prod_{j=1}^{l_1}(z-z_j^-)}, \quad t \in L. \quad (3.2.4)$$

The function $f_1(t)$ is Hölder-continuous and nonvanishing on L. Moreover

$$\operatorname{wind}_L f_1(t) = \chi - k + l = 0. \quad (3.2.5)$$

Taking the logarithm of both sides of (3.2.4) we reduce it to the following boundary value problem

$$\Psi_1^+(t) - \Psi_2^-(t) = \log f_1(t), \ t \in L, \quad (3.2.6)$$

where $\Psi_1^+(z) := \log \Psi^+(z)$, $\Psi_1^-(z) := \log \Psi^-(z)$, with any branch of logarithmic functions in these representations. The unique solution of problem (3.2.6) has the form

$$\Psi_1^\pm(z) = \frac{1}{2\pi i}\int_L \frac{\log f_1(\tau)}{\tau - z}d\tau =: \Gamma_1(z), \ z \in D^\pm. \quad (3.2.7)$$

Correspondingly, the solution of problem (3.2.4) is delivered by the formula

$$\Psi^\pm(z) = C^{\pm 1}\exp\{\pm\Gamma_1(z)\}, \ z \in D^\pm, \quad (3.2.8)$$

where $C \neq 0$ is an arbitrary constant. Finally, the general solution of problem (3.2.1) in the considered class has the form

$$\begin{cases} \Phi^+(z) = C\prod_{j=1}^{k}(z-z_j^+)\exp\{\Gamma_1^+(z)\}, \ z \in D^+, \\ \Phi^-(z) = \frac{1}{Cz^{l_0}}\prod_{j=1}^{l_1}(z-z_j^-)\exp\{\Gamma_1^-(z)\}, \ z \in D^-. \end{cases} \quad (3.2.9)$$

3.3 Entire functions methods for nonlinear conjugation problem of multiplication type.

3.3.1 Multiplication type problem for an open arc.

Let L be a simple bounded smooth open arc $\widehat{ab}$ oriented from a to b; $\tilde{L} = L \setminus \{a,b\}$; $D = \widehat{\mathbb{C}} \setminus L$. We introduce the following classes of functions (using letter $\mathcal{A}$ for analytic, and $\mathcal{B}$ for bounded):

$$\tilde{\mathcal{B}} = \left\{ F \in \mathcal{A}(D) \bigcap \mathcal{B}(G), \text{ for any closed domain } G \subset \widehat{\mathbb{C}} \setminus \{a,b\} \right\};$$

$$\tilde{\mathcal{B}}^0 = \left\{ F \in \tilde{\mathcal{B}} : F(z) \neq 0, \forall z \in D \right\};$$

$$\widehat{\mathcal{B}} = \left\{ F \in \tilde{\mathcal{B}} : \Re F(z) \in \mathcal{B}\left(U(a)\bigcup U(b)\right)\right\},$$

for certain neighbourhoods U(a), U(b) of points z=a, z=b;

$$\mathcal{B} = \mathcal{A}(D) \bigcup \mathcal{B}(D);$$

$$\mathcal{B}^n = \{F \in \mathcal{B} : F \text{ has exactly n zeros in } D\},$$

$n = 0, 1, 2, \ldots$.

Definition 3.1 *A function $F \in \mathcal{A}(D)$ satisfying in D the following condition*

$$\sup |\log |F(z)|| < M_D < +\infty$$

is called log-bounded in D.

Definition 3.2 *A function $F \in \mathcal{A}(\widehat{\mathbb{C}} \setminus \{z_1, \ldots, z_n\})$ single-valued in $\widehat{\mathbb{C}} \setminus \{z_1, \ldots, z_n\}$ is called an entire function with respect to points $z_1, \ldots, z_n$. The class of all such functions is denoted $\Omega_{z_1,\ldots,z_n}(z)$.*

Definition 3.3 *It will be said that a curve (an arc) L has a tangency not smaller than μ ($\mu > 1$) at a point $t \in L$ if*

$$\limsup_{r \to 0} \frac{h(r)}{r^\mu} = c, \ 0 \leq c < \infty, \ c = const,$$

where $h(r) = dist(\tau, l(t))$, $\tau \in L$, $r = |\tau - t|$, $l(t)$ is the tangent line to L at the point t.

We consider here the following problem: given $f \in \mathcal{H}^\alpha(L)$, $f(t) \neq 0$, $t \in L$, find all functions $\Phi \in \mathcal{B}_n$, $n = 0, 1, \ldots$, having one-sided boundary values everywhere on $\tilde{L}$, satisfying nonlinear condition

$$\Phi^+(t)\Phi^-(t) = f(t), \ t \in \tilde{L}, \qquad (3.3.1)$$

where $\Phi^+(t)$ ($\Phi^-(t)$) mean boundary limits from the left (right) according to orientation on L. Let us consider first the model problem: find all functions $\Phi \in \mathcal{B}_0$, satisfying the following boundary condition:

$$\Phi^+(t)\Phi^-(t) = 1, \ t \in \tilde{L}. \qquad (3.3.2)$$

This problem has two "trivial" solutions

$$\Phi_1(z) \equiv 1, \ \Phi_2(z) \equiv -1, z \in D.$$

Let now $\Phi(z)$ ($\neq \Phi_1(z)$, $\neq \Phi_2(z)$) be a solution of (3.3.2). Then all branches of $\log \Phi(z)$ are single-valued in D and bounded in every closed domain $G \subset \hat{\mathbb{C}} \setminus \{a, b\}$. Choosing any such branch one can reduce the boundary condition (3.3.2) to a collection of two conditions:

$$\log \Phi^+(t) + \log \Phi^-(t) = 0; \qquad (3.3.3)$$

$$\log \Phi^+(t) + \log \Phi^-(t) = 2\pi i. \qquad (3.3.4)$$

If $\Phi(z)$ is a solution of (3.3.4), then $-\Phi(z)$ is a solution of (3.3.3). Hence it is enough to consider problem (3.3.3). The latter is equivalent to the simple linear boundary value problem

$$\Psi^+(t) = (-1)\Psi^-(t), \ t \in \tilde{L}, \qquad (3.3.5)$$

with respect to $\Psi(z) := \log \Phi(z)$, which has to be solved in the class $\hat{\mathcal{B}}$. Let us first consider $\tilde{\mathcal{B}}$-solutions of (3.3.5). They have the form

$$\Psi(z) = \Psi_0(z)\,\Omega_{a,b}(z), \qquad (3.3.6)$$

where $\Psi_0(z)$ is a solution of (3.3.5) in $\tilde{\mathcal{B}}_0$, and $\Omega_{a,b}(z)$ is an entire function with respect to points a, b. $\Psi_0(z)$ is in fact the solution of problem

$$\log \Psi^+(t) - \log \Psi^-(t) = \pi i, \ t \in \tilde{L}.$$

Therefore, it is an arbitrary branch of the multi-valued function

$$\Psi_0(z) = \sqrt{\frac{z-b}{z-a}}.$$

3.3. ENTIRE FUNCTIONS METHODS. 77

Hence, the general formula of solutions of (3.3.5) can be rewritten as

$$\Psi(z) = \frac{F_{a,b,\infty}(z)}{\sqrt{(z-a)(z-b)}}, \qquad (3.3.7)$$

where $F_{a,b,\infty}$ is an entire function with respect to points a, b, ∞, having not more than the first order pole at ∞. In order to discover the conditions on $F_{a,b,\infty}$ we have to use the following special type uniqueness result ([103, pp. 52-53]):

Lemma 3.1 *Let $\sqrt{w}$ be any branch of square root in the complex plane cut along closed arc Λ with the end-points $w = 0$, $w = \infty$. Let f be an entire function such that*

$$\sup_w \Re[\sqrt{w}f(w)] < constant.$$

Then $f(w) \equiv constant$.

It follows immediately

Lemma 3.2 *If $\Psi(z)$ in (3.3.7) is a solution of problem (3.3.5) in the class $\widehat{\mathcal{B}}$, then*

$$F_{a,b,\infty}(z) = cz + d.$$

Proof. As $F_{a,b,\infty}$ has not more than a simple pole at infinity, it is sufficient to show that in the following series

$$\frac{F_{a,b,\infty}(z)}{\sqrt{z-b}} = \sum_{n=-\infty}^{\infty} a_n(z-a)^n, \quad \frac{F_{a,b,\infty}(z)}{\sqrt{z-a}} = \sum_{n=-\infty}^{\infty} b_n(z-b)^n$$

all coefficients a_n, b_n with negative indices are equal to zero ($a_n = b_n = 0$, $n = -1, -2, \ldots$). Let us denote

$$\frac{F_{a,b,\infty}(z)}{\sqrt{z-b}} = \sum_{n=-\infty}^{0} a_n(z-a)^n + \sum_{n=1}^{\infty} a_n(z-a)^n =: F_1(z) + F_2(z).$$

As $\Psi(z)$ in (3.3.7) belongs to $\widehat{\mathcal{B}}$ then

$$\sup_{z\in\mathbb{C}\setminus l} \Re \frac{F_1(z)}{\sqrt{z-a}} < const, \qquad (3.3.8)$$

where $\sqrt{z-a}$ is any branch in $\mathbb{C}$ with a cut along an arc l from a to ∞; $F_1(z)$ is single-valued analytic in $0 < |z - a| \leq \infty$. Using the transformation $w = \frac{1}{z-a}$ and denoting $f(w) := F_1(a + \frac{1}{w})$ one gets immediately from (3.3.8)

$$\sup_{w \in \mathbb{C} \setminus \Lambda} \Re[\sqrt{w} f(w)] < const,$$

where Λ is the image of l. Thus by Lemma 1 $f(w) \equiv const$; hence $(a_n = 0, n = -1, -2, \ldots)$. The same can be done for b_n. ∎

Corollary 3.1 *If $\Psi(z)$ in (3.3.7) is a solution of problem (3.3.5) in the class $\widehat{\mathcal{B}}$, then it has a form*

$$\Psi(z) = \frac{c(z - z_0)}{\sqrt{(z-a)(z-b)}}, \qquad (3.3.9)$$

where c, z_0 are constants, and $\sqrt{(z-a)(z-b)}$ is any analytic branch in D.

Let us study the behaviour of $\Psi(z)$, given by formula (3.3.9) near the points $z = a$, $z = b$. It is clear that if $z_0 = a$ ($z_0 = b$), then $\Psi(z)$ is bounded in $U(a)$ (in $U(b)$ respectively). Let l_a (l_b) be a ray beginning at the point $z = a$ ($z = b$), and having there an angle π with L. We draw the cut connecting a point $t \in \widetilde{L}$ with $z = \infty$, and fix any branch of $\arg \sqrt{(z-a)(z-b)}$ in the cut plane (see Figure 3.5). Denote

$$\lim_{l_a \ni z \to a} \arg \sqrt{(z-a)(z-b)} =: \alpha,$$

$$\lim_{l_b \ni z \to b} \arg \sqrt{(z-a)(z-b)} =: \beta.$$

Lemma 3.3 *Let $\Psi(z)$ in (3.3.9) be a solution of problem (3.3.5) in the class $\widehat{\mathcal{B}}$. Then*

$$\begin{cases} \arg c + \arg(a - z_0) = \alpha + \pi, \\ \arg c + \arg(b - z_0) = \beta + \pi. \end{cases} \qquad (3.3.10)$$

Proof. If $z_0 \neq a$, then we have to show that from

$$\sup_{0 < |z-a| < \infty} \Re \Psi(z) < \infty$$

follows the first equality of (3.3.10). If not, i.e., if

$$\arg c + \arg(a - z_0) = \alpha + \pi + \theta$$

3.3. ENTIRE FUNCTIONS METHODS.

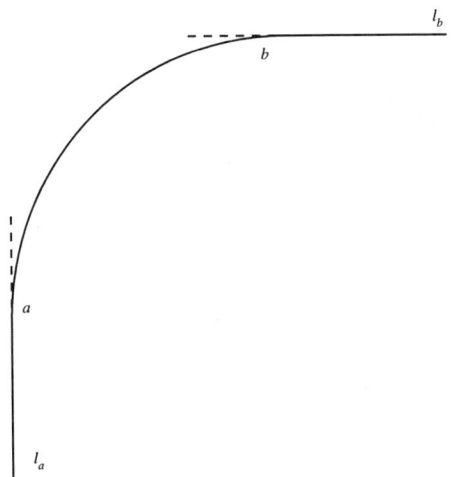

Figure 3.5: The arc $L = (a,b)$ (the angle between L and l_a, (l_b) is equal to π)

for certain $0 < |\theta| \leq \pi$ (without loss of generality we can suppose that $\theta > 0$), then denoting by $l_{a,\theta}$ a ray from $z = a$ such that

$$\arg(z-a)|_{z \in l_{a,\theta}} = \arg(z-a)|_{z \in l_a} - \pi + \theta$$

one gets

$$\arg \Psi(z)|_{z \in l_{a,\theta}} = [\arg c + \arg(\zeta - z_0) - \arg \sqrt{(\zeta-a)(\zeta-b)}$$
$$+ \frac{\pi}{2} - \frac{\theta}{2} + o(1)]|_{\zeta \in l_a} = \frac{3\pi}{2} + \frac{\theta}{2} + o(1), \ z \to a, \ \zeta \to a.$$

Hence
$$\lim_{l_a \ni z \to a} \cos \arg \Psi(z) = \cos\left(\frac{3\pi}{2} + \frac{\theta}{2}\right) = \eta > 0.$$

Therefore
$$\lim_{l_a \ni z \to a} \arg \Psi(z) = +\infty.$$

The contradiction gives us the necessary result. ∎

The same considerations lead to the second equality of (3.3.10). Using Lemma 3.3 one can prove the following

Lemma 3.4 Let c and z_0 ($z_0 \neq a$) (respectively $z_0 \neq b$) satisfy the condition (3.3.10). The real part of Ψ in (3.3.9) is bounded in a neighbourhood of $z = a$ (respectively, of $z = b$), i.e.,

$$\sup_{0<|z-d|<\varepsilon} \Re\Psi(z) < +\infty, \quad d = a \text{ or } b, \qquad (3.3.11)$$

iff the arc L has at the point $z = a$ ($z = b$) the order of tangency not smaller than $\frac{3}{2}$, i.e., $\mu_a \geq \frac{3}{2}$ (respectively $\mu_b \geq \frac{3}{2}$).

Proof. It is enough to get the result only for one point. (Necessity) Let Ψ in (3.3.9) satisfy the inequality (3.3.11). Then for all z, $0 < |z - a| < \varepsilon$, we have

$$\arg \frac{c(z - z_0)}{\sqrt{z - b}} = \Im \log \frac{c(z - z_0)}{\sqrt{z - b}} = \Im[c_0 + c_1(z - a) + \ldots] = \Im c_0 + O(r), \ r \to 0,$$

where $r = |z - a|$. Thus it follows from (3.3.10) that

$$\Im c_0 = \pi + \frac{1}{2} \arg(z - a)_{|z \in l_a},$$

and consequently

$$\arg_{0<|z-a|<\varepsilon} \frac{c(z - z_0)}{\sqrt{z - b}} = \pi + \frac{1}{2} \arg(z - a)_{|z \in l_a} + O(r), \ r \to 0. \qquad (3.3.12)$$

Let t^+ (t^-) be left-sided (right-sided) points on the cut along L; $\omega(r)$ is the angle between vector $\overline{at}$, $t \in L$, and positive in the sense of orientation tangent vector to L at $z = a$. Clearly, $\omega(r) \to 0$, as $r \to 0$. Then

$$\begin{cases} \arg(t^+ - a) = \arg(z - a)_{|z \in l_a} - \pi + \omega(r), \\ \arg(t^- - a) = \arg(z - a)_{|z \in l_a} + \pi + \omega(r) \end{cases} \qquad (3.3.13)$$

holds in a neighbourhood of $z = a$. It follows from (3.3.12), (3.3.13) that

$$\arg \Psi^+(t) = \arg \frac{c(t^+ - z_0)}{\sqrt{t^+ - b}} - \frac{1}{2} \arg(t^+ - a) = \frac{3\pi}{2} - \frac{\omega(r)}{2} + O(r), \ r \to 0.$$

Respectively

$$\arg \Psi^-(t) = \frac{\pi}{2} - \frac{\omega(r)}{2} + O(r), \ r \to 0.$$

Hence

$$\max\{\Re\Psi^+(t), \Re\Psi^-(t)\} = \frac{|c||t - z_0|}{\sqrt{|t - b|}} \frac{1}{\sqrt{r}} \left[\sin \frac{|\omega(r)|}{2} + O(r)\right]$$

$$\geq C_1 \frac{\sin |\omega(r)|}{\sqrt{r}}, \ 0 < C_1 < \infty. \qquad (3.3.14)$$

3.3. ENTIRE FUNCTIONS METHODS.

Using (3.3.11) we get immediately
$$C_1 \frac{\sin|\omega(r)|}{\sqrt{r}} = C_1 \frac{h(r)}{r^{3/2}} \leq C_2 < \infty, \ r \to 0.$$

According to Definition 3.2 it means that L has at $z = a$ the order of tangency not smaller than $\frac{3}{2}$ ($\mu \geq \frac{3}{2}$.)

(Sufficiency) Let us use the following fact: if a function $\Psi(z)$ is analytic in
$$\Delta_\varepsilon := \{z : |z-a| < \varepsilon\} \setminus L, \ \varepsilon > 0,$$
continuous up to the boundary excluding (possibly) the point $z = a$ and satisfies the conditions
$$\begin{cases} \sup_{z \in \Delta_\varepsilon} \Re\Psi(z) < \frac{k}{\sqrt{r}}, \ k = \text{constant}, \ r \to 0, \\ \sup_{z \in l_a} \Re\Psi(z) < k, \\ \sup_{t \in L} \Re\Psi^\pm(t) < k, \end{cases} \quad (3.3.15)$$
then (3.3.11) holds for $\Psi(z)$. This result is a variant of Phragmen–Lindelöf theorem (see, e.g., [82], p. 357; cf. Section 2.10). For the function Ψ of the form (3.3.9) the first two conditions (3.3.15) are of immediate verification. The third one follows from $\mu_a \geq \frac{3}{2}$ because
$$\begin{aligned} \max\{\Re\Psi^+(t), \Re\Psi^-(t)\} &= \frac{|c||t-z_0|}{\sqrt{|t-b|}} \frac{1}{\sqrt{r}} \left[\sin\frac{|\omega(r)|}{2} + O(r) \right] \\ &\leq \tilde{C}_1 \frac{\sin|\omega(r)|}{\sqrt{r}} = \tilde{C}_1 \frac{h(r)}{r^{\frac{3}{2}}}. \end{aligned}$$

It completes the proof. ∎

Remark 3.2 *It is clear that if $z_0 = a$ ($z_0 = b$) then the condition (3.3.11) is valid without any restriction on the behaviour of L at $z = a$ ($z = b$).*

Denote
$$\Psi(z, c, z_0) := \frac{c(z-z_0)}{\sqrt{(z-a)(z-b)}}, \quad (3.3.16)$$
where z_0 is an arbitrary point on the closed arc L. Then
$$\begin{aligned} \arg(a - z_0) - \arg(b - z_0) &= \alpha - \beta, \\ c = \lambda \exp\{i[\alpha + \pi - \arg(a - z_0)]\} &\equiv \lambda \exp\{i[\beta + \pi - \arg(b - z_0)]\}, \\ 0 \leq \lambda < \infty, \end{aligned}$$
where α, β are determined above. The direct consequence of Lemma 3.4 is the following solvability result for problem (3.3.5):

Theorem 3.2 *The general solution of problem (3.3.5) in the class $\mathcal{B}$ is delivered by the formula*

$$\Psi(z) = \begin{cases} \Psi(z, c, z_0), & \text{if } \mu_a \geq \frac{3}{2}, \mu_b \geq \frac{3}{2}, \\ \Psi(z, c, a), & \text{if } \mu_a < \frac{3}{2}, \mu_b \geq \frac{3}{2}, \\ \Psi(z, c, b), & \text{if } \mu_a \geq \frac{3}{2}, \mu_b < \frac{3}{2}, \\ 0, & \text{if } \mu_a < \frac{3}{2}, \mu_b < \frac{3}{2}, \end{cases} \quad (3.3.17)$$

with an arbitrary $z_0 \in L$.

Corollary 3.2 *The general solution of problem (3.3.2) in the class $\mathcal{B}^0$ is given by the formula*

$$\Phi(z) = \begin{cases} \exp\{\Psi(z, c, z_0)\}, & \text{if } \mu_a \geq \frac{3}{2}, \mu_b \geq \frac{3}{2}, \\ \exp\{\Psi(z, c, a)\}, & \text{if } \mu_a < \frac{3}{2}, \mu_b \geq \frac{3}{2}, \\ \exp\{\Psi(z, c, b)\}, & \text{if } \mu_a \geq \frac{3}{2}, \mu_b < \frac{3}{2}, \\ \pm 1, & \text{if } \mu_a < \frac{3}{2}, \mu_b < \frac{3}{2}, \end{cases} \quad (3.3.18)$$

with an arbitrary $z_0 \in L$.

It should be noted that the only two solutions of problem (3.3.2) in the class $\mathcal{B}^0$ are log-bounded, namely, $\Phi(z) = \pm 1$. The conditions $\mu_a \geq \frac{3}{2}$, $\mu_b \geq \frac{3}{2}$ are valid automatically for an arc which is analytic near the end points or is the Lyapunov-arc with the exponent $\lambda \geq \frac{1}{2}$ (see, e.g., [205]). Let us consider now the general problem of multiplication type (3.3.1), i.e.,

$$\Phi^+(t) \cdot \Phi^-(t) = f(t), \ t \in \widetilde{L},$$

$f \in \mathcal{H}$, $f(t) \neq 0$, $t \in L$, in the class $\mathcal{B}^0$. The general solution $\Phi(z)$ of problem (3.3.1) can be represented in the form

$$\Phi(z) = \Phi_0(z)\Phi_*(z), \quad (3.3.19)$$

where $\Phi_0(z)$ is a log-bounded solution of problem (3.3.1), and $\Phi_*(z)$ - the general $\mathcal{B}^0$-solution of problem (3.3.5). As any analytic branch of $\log \Phi_0(z)$ is single-valued in D and bounded in every closed domain $G \subset \widehat{\mathbb{C}} \setminus \{a, b\}$, then $\Phi_0(z)$ can be found as a solution of the following boundary value problem:

$$\log \Phi_0^+(t) = (-1) \log \Phi_0^-(t) + \log f(t), \ t \in \widetilde{L}. \quad (3.3.20)$$

The bounded in D solution of (3.3.20) is given by the formula (see SubSection 2.8.3)

$$\log \Phi_0(z) = \frac{\sqrt{(z-a)(z-b)}}{2\pi i} \int_a^b \frac{\log f(\tau)}{\sqrt{(\tau-a)(\tau-b)}} \frac{d\tau}{\tau - z}, \quad (3.3.21)$$

3.3. ENTIRE FUNCTIONS METHODS.

where $\sqrt{(z-a)(z-b)}$ is any branch of corresponding multi-valued function in D, $\sqrt{(\tau-a)(\tau-b)} := \sqrt{(\tau^+ - a)(\tau^+ - b)}$. It is not hard to see that

$$\Phi_0(z) := \exp\left[\frac{\sqrt{(z-a)(z-b)}}{2\pi i}\int_a^b \frac{\log f(\tau)}{\sqrt{(\tau-a)(\tau-b)}}\frac{d\tau}{\tau-z}\right] \qquad (3.3.22)$$

is log-bounded. Hence, (3.3.19) leads us to the following solvability result for (3.3.1) in $\mathcal{B}^0$:

Theorem 3.3 *The general solution of problem (3.3.1) in the class $\mathcal{B}^0$ is determined by the formula*

$$\Phi(z) = \begin{cases} \pm\Phi_0(z)\exp\{\Psi(z,c,z_0)\}, & \text{if } \mu_a \geq \frac{3}{2}, \mu_b \geq \frac{3}{2}, \\ \pm\Phi_0(z)\exp\{\Psi(z,c,a)\}, & \text{if } \mu_a < \frac{3}{2}, \mu_b \geq \frac{3}{2}, \\ \pm\Phi_0(z)\exp\{\Psi(z,c,b)\}, & \text{if } \mu_a \geq \frac{3}{2}, \mu_b < \frac{3}{2}, \\ \pm\Phi_0(z), & \text{if } \mu_a < \frac{3}{2}, \mu_b < \frac{3}{2}, \end{cases} \qquad (3.3.23)$$

where $\Phi_0(z)$ is given in (3.3.22), $\Psi(z,c,z_0)$ is defined in (3.3.16) with z_0 being an arbitrary point of L.

Let us now consider problem (3.3.1) in the class $\mathcal{B}^n, n = 1, 2, \ldots$. Suppose that the solution $\Phi(z)$ has zeros at $z = z_1, \ldots, z_s$, $s \leq n$ and at $z = \infty$ (the latter of order $n-s$). First of all we construct the "simple" function $N(z) \in \mathcal{B}^n$ having the same zeros as $\Phi(z)$. For this we add to L an arc l so that $L \cup l$ is a simple closed smooth curve, and consider the following boundary value problem

$$X^+(t) = G(t)X^-(t), \ t \in L \cup l, \qquad (3.3.24)$$

where

$$G(t) = \begin{cases} e^{2\pi i n\left(\frac{t-a}{t-b}\right)}, & t \in L, \\ 1, & t \in l. \end{cases} \qquad (3.3.25)$$

Clearly
$$\text{wind}_{L \cup l}\, G(t) = n.$$

Consider the following partial solution of (3.3.24) (see, e.g., [91, p. 488]; cf. Section 2.8):

$$\begin{aligned} X_0(z) &= \frac{1}{e^n(z-a)^n}\exp\left\{\frac{1}{2\pi i}\int_a^b \frac{\log G(\tau)}{\tau-z}d\tau\right\} \\ &= \frac{1}{(z-a)^n}\exp\left\{\frac{n}{b-a}(z-a)\log\frac{b-z}{a-z}\right\}, \end{aligned} \qquad (3.3.26)$$

where the branch of $\log \frac{b-z}{a-z}$ is determined by the condition $\lim_{z\to\infty} \log \frac{b-z}{a-z} = 0$. The function $X_0(z)$ is bounded and has no zero in D besides of nth order zero at infinity. Hence

$$N(z) = X_0(z) \prod_{k=1}^{s} (z - z_k) \qquad (3.3.27)$$

is analytic in D, Hölder-continuous up to its boundary and has the same zeros as $\Phi(z)$. Then the function $\Gamma(z) = \Phi(z)/N(z)$ belongs to the class $\mathcal{B}_0$ and satisfies the boundary condition

$$\Gamma^+(t)\Gamma^-(t) = \frac{f(t)}{N^+(t)N^-(t)}, \ t \in \tilde{L}. \qquad (3.3.28)$$

It is clear that problem (3.3.1) has a solution $\Phi \in \mathcal{B}^n$ iff problem (3.3.28) has a solution $\Gamma \in \mathcal{B}^0$. It gives

Theorem 3.4 *The general solution of problem (3.3.1) in the class $\mathcal{B}^n$ is given by the formula*

$$\Phi(z) = \begin{cases} \pm N(z)\Gamma_0(z)\exp\{\Psi(z,c,z_0)\}, & \text{if } \mu_a \geq \frac{3}{2}, \ \mu_b \geq \frac{3}{2}, \\ \pm N(z)\Gamma_0(z)\exp\{\Psi(z,c,a)\}, & \text{if } \mu_a < \frac{3}{2}, \ \mu_b \geq \frac{3}{2}, \\ \pm N(z)\Gamma_0(z)\exp\{\Psi(z,c,b)\}, & \text{if } \mu_a \geq \frac{3}{2}, \ \mu_b < \frac{3}{2}, \\ \pm N(z)\Gamma_0(z), & \text{if } \mu_a < \frac{3}{2}, \ \mu_b < \frac{3}{2}, \end{cases} \qquad (3.3.29)$$

where

$$\Gamma_0(z) = \exp \frac{\sqrt{(z-a)(z-b)}}{2\pi i} \int_a^b \log \frac{f(\tau)}{N^+(\tau)N^-(\tau)} \frac{1}{\sqrt{(\tau-a)(\tau-b)}} \frac{d\tau}{\tau-z},$$

N, Ψ *are determined by (3.3.27), (3.3.16) respectively.*

Let L be an unbounded simple smooth arc beginning at $z = a$ and ending at $z = \infty$, $D = \mathbb{C} \setminus L$. Without loss of generality $0 \notin L$. By definition the order of tangency of L at $z = \infty$ is equal to the order of tangency of Λ at $w = 0$, where Λ is the image of L under the mapping $w = \frac{1}{z}$. Let us consider in this situation the problem, analogous to problem (3.3.1), namely: find $\Phi \in \mathcal{B}^n$ such that

$$\Phi^+(t)\Phi^-(t) = f(t), \ t \in L, \qquad (3.3.30)$$

$f \in \mathcal{H}^*$ (i.e., Hölder-continuous on any finite part of $\tilde{L}$, possibly with integrable singularities at the end-points of L (see, e.g., [205])), $f(t) \neq 0$, $\forall t \in L$. Suppose that $\Phi(z)$ has zeros at the points $z = z_1, \ldots, z_s$, $s \leq n$ and $(n-s)$th order zero at $z = 0$. The direct consequence of the previous result is the following

3.3. ENTIRE FUNCTIONS METHODS.

Theorem 3.5 *The general solution of problem (3.3.1) in the class $\mathcal{B}^n$ has the form*

$$\Phi(z) = \begin{cases} \pm N(z)\Gamma_0(z)\exp\{\Psi(z,c,z_0)\}, & \text{if } \mu_a \geq \frac{3}{2},\ \mu_b \geq \frac{3}{2}, \\ \pm N(z)\Gamma_0(z)\exp\{\Psi(z,c,a)\}, & \text{if } \mu_a < \frac{3}{2},\ \mu_b \geq \frac{3}{2}, \\ \pm N(z)\Gamma_0(z)\exp\{\Psi(z,c,b)\}, & \text{if } \mu_a \geq \frac{3}{2},\ \mu_b < \frac{3}{2}, \\ \pm N(z)\Gamma_0(z), & \text{if } \mu_a < \frac{3}{2},\ \mu_b < \frac{3}{2}, \end{cases} \quad (3.3.31)$$

where $N(z) = z^n \prod_{k=1}^{s}(1-\frac{z_k}{z})\exp\left\{n(1-\frac{a}{z})\log(1-\frac{z}{a})\right\}$, *the branch of logarithmic function* $\log(1-\frac{z}{a})$ *is determined by condition* $\lim_{z\to 0}\log(1-\frac{z}{a}) = 0$;

$$\Gamma_0(z) = \exp\left\{\frac{\sqrt{a-z}}{2\pi i}\right\}\int_a^\infty \log\frac{f(\tau)}{N^+(\tau)N^-(\tau)}\frac{1}{\sqrt{a-\tau}}\frac{d\tau}{\tau-z};$$

$$\Psi(z,c,z_0) := \frac{c\left(1-\frac{z}{z_0}\right)}{\sqrt{1-\frac{z}{a}}},$$

the point z_0 belongs to the closed ray

$$\arg(a-z_0) = \arg a + \alpha - \beta;\ c = \lambda\exp i\{\beta + \arg z_0\},\ 0 \leq \lambda < \infty,$$

$$\alpha := \lim_{l_a \ni z \to a} \arg \frac{1}{z}\sqrt{1-\frac{z}{a}};\ \beta := \lim_{l_\infty \ni z \to \infty} \arg \frac{1}{z}\sqrt{1-\frac{z}{a}},$$

$$l_\infty := \left\{z \in \mathbb{C} : \arg z = \pi + \lim_{L \ni z \to \infty} \arg z\right\}.$$

3.3.2 Multiplication type problem for a compound contour.

Let L be a piece-wise smooth connected curve consisting of a finite collection of an open arc having no common points besides possibly their end points. Without loss of generality we can suppose that $\infty \notin L$. An end point of at least one arc of L is called *a knot of L*. An order of a knot of L is the number of arcs of L having this knot as an end point. Let us suppose additionally that all knots $\alpha_1,\ldots,\alpha_\nu$, $\nu \in \mathbb{N}$, of the curve L have even orders. As L is connected, then it divides the complex plane $\mathbb{C}$ into the finite number of simply connected domains $D_1,\ldots,D_\mu$. Besides L is an Euler graph. Hence, one can imagine $\widehat{\mathbb{C}} \setminus L$ as a 2-colourable chart such that any two domains with common boundary have different colour (see [284]).

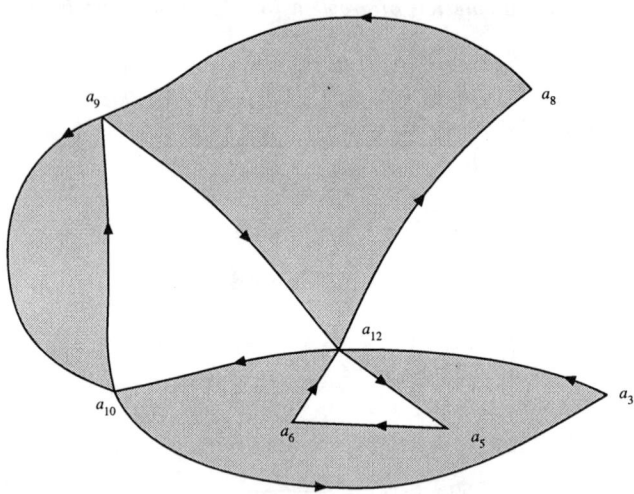

Figure 3.6: The compound contour (2-colourable graph) ($a_{12} = a_7 = a_4 = a_1$; $a_{10} = a_2$; $a_{11} = a_9$)

Denote the union of all domains with the same colour as that one containing ∞ by D^-, but the union of all others by D^+ (see Figure 3.6). Let us also choose the "positive" orientation on L, i.e., D^+ is always to the left of L. We consider the following classes of functions: $\mathcal{B}$ is a collection of all pairs of functions, analytic and bounded on $\mathbb{C} \setminus L$, nonvanishing on $L \setminus \{\alpha_1, \ldots, \alpha_\nu\}$; $\mathcal{B}^{k,l} \subset \mathcal{B}$, $k, l \in \mathbb{N}_0$, a subclass of all functions from $\mathcal{B}$ having exactly k zeros in D^+ and l zeros in D^-; $\mathcal{B}^0 := \mathcal{B}^{0,0}$; $\widetilde{\mathcal{B}}$ is a collection of all pairs of analytic functions on $\mathbb{C} \setminus L$, bounded in every closed domain $G \subset \mathbb{C} \setminus \{a_1, \ldots, a_\nu\}$. Let f be a complex-valued piece-wise continuous on L function, satisfying the Hölder condition on $L \setminus \{\alpha_1, \ldots, \alpha_\nu\}$; $f(t) \neq 0$, $t \in L$. We study problem: find all pairs of analytic functions (Φ^+, Φ^-), continuous up to $L \setminus \{\alpha_1, \ldots, \alpha_\nu\}$, by the following boundary condition:

$$\Phi^+(t)\Phi^-(t) = f(t),\ t \in L \setminus \{\alpha_1, \ldots, \alpha_\nu\}, \qquad (3.3.32)$$

As in previous subsection one can start with the investigation of the model problem:

$$\Phi^+(t)\Phi^-(t) = 1,\ t \in L \setminus \{\alpha_1, \ldots, \alpha_\nu\}, \qquad (3.3.33)$$

3.3. ENTIRE FUNCTIONS METHODS.

in the class $\mathcal{B}^0$. For brevity let us write

$$\Phi(z) = [H(z)]^{\pm}, \qquad (3.3.34)$$

meaning

$$\Phi(z) = \begin{cases} H(z), & z \in D^+, \\ [H(z)]^{-1}, & z \in D^-. \end{cases}$$

After changing unknown function

$$\Psi(z) := [\Phi(z)]^{\pm},$$

we get the following linear boundary condition

$$\Psi^+(t) = \Psi^-(t), \ t \in L \setminus \{\alpha_1, \ldots, \alpha_\nu\}. \qquad (3.3.35)$$

The Analytic Continuation Principle leads us to the following general solution of (3.3.35) in $\widetilde{\mathcal{B}}$

$$\Psi(z) = F_{\alpha_1,\ldots,\alpha_\nu}(z),$$

where $F_{\alpha_1,\ldots,\alpha_\nu}$ is an entire function with respect to points $\alpha_1, \ldots, \alpha_\nu$. It is evident that

$$\Phi(z) = [F_{\alpha_1,\ldots,\alpha_\nu}(z)]^{\pm} \qquad (3.3.36)$$

is the general solution of (3.3.33) in $\widetilde{\mathcal{B}}$ with $F_{\alpha_1,\ldots,\alpha_\nu}$ nonvanishing in D^-. Among them are in particular the solution of the type $\Phi(z) = [c]^{\pm}, c \neq 0$, which will be called *trivial solutions* of problem (3.3.33). Therefore, the question is what are the conditions under which (3.3.33) has nontrivial solutions in $\mathcal{B}^0$?

Lemma 3.5 *If the function Φ in (3.3.36) is a solution of problem (3.3.33) in $\mathcal{B}^0$, then the entirefunction $F_{\alpha_1,\ldots,\alpha_\nu}$ with respect to $\alpha_1, \ldots, \alpha_\nu$ from (3.3.6) can be represented in the form*

$$F_{\alpha_1,\ldots,\alpha_\nu} = \prod_{j=1}^{\nu}(z-\alpha_j)^{m_j} \exp\left\{\sum_{j=1}^{\nu} F_j\left(\frac{1}{z-\alpha_j}\right)\right\}, \qquad (3.3.37)$$

where $\sum_{j=1}^{\nu} m_j = 0$, $m_j \in \mathbb{Z}$, $j = 1,\ldots,\nu$, and F_j are some entire functions.

Proof. It is sufficient to get representation (3.3.37) for two points α_j, say $\alpha_0 = 0$, $\alpha_1 = 1$. Let us fix positive numbers r_0, r_1 in such a way that the circles

$$\mathbb{T}(0,r_0) := \{z : |z| = r_0\}, \ \mathbb{T}(1,r_1) := \{z : |z-1| = r_1\}$$

are separated, i.e., $r_0 + r_1 < 1$. Denote also

$$wind_{\mathbb{T}(0,r_0)} F_{0,1} = m_0, \quad wind_{\mathbb{T}(1,r_1)} F_{0,1} = m_1.$$

Evidently $m_0, m_1 \in \mathbb{Z}$ as $F_{0,1}$ is analytic in $\widehat{\mathbb{C}} \setminus \{0,1\}$. If we take $b > 0$ so large that

$$int\,\mathbb{D}(0,b) := \{z : |z| < b\} \supset \mathbb{T}(0,r_0) \bigcap \mathbb{T}(1,r_1),$$

then by the Argument Principle

$$\begin{cases} wind_{\mathbb{T}(0,b)} F_{0,1} = 0, \\ wind_{\mathbb{T}(0,b)} F_{0,1} = m_0 + m_1. \end{cases}$$

Therefore the function

$$F(z) := F_{0,1}(z) z^{-m_0} (z-1)^{m_1}$$

is the entire function with respect to $z = 0$, $z = 1$, nonvanishing in $\widehat{\mathbb{C}} \setminus \{0,1\}$. Besides,

$$\oint_\gamma d\log F(z) = 0$$

for any closed smooth curve $\gamma \in \widehat{\mathbb{C}} \setminus L$. Hence any analytic branch of the function $\log F(z)$ is an entire function with respect to $z = 0$, $z = 1$. Then (see, e.g., [152]) it can be written in the form

$$\log F(z) = F_0(\frac{1}{z}) + F_1(\frac{1}{z-1}),$$

where F_0, F_1 are entire functions. The last formula gives us the representation (3.3.37), i.e.,

$$\Phi(z) = \left[\prod_{j=1}^{\nu} (z - \alpha_j)^{m_j} \exp\left\{ \sum_{j=1}^{\nu} F_j(\frac{1}{z - \alpha_j}) \right\} \right]^{\pm}. \qquad (3.3.38)$$

It completes the proof. ∎

Lemma 3.5 permits studying solutions of the type (3.4.4) locally, near one of the knots α_j. For definiteness we will do it at the point α_0, and suppose $\alpha_0 := 0$. The following results generalize Lemma 3.5 (cf. [103])

Lemma 3.6 *Let $\Phi(z)$ in (3.3.36) be of the class $\mathcal{B}^0$ in a neighbourhood of the knot $\alpha_0 = 0$. Then F_0 is a polynomial.*

3.3. ENTIRE FUNCTIONS METHODS.

Proof. After changing variable $\zeta = \frac{1}{z}$ we have to study the behaviour of $\Phi(z)$ in a neighbourhood of $\zeta = \infty$. The image L_ζ of contour L coincides locally with m simple smooth arcs met at $\zeta = \infty$. Due to smoothness of L there exists $r_0 > 0$ such that any circle $\mathbb{T}(0, r)$, $r > r_0$, has only one common point with each arc. Let

$$F_0(\zeta) = \sum_{n=0}^{\infty} a_n \zeta^n \qquad (3.3.39)$$

be a *transcendental* (i.e., non-polynomial) entire function, and $a_\nu \zeta^\nu$ be its *maximal term*, $\nu = \nu(r)$ be its *central index*. (see Section 2.10) Remember that a set $E \subset [0, \infty)$ has a *finite logarithmic measure* if

$$\limsup_{r \to \infty} \frac{\log \, mes \, E \cap [0, r)}{\log r} < \infty.$$

The next result is known as *Valiron's estimate* (cf. [265, p. 195], see also Section 2.10): let F_0 be a transcendental entire function (3.3.39), $\zeta(r) = re^{i\phi_0}$ be a point on $\mathbb{T}(0, r)$ such that

$$|F_0(\zeta(r))| = M(r, F_0) := \max_{|\zeta|=r} |F_0(\zeta)|;$$

then for all $r > 0$ except of a set E of finite logarithmic measure the following formula takes place

$$F_0(re^{i\phi}) = e^{i(\phi - \phi_0)\nu(r)} F_0(\zeta(r)) (1 + \chi(r, \phi)), \qquad (3.3.40)$$

where $|\phi - \phi_0| < \nu(r)^{-\frac{15}{16}}$, $|1 + \chi(r, \phi)| < K\nu(r)^{-\frac{1}{16}}$, $K = const$.
Let us fix $r > r_0$, $r \notin E$. Divide the arc

$$\sigma(r) := \left\{ re^{i\phi} \in \mathbb{T}(0, r) : |\phi - \phi_0| < \nu(r)^{-\frac{15}{16}} \right\}$$

onto $(m + 1)$ equal parts. One of these subarcs, say,

$$\sigma_0(r) := \left\{ re^{i\phi} \in \mathbb{T}(0, r) : \phi_1 < \phi < \phi_2, \phi_2 - \phi_1 = \frac{1}{m+1} \nu(r)^{-\frac{15}{16}} \right\},$$

lays between two neighbouring arcs of the contour L_ζ in a domain D_0 (without loss of generality $D_0 \subseteq D^+$). Then for $\zeta \in D_0$ the corresponding component of the solution (3.3.38) contains the factor

$$\zeta^{-m_0} \exp\{F_0(\zeta)\}$$

determining behaviour of $\Phi(z)$ at $z = 0$. It follows from (3.3.40) that

$$\Delta := \arg F_0(re^{i\phi_2}) - \arg F_0(re^{i\phi_1})$$
$$= (\phi_2 - \phi_1)\nu(r) + \arg(1 + \chi(r, \phi_2)) - \arg(1 + \chi(r, \phi_1)).$$

By construction the Valiron's estimate gives

$$\Delta > 2\pi, \text{ for all } r > r_1.$$

Hence there exists a point $\zeta_r \in \sigma_0(r) : F_0(\zeta_r) > 0$, i.e., $\Re F_0(\zeta_r) = |F_0(\zeta_r)|$. Therefore, (3.3.40) leads to

$$\lim_{E \ni r \to +\infty} \Re F_0(\zeta_r) = \lim_{E \ni r \to +\infty} F_0(\zeta_r) = \lim_{r \to +\infty} \max_\phi |F_0(re^{i\phi})| = +\infty.$$

Therefore, $\Phi \notin \mathcal{B}^0$. Hence F_0 is a polynomial. ∎

We suppose in what follows that the knot $\alpha_0 = 0$ is a common point of $2n$ simple smooth arcs of L (all knots are of even order).

Lemma 3.7 *If Φ is a solution of (3.3.33) in $\mathcal{B}^0$, then $\deg F_0 \leq n$.*

Proof. If not, i.e., $\deg F_0 = m \geq n+1$, then simple geometric considerations show that for any ε, $0 < \varepsilon < \frac{\pi}{n}$, there exists a sector

$$V_\varepsilon := \left\{ z = re^{i\phi} : 0 < r < r_\varepsilon, \gamma_1 < \phi < \gamma_2, \gamma_2 - \gamma_1 = \frac{\pi}{n} - \varepsilon \right\},$$

having no common point with L (thus $V_\varepsilon \subset D^+$ or $V_\varepsilon \subset D^-$). For sufficiently small $\varepsilon > 0$ one can find two rays v_1, v_2 in V_ε with the angle between them bigger than $\frac{\pi}{m}$. On these rays the real part of F_0 has different signs, i.e.,

$$\Re F_0(\tfrac{1}{z}) = \tfrac{C_1}{|z|^m} + o(\tfrac{1}{|z|^m}), \ z \in v_1, \ z \to 0,$$
$$\Re F_0(\tfrac{1}{z}) = \tfrac{C_2}{|z|^m} + o(\tfrac{1}{|z|^m}), \ z \in v_2, \ z \to 0,$$

for certain constants C_1, C_2; $C_1 \cdot C_2 < 0$. It shows that Φ could not be bounded in the neighbourhood of $\alpha_0 = 0$. ∎

Up to rotation the polynomial F_0 can be written in the form

$$F_0(z) = aiz^n + c_{n-1}z^{n-1} + \cdots + c_1 z, \ a \in \mathbb{R}, \ c_j \in \mathbb{C}, \ j = 1, \ldots, n-1. \quad (3.3.41)$$

Let us consider first the solvability of (3.3.33) in the case when the arcs of L have no common one-sided tangent line at $\alpha = 0$. It is not hard to see that in this situation $a \neq 0$ in (3.3.41). Thus, if $a = 0$, then $\deg F_0 \leq n-1$. Denote

$$l(\rho, \eta) := \{ z = re^{i\phi} : 0 < r < \rho, \phi = \eta \}$$

3.3. ENTIRE FUNCTIONS METHODS.

a segment with $z = 0$ being one of its end-points. Let ϕ_0 be the minimal angle between two arcs of L at $\alpha = 0$. Then $l(\rho, \tilde{\eta})$, $l(\rho, \tilde{\eta}+\phi_0)$ lay on one-sided tangent lines to two arcs of L. We can suppose that the domain between these arcs belongs to D^+. So, the function $\Re F_0(\frac{1}{z})$ should be upper-bounded on any segment $l(\rho, \eta)$, $\tilde{\eta} < \eta < \tilde{\eta}+\phi_0$, for sufficiently small ρ, and lower-bounded on the segments $l(\rho, \tilde{\eta}-\varepsilon), l(\rho, \tilde{\eta}+\phi_0+\varepsilon)$, $0 < \varepsilon < \phi_0$, for sufficiently small ρ. If $\deg F_0 = n - 1$, then the neighbourhood of $z = 0$ is divided onto $n - 1$ equal sectors

$$S_i := \{z = re^{i\phi} : 0 < r < \rho, \eta_i < \phi < \eta_{i+1}\}, i = 1, \ldots, n-1, \eta_n = \eta_1,$$

such that the maximal subsector on which $\Re F_0(\frac{1}{z})$ is upper-bounded is sufficiently closed (but smaller) to certain S_i. On the same subsector $\Re F_0(\frac{1}{z})$ is unbounded from below. Since $\phi_0 \leq \frac{\pi}{n}$ then this function should be unbounded from below or on $l(\rho, \eta_i - \varepsilon)$, or on $l(\rho, \eta_i + \phi_0 + \varepsilon)$ for small $\varepsilon > 0$, because $\phi_0 + 2\varepsilon < \frac{\pi}{n-1}$. It contradicts the above described behaviour of F_0.

Let $\Phi \in \mathcal{B}^0$ be a solution of problem (3.3.33). Then its boundary values are log-bounded on all arcs of L, i.e., the function $\Re F_0(\frac{1}{z})$ is upper- and lower-bounded there. For a polynomial of the degree n there exist exactly $2n$ rays on which $\Re F_0(\frac{1}{z})$ is upper- and lower-bounded. Therefore we get immediately the following necessary condition for the existence of nontrivial solution of (3.3.33):

Corollary 3.3 *If (3.3.36) is a nontrivial solution in $\mathcal{B}^0$ of the problem (3.3.33), then the arcs L_k are tangent to the system of $2n$ rays with angles $\frac{\pi}{n}$ between them.*

These rays coincide with $l(\rho, \phi_k)$, $\phi_k = \frac{k\pi}{n}$, $k = 0, \ldots, 2n-1$, if F_0 has the form (3.3.41). So, we can assume that $l(\rho, \phi_k)$ are tangent lines to L_k.

Lemma 3.8 *If $a < 0$ and the function*

$$H(z) = \left[z^m \exp\left\{F_0(\frac{1}{z})\right\}\right]^{\pm}, \quad m \in \mathbb{Z}, \tag{3.3.42}$$

is bounded on the arcs L_k, $k = 0, \ldots, 2n-1$, then $\Phi \in \mathcal{B}_0$ in a neighbourhood of $\alpha_0 = 0$.

First we note that for small $\rho > 0$ the segments $l(\rho, \frac{\pi(4k+1)}{2n})$, $k = 0, \ldots, n-1$, belong to D^+, and the segments $l(\rho, \frac{\pi(4k-1)}{2n})$, $k = 1, \ldots, n$, belong to D^-. On all these segments we have

$$|H(z)| = \exp\left\{ar^{-n} + O(r^{-n+1})\right\}, \quad z = re^{i\phi}, r \to 0.$$

Consider now any domain D between the arc L_k and one of such segments in a neighbourhood of $z = 0$. After changing variable $\zeta := \frac{1}{z}$ the domain D becomes G_0. Applying the Phragmen–Lindelöf theorem (see, e.g., [82, p. 357]; cf. Section 2.10) to $H(\frac{1}{\zeta})$ in G_0 one can get its boundedness.

Remark 3.3 *If Φ is a solution of (3.3.33) of the class B_0 in a neighbourhood of $z = 0$, then $a < 0$.*

It follows from the fact that the main part of Φ is the function H given by formula (3.3.42).

Let us now study an asymptotic behaviour of the polynomial F_0 in (3.3.41) at a neighbourhood of the knot $\alpha_0 = 0$. Suppose that the arcs L_k, $k = 0, \ldots, 2n - 1$, have the equations $\phi = \phi_k(r)$ in polar coordinate $z = re^{i\phi}$ near $\alpha_0 = 0$. It follows from the Corollary 3.3 that

$$\phi_k(r) \to \frac{\pi k}{n}, \quad k = 0, \ldots, 2n - 1.$$

Lemma 3.8 shows that $H(z)$ in (3.3.42) is bounded iff it is bounded on arcs L_k. Hence

$$-\frac{\sin n\phi_k(r)}{r^n} + \sum_{j=1}^{n-1} \frac{a_j \cos j\phi_k(r) + b_j \sin j\phi_k(r)}{r^j} + \frac{m}{a} \log r = O(1), \quad r \to 0, \quad (3.3.43)$$

or in terms of $\psi_k(r) := \phi_k(r) - \frac{\pi k}{n}$, $k = 0, \ldots, 2n - 1$,

$$(-1)^k \sin n\psi_k(r) - \sum_{j=1}^{n-1}(A_{j,k} \cos j\psi_k(r) + B_{j,k} \sin j\psi_k(r))r^{n-j}$$
$$- \frac{m}{a} r^n \log r = O(r^n), r \to 0,$$

where $A_{j,k} := a_j \cos \frac{jk\pi}{n} + b_j \sin \frac{jk\pi}{n}$, $B_{j,k} := -a_j \sin \frac{jk\pi}{n} + b_j \cos \frac{jk\pi}{n}$. Using the following identities for trigonometric functions (see, e.g., [2])

$$\begin{cases} \cos kz := {}_2F_1(\frac{k}{2n}, -\frac{k}{2n}; \frac{1}{2}; \sin^2 nz), \\ \sin kz := \frac{k}{n} \sin nz \, {}_2F_1(\frac{1}{2} + \frac{k}{2n}, \frac{1}{2} - \frac{k}{2n}; \frac{3}{2}; \sin^2 nz), \end{cases} \quad (3.3.44)$$

where ${}_2F_1$ is the Gauss hypergeometric function (cf. [80]), one can expand $\cos kz, \sin kz$ in a power series with respect to $\sin nz$. Besides, the first expansion contains only even powers, but the second one contains only odd ones. We need now in the following technical result:

3.3. ENTIRE FUNCTIONS METHODS.

Lemma 3.9 *([115]) Let the functions $f : [0, r_0] \to \mathbb{R}$,*

$$f(r) = o(1), \ r \to 0, f_j(r) = \sum_{k=0}^{\infty} a_{k,j}(f(r))^k, \ j = 1, \ldots, n-1, a_{k,j} \in \mathbb{R}, |a_{k,j}| < \frac{c}{k^{1+\varepsilon}},$$

be such that the following asymptotic equality takes place

$$f(r) + \sum_{k=0}^{\infty} f_j(r) r^{n-j} + \theta r^n \log r = O(r^n), \ r \to 0, \ \theta \in \mathbb{R}. \tag{3.3.45}$$

Then f satisfies the asymptotic representation

$$f(r) = \sum_{k=0}^{\infty} b_k r^k - \theta r^n \log r + O(r^n), \ r \to 0, \tag{3.3.46}$$

where b_k are real constants.

The result of Lemma 3.9 means that the asymptotic relation (3.3.45) is in fact uniform, i.e., the coefficients of the main part of asymptotics (the sum in (3.3.45)) contain only pure power terms with constant coefficients. A straightforward consequence of Lemma 3.9 and the properties of hypergeometric function (see [80, p. 70]) is the next asymptotic description of the behaviour of arcs L_k at $\alpha_0 = 0$ in the case of the existence of a nontrivial solution of (3.3.33) in $\mathcal{B}^0$.

Lemma 3.10 *Let $\phi_k(r) = \psi_k(r) + \frac{\pi k}{n}$, $k = 0, \ldots, 2n-1$, be the equations of arcs L_k in polar co-ordinates, and the asymptotic formulae (3.3.43) are valid. Then*

$$\sin n\psi_k(r) = \sum_{j=1}^{n-1} d_{j,k} r^j + (-1)^k \frac{m}{a} r^n \log r + O(r^n), \ r \to 0. \tag{3.3.47}$$

Let now some of arcs L_k have a common one-sided tangent line at a knot α. The necessary condition for the existence of nontrivial solution of (3.3.33) is given in the following

Lemma 3.11 *If problem (3.3.33) has a nontrivial solution in $\mathcal{B}^0$, then necessarily only odd number of arcs L_k can have a common one-sided tangent line at a knot α.*

Proof. Without loss of generality we can suppose $\alpha = \alpha_0 = 0$. It was already shown that if (3.3.33) has a nontrivial solution, then $deg\, F_0 \leq n$, where $2n$ is the number of simple arcs L_k at the knot α_0. As $\Phi^{\pm}(t)$ are bounded on L, then $\exp F_0(\frac{1}{z})$ is log-bounded on L_k, $k = 0, \ldots, 2n - 1$. Let us suppose that an *even* number of L_k, namely $L_0, \ldots, L_{2k_0-1}$, $k_0 \leq n$, has a common one-sided tangent line $\phi = 0$. Suppose also that there exists a polynomial F_0, $deg\, F_0 = n_0 \leq n$, such that $H(z)$ in (3.3.42) is bounded. It follows from Lemma 3.7 that

$$F_0(z) = aiz^{n_0} + C_{n_0-1}z^{n_0-1} + \cdots + c_1 z, a \in \mathbb{R} \setminus \{0\}, c_k \in \mathbb{C}.$$

Besides the arcs L_j, $j = 2k_0, \ldots, 2n - 1$, have one of the following tangent lines: $\phi = \frac{\pi k}{n_0}$, $k = 1, \ldots, 2n_0 - 1$. One can note that the region between L_{2k_0-1} and L_{2k_0}, as well as that between L_{2n-1} and L_0 lies in the domains of the same type, say D^+ for certainty. As L_{2k_0} is tangent to the ray $\phi = \frac{\pi}{n_0}$ and L_{2n-1} is tangent to the ray $\phi = 2\pi - \frac{\pi}{n_0}$. Hence the rays $\phi = \frac{\pi}{2n_0}$ and $\phi = 2\pi - \frac{\pi}{2n_0}$ are inside of D^+. Hence the function $\Re F_0(\frac{1}{z})$ should be bounded on these rays. This is impossible because the following asymptotic relations

$$\Re F_0(r^{-1}e^{-\frac{i\pi}{2n_0}}) \sim \frac{a}{r^{n_0}}, \qquad r \to 0,$$
$$\Re F_0(r^{-1}e^{-i(2\pi-\frac{\pi}{2n_0})}) \sim -\frac{a}{r^{n_0}}, \qquad r \to 0$$

lead to a contradiction. ∎

So, only the odd number of arcs L_k can have a common one-sided tangent line. Choose one representative corresponding to each family of arcs L_k with common one-sided tangent line, and denote them $\widetilde{L_0}, \ldots, \widetilde{L_{2m_0-1}}$ (it is evident that there is an even quantity of such families). As it was already seen $H(z)$ in (3.3.42) is a nontrivial solution of (3.3.33) if

$$deg\, F_0(z) = m_0.$$

Up to rotation $\widetilde{L_0}$ is tangent to $\phi = 0$. Then the same considerations as above lead us to the following representation for F_0:

$$F_0(z) = \tilde{a}\left(-iz^{m_0} + \sum_{j=1}^{m_0-1}(\tilde{a}_k + i\tilde{b}_k)z^k\right), \tilde{a} > 0, \tilde{a}_k, \tilde{b}_k \in \mathbb{R}.$$

Therefore $\widetilde{L_k}$ are tangent to $\phi = \frac{2\pi k}{m_0}$, $k = 0, \ldots, 2m_0 - 1$. As $\Re F_0(\frac{1}{z})$ is bounded on $\widetilde{L_k}$, then one can get the asymptotic formula for the equation $\phi = \tilde{\phi}_k(r) = \tilde{\psi}_k(r) + \frac{\pi k}{m_0}$

3.3. ENTIRE FUNCTIONS METHODS.

of the arcs $\widetilde{L_k}$ at $\alpha_0 = 0$

$$(-1)^k \sin m_0 \widetilde{\psi}_k(r) - \sum_{j=1}^{m_0-1}(\widetilde{A}_{j,k} \cos j\widetilde{\psi}_k(r) \\ + \widetilde{B}_{j,k} \sin j\widetilde{\psi}_k(r))r^{m_0-j} - \tfrac{m_0}{\tilde{a}}r^{m_0} \log r = O(r^{m_0}), r \to 0, \qquad (3.3.48)$$

where $\widetilde{A}_{j,k} = \tilde{a}_j \cos \tfrac{jk\pi}{m_0} + \tilde{b}_j \sin \tfrac{jk\pi}{m_0}$; $\widetilde{B}_{j,k} = -\tilde{a}_j \sin \tfrac{jk\pi}{m_0} + \tilde{b}_j \cos \tfrac{jk\pi}{m_0}$. Besides, for the arcs L_k with common tangent line the relation (3.3.48) takes place with the same parameter k. Hence

$$\sin m_0 \widetilde{\psi}_{k_1}(r) - \sin m_0 \widetilde{\psi}_{k_2}(r) = O(r^{m_0}), \ r \to 0, \qquad (3.3.49)$$

for every two arcs L_{k_1}, L_{k_2} with common tangent line.

Now we have to construct polynomials F_j, $j = 1, \ldots, \nu$, for which the formula (3.3.38) gives nontrivial solution of the model problem (3.3.33) in $\mathcal{B}^0$, as well as to formulate the conditions for their existence.

Definition 3.4 *The curve L is called regular at the knot α_l if any two neighbouring one-sided tangent lines to arcs L_k at α_l form the same angles, and any one-sided tangent line touches an odd number of arcs L_k.*

As it was already mentioned, if problem (3.3.33) has a nontrivial solution in $\mathcal{B}^0$, then the contour L is regular at any knot α_l. Let us fix the knot α_l and choose one arc $L_{l,j}$ from every collection of arcs of L tangent to a certain ray at $z = \alpha_{l_i}$. It was already shown that the equation of $L_{l,j}$ with respect to polar coordinates (ϕ, r) at α_l

$$\phi = \phi(r) := \psi_{l,j} + \beta_l + j\frac{\pi}{n_l}$$

has to have the following asymptotics

$$\sin n_l \psi_{l,j}(r) = \sum_{k=1}^{n_l-1} d_{l,j,k} r^k + (-1)^j \theta_l r^{n_l} \log r + O(r^{n_l}), \ r \to 0. \qquad (3.3.50)$$

The same considerations as before lead us to the necessary condition for the existence of a nontrivial solution of (3.3.33) at α_l:

$$(-1)^j \sin n_l \psi_{l,j}(r) - \sum_{k=1}^{n_l-1}(A_{l,j,k} \cos k\psi_{l,j}(r) + B_{l,j,k} \sin k\psi_{l,j}(r))r^{n_l-k} \\ -\theta_j r^{n_l} \log r = O(r^{n_l}), \ r \to 0, \ l = 1,\ldots,\nu, \ j = 0,\ldots,2n_l-1, \ \theta_j \in \mathbb{R}, \qquad (3.3.51)$$

where $A_{l,j,k} = a_k^{(l)} \cos \frac{jk\pi}{n_l} + b_k^{(l)} \sin \frac{jk\pi}{n_l}$; $\tilde{A}_{l,j,k} = -a_k^{(l)} \sin \frac{jk\pi}{n_l} + b_k^{(l)} \cos \frac{jk\pi}{n_l}$. Besides, if two simple arcs

$$\phi = \phi(r) := \psi_l(r) + \beta_l + \frac{\pi k}{n_l}, \quad \tilde{\phi} = \tilde{\phi}(r) := \tilde{\psi}_l(r) + \beta_l + \frac{\pi k}{n_l}$$

have a common tangent line, then

$$\sin n_l \psi_l(r) - \sin n_l \tilde{\psi}_l(r) = O(r^{n_l}), \quad r \to 0. \tag{3.3.52}$$

Applying representation (3.3.44) to asymptotic formula (3.3.51), and comparing this formula with (3.3.50) one can get the following system of equations with respect to unknown $a_k^{(l)}, b_k^{(l)}$:

$$(-1)^j d_{l,j,k} - A_{l,j,n_l-k} - P_{l,j,k} = 0, \tag{3.3.53}$$

where $d_{l,j,k}$ are the same as in (3.3.50), $A_{l,j,k}$ are given in (3.3.51), and $P_{l,j,k}$ are linearly depending on $a_k^{(l)}, b_k^{(l)}$. The linear algebraic system (3.3.53) consists of $2n_l(n_l-1)$ equations with respect to $2n_l - 2$ unknowns. It is not hard to see that the matrix of (3.3.53) has a maximal rang equal to $2n_l - 2$. Therefore, the subsystem of equations corresponding to $j = 0$ and $j = 1$ has the form

$$\begin{cases} d_{l,0,1} & = a_{n_l-1}^{(l)}, \\ -d_{l,1,1}, & = a_{n_l-1}^{(l)} \cos \frac{\pi}{n_l} + b_{n_l-1}^{(l)} \sin \frac{\pi}{n_l}, \\ \mathcal{L}_{0,1}(a_{n_l-1}^{(l)}, b_{n_l-1}^{(l)}) & = a_{n_l-2}^{(l)}, \\ \mathcal{L}_{0,1}(a_{n_l-1}^{(l)}, b_{n_l-1}^{(l)}) & = -a_{n_l-2}^{(l)} \cos \frac{2\pi}{n_l} + b_{n_l-2}^{(l)} \sin \frac{2\pi}{n_l}, \\ \ldots & \ldots \\ \mathcal{L}_{0,n_l-2}(a_{n_l-1}^{(l)}, b_{n_l-1}^{(l)}, \ldots, a_2^{(l)}, b_2^{(l)}) & = a_1^{(l)}, \\ \mathcal{L}_{0,n_l-2}(a_{n_l-1}^{(l)}, b_{n_l-1}^{(l)}, \ldots, a_2^{(l)}, b_2^{(l)}) & = -a_1^{(l)} \cos \frac{(n_l-1)\pi}{n_l} + b_1^{(l)} \sin \frac{(n_l-1)\pi}{n_l}, \end{cases}$$

where $\mathcal{L}_{j,k}(x_1, \ldots, x_{2k})$, $j = 0, 1$, $k = 1, \ldots, n_l - 1$, are linear functions. The determinant of this system is equal to

$$\mathcal{J} = \prod_{k=1}^{n_l-1} \sin \frac{k\pi}{n_l} \neq 0.$$

So, the system (3.3.53) has a solution iff it contains exactly $2n_l - 2$ linear independent equations. Now one can describe the result concerning nontrivial solutions of (3.3.33) algorithmically.

Algorithm for determining of nontrivial solutions of (3.3.33) in $\mathcal{B}^0$:

3.3. ENTIRE FUNCTIONS METHODS.

$1.^0$ Find knots with respect to which the curve L is regular. If any, then go to the step $2.^0$ If not, problem (3.3.33) has only trivial solution in $\mathcal{B}^0$.

$2.^0$ See for knots satisfying (3.3.50), (3.3.52), where $2n_l$ is a number of different one-sided tangent lines, and θ_l, $d_{l,j,k}$ are real numbers. If these asymptotic relations take place for at least one knot and all arcs end at it, then go to the step $3.^0$. If not, then problem (3.3.33) has only trivial solution in $\mathcal{B}^0$.

$3.^0$ Taking $d_{l,j,k}$ from (3.3.50) consider the system (3.3.53) with respect to $a_k^{(l)}$, $b_k^{(l)}$. If for at least one knot this system has a solution, then go to step $4.^0$. If not, then problem (3.3.33) has only trivial solution in $\mathcal{B}^0$.

$4.^0$ Let $\{\alpha_l \,|\, l \in I\}$ be the family of knots for which the answers of $1.^0 - 3.^0$ are positive, and $\{\theta_l \,|\, l \in I\}$ are real numbers from the asymptotic representation (3.3.50). Problem (3.3.33) has nontrivial solutions in $\mathcal{B}^0$ if either some of θ_l have different signs, or some of them are equal to zero.

Let us introduce some notations: denote m_l as any integer number for which $\text{sign } m_l = \text{sign } \theta_l$; $a_l := \frac{m_l}{\theta_l}$, if $\theta_l \neq 0$, and a_l is an arbitrary positive real number, if $\theta_l = 0$;

$$F_l := P_l(ze^{i\beta_l}), \tag{3.3.54}$$

where

$$P_l(z) := a_l\left(-iz^{n_l} + \sum_{k=1}^{n_l-1}(a_k^{(l)} + ib_k^{(l)})z^k\right), l \in I, \tag{3.3.55}$$

n_l is a number of arcs $L_{l,j}$ with the end-point α_l, β_l is a polar angle to the tangent line $L_{l,0}$, $a_k^{(l)}, b_k^{(l)}$ are determined from the system (3.3.53). From combining the above one can formulate the main result on solvability of problem (3.3.33) in $\mathcal{B}^0$.

Theorem 3.6 *If the answer on $1.^0 - 3.^0$ in the Algorithm is positive, then problem (3.3.33) has nontrivial solutions in $\mathcal{B}^0$, represented in the above notations by the formulae*

$$\Phi(z) := \left[c\prod_{l \in I_1}(z-\alpha_l)^{m_l}\exp\left\{\sum_{l \in I_1}F_l(\frac{1}{z-\alpha_l})\right\}\right]^{\pm}, \tag{3.3.56}$$

where $I_1 \subset I$ is any subset of I, and

$$\sum_{l \in I_1} m_l = 0. \tag{3.3.57}$$

Let us now consider problem (3.3.32), i.e.,

$$\Phi^+(t)\Phi^-(t) = f(t), \, t \in L \setminus \{\alpha_1, \ldots, \alpha_\nu\},$$

in the class $\mathcal{B}^{k,l} := \mathcal{B}^k(D^+) \times \mathcal{B}^l(D^-)$. Suppose that $\Phi^+(z)$ has zeros at the points $z_j^+ \in D^+, j = 1, \ldots, k$, and $\Phi^-(z)$ has zeros at the points $z_j^- \in D^-, j = 1, \ldots, l_1$. Let $z = 0$ belong to D^+, and $l := l_0 + l_1$, where l_0 is an order of $\Phi(z)$ at $z = \infty$. We change the unknown function using the formulas

$$\Phi_1^+(z) := \Phi^+(z) \prod_{j=1}^{k} \left(1 - \frac{z}{z_j^+}\right)^{-1}, \quad z \in D^+,$$

$$\Phi_1^-(z) := \Phi^-(z) z^{l_0} \prod_{j=k+1}^{k+l_1} \left(1 - \frac{z_j^-}{z}\right)^{-1}, \quad z \in D^-.$$

The functions $\Phi_1^+(z), \Phi_1^-(z)$ are analytic in the domains D^+, D^- respectively, having no zeros there. These functions do satisfy the boundary conditions

$$\Phi_1^+(t)\Phi_1^-(t) = f_1(t), \quad t \in L \setminus \{\alpha_1, \ldots, \alpha_\nu\}, \tag{3.3.58}$$

where

$$f_1(t) := f(t) t^{-l_0} \prod_{j=1}^{k} \left(1 - \frac{t}{z_j^+}\right) \prod_{j=1}^{l_1} \left(1 - \frac{z_j^-}{z}\right). \tag{3.3.59}$$

The set $L \setminus \{\alpha_1, \ldots, \alpha_\nu\}$ is decomposed onto a finite collection of open smooth arcs L_j. On every arc L_j one can choose a single-valued branch of the following function

$$\log f_1(z) := \log f(z) + \log \left[z^{-l_0} \prod_{j=1}^{k} \left(1 - \frac{z}{z_j^+}\right) \prod_{j=1}^{l_1} \left(1 - \frac{z_j^-}{z}\right) \right].$$

Moreover,

$$\frac{1}{2\pi} \left(\sum_j{}'' \arg_{L_j''} f(\alpha_k) - \sum_j{}' \arg_{L_j'} f(\alpha_k) \right) = \\ \frac{1}{2\pi} \left(\sum_j{}'' \arg_{L_j''} f_1(\alpha_k) - \sum_j{}' \arg_{L_j'} f_1(\alpha_k) \right) =: \chi_k, \tag{3.3.60}$$

$k = 1, \ldots, \nu$, where the sum $\sum_j{}'$ ($\sum_j{}''$) is taken over all simple arcs L_j' (L_j'' respectively) which begin (respectively end) at α_k; $\arg_{L_j'} f(\alpha_k)$ ($\arg_{L_j''} f(\alpha_k)$) are limit values of f at α_k taken along the arcs L_j' (L_j'') respectively). Introducing the pair of functions

$$\Phi_0^\pm := [\exp\{\Gamma_0(z)\}]^\pm, z \in D^\pm, \tag{3.3.61}$$

$$\Gamma_0(z) := \frac{1}{2\pi i} \int_L \frac{\log f_1(\tau)}{\tau - z} d\tau,$$

3.3. ENTIRE FUNCTIONS METHODS.

we reduce problem (3.3.32) in $\mathcal{B}^{k,l}$ to problem

$$\Psi^+(t)\Psi^-(t) = 1, \ t \in L \setminus \{\alpha_1, \ldots, \alpha_\nu\},$$

in the class $\mathcal{B}^0$ with respect to the functions

$$\Psi^\pm(z) := \Phi_1^\pm \left[\Phi_0^\pm\right]^{\mp 1}, \ z \in D^\pm. \tag{3.3.62}$$

The latter problem is already investigated. Therefore, one can formulate the final result about solvability of problem (3.3.32) in $\mathcal{B}^{k,l}$. To do it we introduce the following notations:

$$I_1 := \{l \in I : \chi_l = [\chi_l] \wedge \theta_l = 0\}, I_2 := \{l \in I : \theta_l > 0\},$$

$$I_3 := \{l \in I : \theta_l < 0\}, I_4 := \{l \in \mathbb{N} \setminus I : \chi_l = [\chi_l]\};$$

$$I_0 := I_1 \bigcup I_2 \bigcup I_3 \bigcup I_4;$$

$$M_l = \begin{cases} \{-\chi_l\} & , \text{if } l \in I_1 \bigcup I_4; \\ m \in \mathbb{Z}, m \geq -[\chi_l], & \text{if } l \in I_2; \\ m \in \mathbb{Z}, m \leq [-\chi_l], & \text{if } l \in I_3; \end{cases}$$

$$a_l = \begin{cases} \text{any positive real number} & , \text{if } l \in I_1; \\ (m_l + \chi_l)\theta_l^{-1} & , \text{if } l \in I_0 \setminus I_1. \end{cases}$$

Theorem 3.7 *Problem (3.3.32) has nontrivial solutions in $\mathcal{B}^{k,l}$ iff $I_0 \neq \emptyset$ and there exist $m_l \in M_l$, $l \in I_0$, such that*

$$\sum_{l \in I_0} m_l = 0. \tag{3.3.63}$$

The general solution is represented by the following formula ($z \in D^\pm$)

$$\Phi^+(z) = \prod_{j=1}^{k}\left(1 - \frac{z}{z_j^+}\right) \exp\{\Gamma_0(z)\} \prod_{l \in I_0}(z - \alpha_l)^{m_l} \exp\left(\sum_{l \in I_0} F_l(\frac{1}{z - \alpha_l})\right),$$

$$\Phi^-(z) = z^{-l_0} \prod_{j=1}^{l_1}\left(1 - \frac{z_j^-}{z}\right) \exp\{-\Gamma_0(z)\} \prod_{l \in I_0}(z - \alpha_l)^{-m_l} \exp\left(-\sum_{l \in I_0} F_l(\frac{1}{z - \alpha_l})\right),$$

where the functions F_l are described in (3.3.54), (3.3.55).

3.4 General Riemann–Hilbert problem of power type.

Let us consider the nonlinear Riemann–Hilbert problem of power type with constant exponent on the unit disc $\mathbb{U}$. It consists of searching for a function w analytic in $\mathbb{U}$, continuous up to the boundary ($w(z) \in \mathcal{C}_A(\mathbb{U})$), satisfying the following boundary condition

$$\Re\left\{\overline{\lambda(s)}w^p(s)\right\} = f(s), \quad s \in [0, 2\pi). \tag{3.4.1}$$

The given function $\lambda(s) = a(s) + ib(s) \neq 0$, $f(s)$ are supposed to be Hölder-continuous on $\mathbb{T}$, 2π-periodic with respect to s. As usual (see, e.g., [91]; cf. also Section 2.9) one can assume that f is real-valued and $|\lambda(s)| \equiv 1$, $s \in [0, 2\pi]$. Lastly, the parameter p is a (real or complex) constant. The simplest method is to solve problem (3.4.1) in a subclass of functions $w(z)$, for which $w^p(z)$ is univalent on $\mathbb{U}$. Under this condition the nonlinear problem (3.4.1) is equivalent to a linear one

$$\Re\left\{\overline{\lambda(s)}\phi(s)\right\} = f(s), \quad s \in [0, 2\pi), \tag{3.4.2}$$

where $\phi(z) = w^p(s)$. In particular, it is the case of the solution $w(z)$ having no zero in $\mathbb{U}$. Hence, our main problem is to describe the zero distribution of the solution. It should be noted (see, e.g., [213]) that there are three characteristic cases of problem (3.4.1):

(a) p is positive integer ($p \in \mathbb{N}$),
(b) p is positive irrational ($p \in \mathbb{R}_+ \setminus \mathbb{Q}$),
(c) p is complex ($p = \alpha + i\beta$) with positive integer or positive irrational $\alpha = \Re p$.

In order to see that the other situations are not too essential it is enough to consider the homogeneous problem, corresponding to (3.4.1)

$$\Re\left\{\overline{\lambda(s)}w^p(s)\right\} = 0, \quad s \in [0, 2\pi). \tag{3.4.3}$$

In particular, if p is negative then one can rewrite the boundary condition (3.4.3) in the form

$$\Re\left\{\lambda(s)w^{-p}(s)\right\} = 0, \quad s \in [0, 2\pi). \tag{3.4.4}$$

If $p = \frac{m}{n}$ is rational one can consider instead of (3.4.3) the problem

$$\Re\left\{i^{n+1}\overline{\lambda(s)}^n w^m(s)\right\} = 0, \quad s \in [0, 2\pi). \tag{3.4.5}$$

3.4. GENERAL RIEMANN–HILBERT PROBLEM OF POWER TYPE.

3.4.1 Homogeneous problem with positive integer exponent.

Let us study first the homogeneous problem (3.4.3) corresponding to problem (3.4.1). If p is positive integer, then the changing of unknown functions $\phi(z) = w^p(z)$ led us again to the boundary condition

$$\Re\left\{\overline{\lambda(s)}\phi(s)\right\} = 0, \ s \in [0, 2\pi). \tag{3.4.6}$$

Following [213] introduce the subclass

$$\mathcal{A}_p(\mathbb{U}) := \{w \in \mathcal{A}(\mathbb{U}) : p \text{ is a divisor of multiplicity of any zero of } w\},$$

$$\mathcal{C}_{\mathcal{A},p}(\mathbb{U}) := \mathcal{A}_p(\mathbb{U}) \cap \mathcal{C}(cl\mathbb{U}).$$

The following evident result plays a key role in the description of the solvability picture for problem (3.4.3).

Lemma 3.12 *Boundary value problem (3.4.3) has a solution in $\mathcal{C}_\mathcal{A}(\mathbb{U})$ iff problem (3.4.4) has a solution in $\mathcal{C}_{\mathcal{A},p}(\mathbb{U})$. To each solution $\phi(z)$ of (3.4.4) in $\mathcal{C}_{\mathcal{A},p}(\mathbb{U})$ corresponds exactly p different linear independent solutions of (3.4.3) in $\mathcal{C}_\mathcal{A}(\mathbb{U})$. The latter are all different analytic branches of the function $\phi^{1/p}(z)$.*

Thus, we need only to solve the linear boundary value problem (3.4.4) in the special class $\mathcal{C}_{\mathcal{A},p}(\mathbb{U})$. For this purpose it is enough to use the standard scheme (see, e.g., [91]) taking into account zero distribution of the solutions. As usual the solvability of (3.4.4) depends on the winding number of the coefficient $\lambda(s)$:

$$\chi := wind_\mathbb{T}\lambda(s). \tag{3.4.7}$$

i) $\chi < 0$. In this case problem (3.4.4) has no solution in $\mathcal{C}_\mathcal{A}(\mathbb{U})$; hence in any subclass $\mathcal{C}_{\mathcal{A},p}(\mathbb{U})$, $p \in \mathbb{N}$.

ii) $\chi = 0$. The unique solution of (3.4.4) in $\mathcal{C}_\mathcal{A}(\mathbb{U})$ can be presented in the form ([91]):

$$\phi(z) = iC \exp\left\{i\frac{1}{2\pi}\int_0^{2\pi} \arg \lambda(\sigma)\frac{e^{i\sigma}+z}{e^{i\sigma}-z}d\sigma\right\} =: iC\exp\{i\mathbf{T}(\arg \lambda(\sigma))\}, \tag{3.4.8}$$

where $\mathbf{T}$ is the *Schwarz operator for unit disc* (see Section 2.7), and C is an arbitrary positive constant. Solution (3.4.4) has no zero in $\mathbb{U}$ for any $C \neq 0$. Thus the complete collection of the solutions of (3.4.4) in $\mathcal{C}_{A,p}(\mathbb{U})$ ($\supset \mathcal{C}_A^0(\mathbb{U})$) contains only function (3.4.8). Besides this solution has no zero in $cl\,\mathbb{U}$.

 iii) $\chi > 0$. In order to describe all the solutions in this case let us introduce the subclasses:
$$\mathcal{A}_p^l(\mathbb{U}) := \{\phi \in \mathcal{A}_p(\mathbb{U}) : n_\mathbb{U}(\phi) = pl\},$$
$$\mathcal{C}_{A,p}^l(\mathbb{U}) := \mathcal{A}_p^l(\mathbb{U}) \cap C(cl\mathbb{U}), \quad \mathcal{C}_A^0 := \mathcal{C}_{A,1}^0, \qquad (3.4.9)$$

where $n_\mathbb{U}(\phi)$ is the quantity of zeros of ϕ in $\mathbb{U}$ (with their multiplicities). The following properties can be checked directly:

a) $\mathcal{A}_p^0(\mathbb{U}) = \mathcal{A}_0(\mathbb{U})$,
b) $\mathcal{A}_p(\mathbb{U}) = \bigcup_{l \geq 0} \mathcal{A}_p^l(\mathbb{U})$,
c) $\mathcal{A}_p^{l_1}(\mathbb{U}) \cap \mathcal{A}_p^{l_2}(\mathbb{U}) = \emptyset, l_1 \neq l_2$.

Therefore, we need only to describe the solution of (3.4.4) in a precise subclass $\mathcal{C}_{A,p}^l(\mathbb{U})$. Any function from this subclass has the following representation:

$$\phi(z) = \prod_{k=1}^{l} \left(\frac{z - z_k}{1 - \overline{z_k}z}\right)^p \phi^*(z) =: B^p(z)\phi^*(z), \qquad (3.4.10)$$

where z_k are arbitrary but fixed points in $\mathbb{U}$ (not necessarily different), $\phi^*(z) \in \mathcal{C}_A^0(\mathbb{U})$. Substituting (3.4.10) into equality (3.4.4) one gets the linear homogeneous boundary value problem

$$\Re\{e^{-i\chi_0 s}\phi_0(s)\} = 0, \ s \in [0, 2\pi), \qquad (3.4.11)$$

with respect to unknown function $\phi_0 \in \mathcal{C}_A^0(\mathbb{U})$, where

$$\chi_0 := \chi - pl, \qquad (3.4.12)$$

$$\phi_0(z) := \phi^*(z)\exp\{i\mathbf{T}(\arg \lambda(s) - p\arg B(s) - \chi_0 s)\}, \qquad (3.4.13)$$

$\mathbf{T}$ is the Schwarz operator for unit disc. If $\chi_0 = \chi - pl < 0$ then problem (3.4.11) has no solution in $\mathcal{C}_A(\mathbb{U})$, hence in $\mathcal{C}_{A,p}(\mathbb{U})$. Thus there is no solution of (3.4.4) in any subclass $\mathcal{C}_{A,p}^l(\mathbb{U})$ for $l > \left[\frac{\chi}{p}\right]$. If $\chi_0 \geq 0$ (i.e., $0 \leq l \leq \left[\frac{\chi}{p}\right]$), then the general solution of (3.4.11) in $\mathcal{C}_A(\mathbb{U})$ is given by the polynomial (see, e.g., [91]):

$$Q_{2\chi_0}(z) = z^{\chi_0}\left(P_{\chi_0}(z) - \overline{P}_{\chi_0}(\frac{1}{z})\right),$$

where P_{χ_0} is an arbitrary polynomial, $\deg P_{\chi_0} = \chi_0$. As the following implication is valid: $\left(Q_{2\chi_0}(\frac{1}{\overline{z}}) = 0 \Leftrightarrow Q_{2\chi_0}(z) = 0\right)$, the polynomial $Q_{2\chi_0}$ has no zero in $\mathbb{U}$ if all its

3.4. GENERAL RIEMANN–HILBERT PROBLEM OF POWER TYPE.

zeros lie on the unit circle $\mathbb{T}$, i.e.,

$$\phi_0(z) = C_{\chi_0} \prod_{k=1}^{2\chi_0}(z - t_k), \quad t_k = e^{is_k}, \; k = 1, \ldots, 2\chi_0, \; C_{\chi_0} = re^{i\theta}.$$

Substituting it into (3.4.8) one can get

$$\Re\left\{ \bar{t}^{\chi_0} re^{i\theta} \prod_{k=1}^{2\chi_0}(z - e^{is_k}) \right\} = r2^{2\chi_0}(-1)^{\chi_0} \cos\left(\theta + \frac{1}{2}\sum_{k=1}^{2\chi_0} s_k\right) \prod_{k=1}^{2\chi_0} \sin\frac{s - s_k}{2} \equiv 0.$$

The latter identity determines the value of θ, namely

$$\theta = -\frac{1}{2}\sum_{k=1}^{2\chi_0} s_k + \left(n + \frac{1}{2}\right)\pi,$$

with an arbitrary $n \in \mathbb{Z}$. Hence, the general solution of (3.4.11) in $\mathcal{C}_A^0(\mathbb{U})$ has the following form:

$$\phi_0(z) = iC \prod_{k=1}^{2\chi_0}(z - t_k)\bar{t}_k^{-1/2},$$

where $C = \pm r$ is a real parameter, $t_k \in \mathbb{T}$ ($k = 1, \ldots, 2\chi_0$) is a collection of arbitrary points (not necessary different) on the unit circle $\mathbb{T}$. Denoting $\gamma(z) = \mathbf{T}(\arg \lambda(s) - \chi s)$ and taking into account the identity

$$\mathbf{T}(\arg \lambda(s) - p\arg B(s) - \chi_0 s) = \mathbf{T}(\arg \lambda(s) - \chi s) - 2ip\log \prod_{k=1}^{l}(1 - \bar{z}_k z),$$

one can represent the function $\phi^*(z)$ from (3.4.10) in the following form:

$$\phi^*(z) = iC \exp(i\gamma(z)) \prod_{k=1}^{l}(1 - \bar{z}_k z)^{2p} \prod_{k=1}^{2\chi_0}(z - t_k)\bar{t}_k^{-1/2}.$$

Finally, the general solution of problem (3.4.3) in the subclass $\mathcal{C}_{A,p}^l(\mathbb{U})$ can be given by the formula:

$$w(z) = \rho \exp\left\{\frac{4j \pm 1}{2p}\pi i\right\} \exp\left\{\frac{i\gamma(z)}{p}\right\} \prod_{k=1}^{l}(z - z_k)(1 - \bar{z}_k z) \prod_{k=1}^{2\chi_0}(z - t_k)^{1/p}\bar{t}_k^{-1/2p},$$

(3.4.14)

where ρ is a non-negative constant, $j = 0, \ldots, p - 1$. The branches of multi-valued functions in the latter formula can be chosen arbitrarily. Thus the following solvability result is now proved,

Theorem 3.8 *The homogeneous Riemann–Hilbert boundary value problem of power type (3.4.3) with positive integer exponent p has no solution in $\mathcal{C}_A(\mathbb{U})$ whenever $\chi < 0$. If $\chi \geq 0$ then the general solution of (3.5.3) in $\mathcal{C}_A(\mathbb{U})$ is presented by the formula (3.4.14) with $l \in \left\{0, \cdots, \left[\frac{\chi}{p}\right]\right\}$, $\chi_0 = \chi - pl \geq 0$.*

3.4.2 Homogeneous problem with real exponent.

Let us consider problem (3.4.3) with the exponent p being a real number. As it was already remarked the only interesting case is that of p positive irrational. It should be noted first that putting in (3.4.14) $l = 0$, $j = 0, \pm 1, \ldots$, one can get for any $\chi \geq 0$ the solution of (3.4.3) from the subclass $\mathcal{C}_A^0(\mathbb{U})$. Hence, we need to discover the solutions having zeros into the unit disc $\mathbb{U}$. The latter can be written in the form:

$$\omega(z) = \prod_{k=1}^{l} \frac{z - z_k}{1 - \bar{z}_k z} \omega^*(z) := B(z)\omega^*(z), \qquad (3.4.15)$$

where $\omega^*(z) \in \mathcal{C}_A^0(\mathbb{U})$ are new unknown functions, $|z_k| < 1$, $k = 1, \ldots, l$ ($l \geq 1$). Then for a fixed branch of a function $\phi^*(z) = \omega^{*p}(z) \in \mathcal{C}_A^0(\mathbb{U})$ problem (3.4.3) is reformulated as the linear boundary value problem

$$\Re\left[\overline{\lambda^*(s)}\phi^*(s)\right] = 0, \qquad (3.4.16)$$

with $\lambda^*(s) = \lambda(s)[B(s)]^p$ (evidently $|\lambda^*(s)| \equiv 1$, $s \in [0, 2\pi]$). There exists a point $t_0 = e^{is_0}$ at which the function $\arg \lambda^*(s) := \theta^*(s)$ has a jump discontinuity:

$$\theta^*(s_0 + 2\pi) - \theta^*(s_0) = 2\pi(\chi - pl),$$

and the number $\chi - pl$ is irrational. Let us reduce problem (3.4.16) to the equivalent boundary value problem with continuous coefficient. For this purpose we introduce an auxiliary function

$$\xi^\delta(z) = (z - t_0)^\delta (1 - \bar{t}_0 z)^\delta.$$

An analytic branch of corresponding multi-valued function is fixed under condition $\xi^\delta(t) \equiv t^\delta |t - t_0|^{2\delta}$. Choosing $\delta = lp - [lp]$ we arrive at the boundary condition

$$\Re\left\{e^{-i\chi_0 s}\phi_0(s)\right\} = 0, \quad s \in [0, 2\pi), \qquad (3.4.17)$$

$\chi_0 = \chi - [lp]$, with respect to unknown function

$$\phi_0(z) = \phi^*(z)\xi^\delta(z) \exp i\gamma_0(z). \qquad (3.4.18)$$

3.4. GENERAL RIEMANN–HILBERT PROBLEM OF POWER TYPE.

Here $\gamma_0(z) = \mathbf{T}(\arg \lambda_0(s) - \chi_0 s)$ is the Schwarz integral with continuous density, because
$$\lambda_0(s) = \lambda^*(s)e^{i\delta s}$$
is continuous everywhere on $\mathbb{T}$, and
$$wind_\mathbb{T} \lambda_0(s) = \chi_0.$$

Problem (3.4.17) has to be solved in the subclass $\mathcal{C}_\mathcal{A}^0(\mathbb{U})$ with additional condition $\frac{\phi_0(t)}{(t-t_0)^{2\lambda}} \in C(\mathbb{T})$. Therefore, if $\chi_0 \leq 0$, then problem (3.4.17) has no solution of such a type (for $\chi_0 < 0$ there is no solution of (3.4.17) even in $\mathcal{C}_\mathcal{A}(\mathbb{U})$, but for $\chi_0 = 0$ the only analytic solution is an imaginary constant). If $\chi_0 = \chi - [pl] > 0$, then $\chi > pl$. Hence, the number l of zeros of the function $w(z)$ in $\mathbb{U}$ is bounded, namely
$$l \leq \left[\frac{\chi}{p}\right].$$

Applying the same approach as in the previous subsection and taking into account the additional condition for problem (3.4.17), one can get the following formulae of general solution of problem (3.4.3) in the considered case:

$$w(z) = \begin{cases} \rho \exp\left\{\frac{4j\pm 1}{p}\pi i\right\} (z-t_0)^{\frac{1-2\delta}{p}} t_0^{\frac{2\delta-1}{p}} \prod_{k=1}^{l} (z-z_k)(1-\bar{z}_k z) \\ \times \prod_{k=1}^{2\chi_0 - 1} (z-t_k)^{\frac{1}{p}} t_k^{\frac{1}{2p}} \exp\frac{i\gamma(z)}{p}, \quad \text{for } 0 < \delta < 1/2, \\ \\ \rho \exp\left\{\frac{4j\pm 1}{p}\pi i\right\} (z-t_0)^{\frac{2-2\delta}{p}} t_0^{\frac{\delta-1}{p}} \prod_{k=1}^{l} (z-z_k)(1-\bar{z}_k z) \\ \times \prod_{k=1}^{2\chi_0 - 2} (z-t_k)^{\frac{1}{p}} t_k^{\frac{1}{2p}} \exp\frac{i\gamma(z)}{p}, \quad \text{for } 1/2 < \delta < 1, \end{cases} \quad (3.4.19)$$

where $\rho \geq 0$, $j = 0, \pm 1, \pm 2, \ldots$, $l = 1, \ldots, \left[\frac{\chi}{p}\right]$, $\gamma(z) = \mathbf{T}(\arg \lambda(s) - \chi s)$, and the branches of multi-valued functions are chosen arbitrarily. The obtained results can be gathered in the following

Theorem 3.9 *The homogeneous nonlinear Riemann–Hilbert boundary value problem of power type (3.4.3) with irrational exponent p has no solution for $\chi < 0$. The solutions of this problem in the case $\chi > 0$ can be found by one of the formulas (3.4.17). These solutions can have l zeros at some points $z_k \in \mathbb{U}$, $k = 1, \ldots, l$, with a possible value of $l: 1 \leq l \leq \left[\frac{\chi}{p}\right]$. The nonvanishing solutions of (3.4.3) in the case $\chi \geq 0$ are given by formula (3.4.14) in which $l = 0$, $j = 0, \pm 1, \pm 2, \cdots$.*

3.4.3 Homogeneous problem with complex exponent.

Let us consider the boundary value problem (3.4.3) with

$$p = \alpha + i\beta, \quad \beta \neq 0. \tag{3.4.20}$$

The same arguments as before show us that the only two cases are essential: $\alpha \geq 0$, α, an integer; $\alpha \geq 0$, α, irrational. Let us begin with $\alpha > 0$ (and of course α is an integer or irrational). Denote by z_k internal zeros of the solution $w(z)$ in $\mathbb{U}$, and isolate the corresponding terms, i.e., represent w in the form

$$w(z) = \prod_{k=1}^{l} \frac{z - z_k}{1 - \bar{z}_k z} w^*(z) =: B(z) w^*(z). \tag{3.4.21}$$

Thus, we reduce problem (3.4.3) to linear boundary value problem (3.4.16) with respect to an arbitrary fixed analytic branch of a function

$$\phi^*(z) := (w^*(z))^{\alpha + i\beta}, \quad z \in \mathbb{U}.$$

The coefficient $\lambda^*(s)$ in (3.4.16) can be taken in the form

$$\lambda^*(s) = \lambda(s) \overline{B(s)}^\alpha, \quad s \in [0, 2\pi], \tag{3.4.22}$$

because $\{B(s)\}^{i\beta} = exp\{-\beta \arg B(s)\}$ is positive for any $s \in [0, 2\pi]$. Hence $\phi^*(z)$ is a solution of the following linear Riemann–Hilbert problem

$$\Re\left[\overline{\lambda^*(s)} \phi^*(s)\right] = 0, \quad s \in [0, 2\pi), \tag{3.4.23}$$

where $\lambda^*(s)$ is determined in (3.4.22). Important are the following implications:

$$w^*(z) \in \mathcal{C}_\mathcal{A}^0(\mathbb{U}) \Leftrightarrow \phi^*(z) = (w^*(z))^{\alpha + i\beta} \in \mathcal{C}_\mathcal{A}^0(\mathbb{U}),$$

$$\phi^*(z) \in \mathcal{C}_\mathcal{A}^0(\mathbb{U}) \Leftrightarrow w^*(z) = (\phi^*(z))^{\frac{1}{\alpha + i\beta}} \in \mathcal{C}_\mathcal{A}^0(\mathbb{U}).$$

They give the equivalence of problems (3.4.23) in $\mathcal{C}_\mathcal{A}^0(\mathbb{U})$ and (3.4.3) in $\mathcal{C}_\mathcal{A}(\mathbb{U})$ with a prescribed *finite* collection of internal zeros. Therefore, the solvability picture is essentially the same as in the case of the real exponent.

Namely, if $wind_\mathbb{T} \lambda(s) < 0$, then problem (3.4.3), (3.4.20) has no solution analytic in $\mathbb{U}$ and continuous up to the boundary $\mathbb{T}$.

If $wind_\mathbb{T} \lambda(s) > 0$, then in the case of positive, integer α formula of general solution of (3.4.3), (3.4.20) coincides with formula (3.4.14) with $0 \geq l \geq \left[\frac{x}{p}\right]$, but in the case

3.4. GENERAL RIEMANN–HILBERT PROBLEM OF POWER TYPE.

of positive irrational α the general solution of (3.4.3), (3.4.20) can be found by one of the formulae (3.4.19) in which $\lambda = \alpha l - [\alpha l]$.

If $wind_T\lambda(s) \geq 0$, then in both cases the nonvanishing solutions of problem (3.4.3), (3.4.20) are given by formula (3.4.14) with $l = 0$, $j = 0, \pm 1, \pm 2, \ldots$.

Let us consider now problem (3.4.3) with the pure imaginary exponent, i.e., for $\alpha = 0$. There are in principle two possible situations here. First, we again can look for a solution with a finite collection of internal zeros. Second, it is possible to consider solutions with an infinite collection of internal zeros. In the first case we represent the unknown function $w(z)$ by formula (3.4.21) and introduce the new unknown function

$$\phi^*(z) := (w^*(z))^{i\beta}, \ z \in \mathbb{U}$$

in order to solve the linear boundary value problem for the latter. But now, if $w^*(z)$ is zero free in $\mathbb{U}$ and continuous up to the boundary, then it is not necessarily the same for $\phi^*(z)$. The latter has discontinuous points at all boundary zeros of $w^*(z)$. Moreover the function $w^*(z) = (\phi^*(z))^{-i/\beta}$, $\phi^* \in C_\mathcal{A}^0(\mathbb{U})$ belongs to $C_\mathcal{A}^0(\mathbb{U})$ iff $\chi = 0$. If $\chi > 0$, then $w^* \in C_\mathcal{A}^0(\mathbb{U}) \cap \{B\,(cl\,\mathbb{U})\} \setminus C\,(cl\,\mathbb{U})$, where B means bounded. If we suppose that the solution $w(z)$ of (3.4.3) has an infinite quantity of zeros then we also can represent $w(z)$ in the form (3.4.21). The Blaschke product $B(z)$ becomes then an infinite one. It converges iff (see Section 2.10)

$$\sum_{k=1}^{\infty}(1-|z_k|) < \infty. \tag{3.4.24}$$

Under this condition $B(z)$ has a boundary limit for almost all $t \in \mathbb{T}$. Is it possible to have such a limit everywhere on $\mathbb{T}$? If z tends to the boundary points in the radial way, i.e., $z = \rho e^{i\theta}, \rho \to 1$, for any fixed $\theta \in [0, 2\pi)$, then we can apply the following condition

$$\sum_{k=1}^{\infty}\frac{1-|z_k|}{|e^{i\theta_j} - z_k|} < \infty, \tag{3.4.25}$$

where $e^{i\theta}$ are all accumulation points of the sequence (z_k). Under this condition $B(z)$ has radial boundary function $\widehat{B}(s)$, and $\widehat{B}(s)^{i\beta}$ is positive for any $s \in [0, 2\pi)$. This is true for radial boundary function $\widehat{B}(s)$, but not for $B(s)$. The problem is that $B(s)$ must be determined only a.e. on $\mathbb{T}$. Moreover of it the asymptotic values of $|B(z)|$ at the points $e^{i\theta}$ almost fill in the interval $[0,1]$. Condition (3.4.26) guarantees only that the sequence (z_k) approaches $e^{i\theta}$ by a tangent way. Hence, the function

$$\lambda^*(s) = \lambda(s)\overline{B(s)}^{-i\beta}, \ \beta \in [0, 2\pi) \tag{3.4.26}$$

can be continuous at $e^{i\theta}$ iff $\lambda(e^{i\theta}) = 0$. The latter is out of consideration from the very beginning of this section. Therefore, if we suppose that a solution of (3.4.3) has an infinite collection of zeros then the problem is reduced to the linear Riemann–Hilbert problem with discontinuous coefficient λ^* determined in (3.4.26). This coefficient can be continuous iff the initial problem (3.4.3) is of exceptional type, i.e., $\lambda(e^{i\theta}) = 0$ at some points $\theta_j \in [0, 2\pi)$. The points $e^{i\theta_j}$ should be actually accumulation points of the sequence of internal zeros of the solution. Taking into account the above discussion we arrive at the following conclusion: *Problem (3.4.3) with a pure imaginary exponent has a solution in $\mathcal{C}_\mathcal{A}(\mathbb{U})$. In the later case this solution is given by the formula:*

$$w(z) = B(z)(\phi^*(z))^{-i/\beta}, \qquad (3.4.27)$$

where $\phi^*(z) = iC \exp\{i\gamma(z)\}$ *is a solution of problem (3.4.23), and $B(z)$ is a finite Blaschke product.*

3.5 The modulus problem and its generalization.

3.5.1 Simple problem on modulus.

Let us consider the following problem: let L be a simple closed smooth curve encircling a domain $D = \text{int} L \not\ni \infty$; find an analytic in D function w via its given modulus on the boundary L, i.e.,

$$|w(t)| = a(t), \ t \in L, \qquad (3.5.1)$$

where $a \geq 0$ is Hölder-continuous on L. We study this problem in two situations: *i)* $a(t) > 0, t \in L$; and *ii)* $a(t) \geq 0$, vanishing at some points of L.

It should be noted that problem (3.5.1) is conformal invariant, i.e., does not change its type under conformal mapping of the domain D onto another simply connected domain on the complex plane $\mathbb{C}$. Hence, without loss of generality one can suppose that $D \ni 0$. Additionally problem (3.5.1) is equivalent to the following one:

$$|w(\omega^{-1}(\tau))| = a(\omega^{-1}(\tau)), \ \tau \in \mathbb{T},$$

where ω is the Riemann map of the domain D onto the unit disc $\mathbb{U}$. Thus, problem (3.5.1) can be considered only for the case of unit disc. We give the results for the general situation with corresponding comments concerning the case of the unit disc.

i) $a(t) > 0, t \in L$. The Uniqueness Theorem for analytic functions shows us that the unknown function w has not more than a finite number of zeros in D. Therefore, we have to consider problem (3.5.1) in one of the classes $\mathcal{A}^k(D)$, where k is a prescribed

3.5. THE MODULUS PROBLEM AND ITS GENERALIZATION.

number of zeros of the solution in D. Let us fix some (not necessarily different) points $z_j \in D$, $j = 1, \ldots, k$, supposing that they are zeros of the solution w (for simplicity suppose additionally that $\frac{1}{\overline{z_j}} \notin cl(D)$). Any function $w \in \mathcal{A}^k(D)$ with the given zeros can be represented in the form:

$$w(z) = \prod_{j=1}^{k} \frac{|z_j|}{z_j} \frac{z_j - z}{1 - \overline{z_j}z} w_0(z), \tag{3.5.2}$$

where $w_0 \in \mathcal{A}^0(D)$, i.e., is analytic and nonvanishing in D. Then the boundary condition (3.5.1) can be rewritten as:

$$|w_0(t)| = a(t) \prod_{j=1}^{k} \left| \frac{1 - \overline{z_j}t}{z_j - t} \right| =: b(t), \ t \in L. \tag{3.5.3}$$

It is evident that in the case of the unit disc the product in the right-hand side of (3.5.3) is identically equal to 1, hence $a(t) \equiv b(t)$. But it is in general not the case for an arbitrary domain D, although b always remains positive and Hölder-continuous on L. Taking logarithms of both sides of (3.5.3) we obtain the following boundary condition equivalent to (3.5.3) (for any choice of the branch of logarithmic function in D):

$$\log |w_0(t)| = \log b(t), \ t \in L. \tag{3.5.4}$$

As every branch of logarithmic function $\log w_0(z)$ is analytic in D, so one can consider (3.5.4) as the boundary condition of the Schwarz problem for $\log w_0(z)$. Its solution has the form:

$$\log w_0(z) = \mathbf{T}\{\log b(t)\}(z), \ z \in D,$$

where $\mathbf{T}$ is the Schwarz operator for the domain D. Hence the solution of the starting problem (3.5.1) in the class $w \in \mathcal{A}^k(D)$ is given by the formula

$$w(z) = \prod_{j=1}^{k} \frac{|z_j|}{z_j} \frac{z_j - z}{1 - \overline{z_j}z} \exp\left\{\mathbf{T}\left\{\log\left(a(t) \prod_{j=1}^{k} \left|\frac{1 - \overline{z_j}t}{z_j - t}\right|\right)\right\}(z)\right\}, \ z \in D. \tag{3.5.5}$$

ii) $a(t) \geq 0$, vanishing at some points of L. Suppose additionally that a has a finite collection of zeros on L. Let

$$a(t) = \prod_{s=1}^{m} (t - t_s)^{d_s} a_0(t); \ t_s \in L, \ d_s \in \mathbb{R}_+; \ a_0(t) \neq 0, \ t \in L. \tag{3.5.6}$$

We can consider problem (3.5.1) under condition (3.5.6) or in the class $\widetilde{\mathcal{A}}^k(D)$ (of analytic functions in D with k zeros there and with admissible zeros on L), or in the class $\widetilde{\mathcal{A}}^\infty(D)$. Besides, we can note that problem (3.5.1), (3.5.6) can be studied "locally", taking into account an influence of every point t_s independently. It means that one can investigate the model problem

$$|w_s(t)| = |t - t_s|^{d_s}, \ s \in \{1, \ldots, m\}, \ t \in L, \qquad (3.5.7)$$

and construct then the general solution of (3.5.1), (3.5.6) as the product of solutions of problem (3.5.7), and solution (3.5.5) with corresponding changing of data. Let us consider problem (3.5.7) for different values of d_s. Let first

$$0 < d_s < 1.$$

Then problem (3.5.7) has solutions in any class $\widetilde{\mathcal{A}}^k(D)$, but not in $\widetilde{\mathcal{A}}^\infty(D)$. It follows from the Uniqueness Theorem for analytic functions, and asymptotic behaviour of such functions near the boundary L. Any solution of the class $\widetilde{\mathcal{A}}^k(D)$ is delivered by the formula

$$w_s(z) = (z - t_s)^{d_s} \prod_{j=1}^{k} \frac{|z_j|}{z_j} \frac{z_j - z}{1 - \overline{z_j} z} w_{0,s}(z) \qquad (3.5.8)$$

where $w_{0,s}$ is analytic and nonvanishing in D; the points z_j are arbitrarily fixed points in D. The branch of multi-valued function $(z - t_s)^{d_s}$ is chosen arbitrary in $\widehat{\mathbb{C}} \setminus L_s$, where L_s is a smooth arc connecting the point $z = t_s$ and $z = \infty$, $L_s \cap L = t_s$. Substituting (3.5.8) into boundary condition (3.5.7) one gets the following problem to be solved with respect to $w_{0,s}$ in $\widetilde{\mathcal{A}}^0(D)$:

$$|w_{0,s}| = \prod_{j=1}^{k} \left| \frac{1 - \overline{z_j} t}{z_j - t} \right|, \ t \in L. \qquad (3.5.9)$$

Hence, the solution of problem (3.5.7) in $\widetilde{\mathcal{A}}^k(D)$ has the form

$$w_s(z) = (z - t_s)^{d_s} \prod_{j=1}^{k} \frac{|z_j|}{z_j} \frac{z_j - z}{1 - \overline{z_j} z} \exp\left\{ \mathbf{T}\left\{ \log\left(\prod_{j=1}^{k} \left| \frac{1 - \overline{z_j} t}{z_j - t} \right|\right)\right\}(z)\right\}, z \in D. \qquad (3.5.10)$$

If the exponent d_s is a positive integer, i.e.,

$$d_s \in \mathbb{N},$$

3.5. THE MODULUS PROBLEM AND ITS GENERALIZATION.

then problem (3.5.7) can have the solution as in any class $\widetilde{\mathcal{A}}^k(D)$ (delivered by formula (3.5.10), in which the first factor is now single-valued), as in the class $\widetilde{\mathcal{A}}^\infty(D)$. Let us show it only in the case of unit disc $\mathbb{U}$. If we choose the points $z_j \in \mathbb{U}$ such that

$$\sum_{j=1}^{\infty} \frac{1-|z_j|}{|t_s - z_j|} < \infty, \qquad (3.5.11)$$

then (see Section 2.10) the Blaschke product

$$B(z) := \prod_{j=1}^{\infty} \left(\frac{|z_j|}{z_j} \frac{z_j - z}{1 - \overline{z_j}z} \right)^{d_s} \qquad (3.5.12)$$

has the radial (and even nontangential) limit at any point of $\mathbb{T}$ (but $B(z)$ does not have the limit at t_s anyway). Besides, the asymptotic values $\widetilde{B}(t)$ of $B(z)$ are such that $|\widetilde{B}(t)| = 1$ for $t \in \mathbb{T} \setminus \{t_s\}$, and $0 \leq |B(t)| \leq 1$, for all $t \in \mathbb{B}bb{T}$. Therefore the solution of problem (3.5.7) in $\mathcal{A}^\infty(\mathbb{U})$ has the following form

$$w_s(z) = (z-t_s)^{d_s} \prod_{p=1}^{\infty} \left(\frac{|z_p|}{z_p} \frac{z_p - z}{1 - \overline{z_p}z} \right)^{d_s} \exp\left\{ \mathbb{T} \left\{ \log \left(\prod_{p=1}^{\infty} |(\frac{1-\overline{z_p}t}{z_p - t})|^{d_s} \right) \right\}(z) \right\}, z \in D.$$

(3.5.13)

If at last $d_s > 1$ is an arbitrary non-integer positive real number then problem (3.5.7) can have solutions as in the class $\widetilde{\mathcal{A}}^k(D)$ (delivered by the formula of type (3.5.10)), as in the class $\widetilde{\mathcal{A}}^\infty(D)$. In the last case one has to represent first the number d_s in the form

$$d_s = d_{s,1} + d_{s,2},$$

where $d_{s,1} \in \mathbb{N}_0$, $d_{s,2} \in \mathbb{R}_+ \setminus \mathbb{N}$. Problem (3.5.7) is reduced then to two problems of the same type but with $d_{s,1}$, $d_{s,2}$ instead of d_s. Both problems were solved before. The general solution is then the product of the solutions of corresponding problems.

3.5.2 Generalized problem on modulus of an analytic function.

We consider here the boundary value problem on determination of an analytic function $w(z)$, $z \in \mathbb{U}$, via the following boundary condition

$$|w(t) - A(t)| = a(t), \ t \in \mathbb{T}, \qquad (3.5.14)$$

where A, a are given Hölder-continuous functions on $\mathbb{T}$, $a(t) > 0$. This problem is studied in the class of meromorphic functions continuous up to the boundary. Determing an analytic function $w_0(z)$ in $\mathbb{U}$ via the boundary condition

$$|w_0(t)| = a(t), \ t \in \mathbb{T},$$

we reduce the starting problem to the following one:

$$|\widetilde{w}(t) - \alpha(t)| = 1, \ t \in \mathbb{T}, \tag{3.5.15}$$

with $\alpha(t) \in \mathcal{H}^\mu(\mathbb{T})$, and $\widetilde{w} := \frac{w}{w_0}$ being the new unknown meromorphic function. Introducing a pair of meromorphic functions by the formula

$$\Phi(z) := \begin{cases} \widetilde{w}(z), & z \in \mathbb{U}, \\ \overline{\widetilde{w}(\frac{1}{\bar{z}})}, & z \in \mathbb{C} \setminus cl(\mathbb{U}), \end{cases} \tag{3.5.16}$$

one gets the following boundary value problem equivalent to problem (3.5.15)

$$\Phi^+(t)\Phi^-(t) - \alpha(t)\left\{\Phi^+(t) + \Phi^-(t)\right\} = 1 - \alpha^2(t), \ t \in \mathbb{T}, \tag{3.5.17}$$

with additional symmetry condition

$$\Phi(z) = \overline{\Phi(\frac{1}{\bar{z}})}. \tag{3.5.18}$$

The function $\frac{1}{\Phi^+(t) - \alpha(t)}$ is Hölder-continuous on $\mathbb{T}$. Hence it can be factorized as follows:

$$\frac{1}{\Phi^+(t) - \alpha(t)} := \frac{\Phi_2^+(t)}{\Phi_2^-(t)}, \ t \in \mathbb{T}, \tag{3.5.19}$$

where $\Phi_2^\pm(z)$ are (in general) meromorphic functions in corresponding domains. Denoting then $\Phi_1^\pm(z) := \Phi^\pm(z)\Phi_2^\pm(z)$ we reduce problem (3.5.17) to the system of linear boundary value problems

$$\begin{cases} \Phi_1^+(t) - \Phi_1^-(t) = [\alpha(t) - 1]\left[\Phi_2^+(t) - \Phi_2^-(t)\right], \\ \Phi_1^+(t) - \Phi_2^+(t) = [\alpha(t) + 1]\left[\Phi_2^+(t) + \Phi_2^-(t)\right]. \end{cases} \tag{3.5.20}$$

Let us represent the function $\Phi_2(z)$ in the form of Schwarz integral with unknown density

$$\Phi_2(z) := \frac{1}{4\pi i} \int_\mathbb{T} \omega(\tau)\frac{\tau + z}{\tau - z}\frac{d\tau}{\tau} - \frac{c}{\pi}$$

3.6. LINEAR FRACTIONAL PROBLEM.

up to an additive real constant c. Then the first equation of (3.5.20) gives immediately the integral representation for Φ_1

$$\Phi_1(z) := \frac{1}{4\pi i} \int_{\mathbb{T}} [\alpha(\tau) - 1] \omega(\tau) \frac{\tau + z}{\tau - z} \frac{d\tau}{\tau},$$

and the second one leads to the following integral equation with respect to ω:

$$(K\omega)(t) := \int_0^{2\pi} \omega(\tau) \left\{ 1 + \frac{\alpha(\tau) - \alpha(t)}{2} \right\} \cot \frac{\sigma - s}{2} d\sigma = ic[\alpha(t) + 1], \tag{3.5.21}$$

$$t = e^{is}, \tau = e^{i\sigma}.$$

This integral equation is a *general singular integral equation with Hilbert kernel* (see, e.g., [91, p. 298]). Its index is equal to zero. Hence or the homogeneous integral equation (i.e., with $c = 0$) has a nontrivial solution, or the inhomogeneous equation is unconditionally solvable for any $c \neq 0$. In the last case the solution takes only purely imaginary values. The solution of the homogeneous integral equation (if exists) can be also taken of the same type. For such solutions both functions Φ_1 and Φ_2 do satisfy the symmetry condition (3.5.18). Introducing then the new unknown function

$$\phi(z) := \frac{1}{\tilde{w}(z) - w_0(z)},$$

where $\tilde{w}(z)$ is an arbitrary solution of (3.5.15), and $w_0(z) := \frac{\Phi_1(z)}{\Phi_2(z)}$ is a special solution determined from (3.5.21), one gets the linear Riemann–Hilbert problem for ϕ

$$[\alpha(t) - u_0(t)]\Re\phi(t) + v_0(t)\Im\phi(t) = \frac{1}{2},\ t \in \mathbb{T},$$

where $w_0(t) := u_0(t) + iv_0(t)$. The solution of the latter is known (see Section 2.9, cf. also [91], Section 29).

3.6 Linear fractional conjugation boundary value problem.

Let L be a simple closed smooth curve dividing the complex plane $\mathbb{C}$ into two domains D^+, D^- ($D^- \ni \infty$); $b(t), c(t)$ are given Hölder continuous functions on L, $1 - b(t)c(t) \neq 0, t \in L$. Consider the following nonlinear boundary value problem

$$\Phi^+(t) + \Phi^-(t) - b(t)\Phi^+(t)\Phi^-(t) = c(t), t \in L, \tag{3.6.1}$$

in the class of piece-wise meromorphic functions $\Phi^\pm(z)$, Hölder continuous up to the boundary $\partial D^\pm$, $\Phi^\pm(t) \neq 0, t \in L$. Introducing new functions by the relations

$$1 - b(t)\Phi^+(t) := -\frac{\phi^-(t)}{\phi^+(t)}, \psi^\pm(t) := \pm\Phi^\pm(t)\phi^\pm(t), t \in L, \qquad (3.6.2)$$

one can reduce (3.6.1) to the system of equations

$$\phi^+(t) + \phi^-(t) = b(t)\psi^+(t), \psi^+(t) + \psi^-(t) = c(t)\phi^-(t), t \in L, \qquad (3.6.3)$$

with respect to $\phi^\pm(t), \psi^\pm(t)$. Denoting $X^\pm(z)$ *the canonical function* of the $\mathbb{C}$-linear conjugation problem with coefficient $1 - b(t)c(t)$ (cf. Section 2.9; see also [91, p. 109]) we obtain the following vector-matrix problem

$$\begin{pmatrix} \phi^+ \\ \psi^+ \end{pmatrix} = \begin{pmatrix} -\frac{X^-}{X^+} & -b\frac{X^-}{X^+} \\ -c\frac{X^-}{X^+} & -\frac{X^-}{X^+} \end{pmatrix} \begin{pmatrix} \phi^- \\ \psi^- \end{pmatrix}, t \in L. \qquad (3.6.4)$$

The latter problem can be reduced by suitable changing of variables to the equivalent one with the matrix $\begin{pmatrix} 1 & b(t) \\ c(t) & 1 \end{pmatrix}$. What are connections between the solutions of (3.6.1) and (3.6.4)? The next result is of immediate verification:

Lemma 3.13 *([123, p. 23-24]) If $\Phi^\pm(z)$ is a piece-wise meromorphic solution of problem (3.6.1), then only those solutions of (3.6.4) correspond to it for which the ratios $\frac{\phi^+}{\phi^-}, \frac{\psi^+}{\psi^-}$ are known, namely*

$$\frac{\phi^+(t)}{\phi^-(t)} = \frac{1}{b(t)\Phi^+(t) - 1} =: \lambda(t), \frac{\psi^+(t)}{\psi^-(t)} = \frac{\Phi^+(t)}{c(t) - \Phi^+(t)} =: \mu(t); \qquad (3.6.5)$$

besides, the functions λ, μ have to satisfy the following relation:

$$\lambda(t) + \mu(t) + (1 - b(t)c(t))\lambda(t)\mu(t) = -1, t \in L. \qquad (3.6.6)$$

If $(\phi^\pm, \psi^\pm)$ is a pair of piece-wise meromorphic solutions of problem (3.6.4), then the piece-wise meromorphic function

$$\Phi^\pm(z) = \pm\frac{\psi^\pm(z)}{\phi^\pm(z)}, z \in D^\pm, \qquad (3.6.7)$$

does satisfy boundary condition (3.6.1). Moreover, formulae (3.6.5) are valid with the functions λ, μ related to each other by (3.6.6).

3.6. LINEAR FRACTIONAL PROBLEM.

Let $\Phi^{\pm}(z)$ be a piece-wise meromorphic solution of (3.6.1), satisfying the additional conditions

$$(\Phi^+(t) - \frac{1}{b(t)})(\Phi^+(t) - c(t)) \neq 0, t \in L \quad (3.6.8)$$

(hence λ, μ are continuous and nonvanishing on L). Among the solutions of problem (3.6.4) we find now a special one which is piece-wise analytic in $\mathbb{C}$ (possibly with pole at infinity). To do it let us denote by a_k, b_j respectively zeros and poles on $\mathbb{C}$ of the function $\Phi^{\pm}(z)$, determined in (3.6.7), and introduce the polynomials

$$T_n(z) := \prod_k (z - a_k), \quad T_p(z) := \prod_j (z - b_j). \quad (3.6.9)$$

Factorizing the functions $\lambda(t), \mu(t)$ (cf. Sec 2.9, see also [91, p. 109])

$$\lambda(t) := \lambda^+(t)[\lambda^-(t)]^{-1}, \quad \mu(t) := \mu^+(t)[\mu^-(t)]^{-1},$$

we obtain

$$\phi^{\pm}(z) = \lambda^{\pm}(z) T_p(z), \quad \psi^{\pm}(z) = \mu^{\pm}(z) T_n(z). \quad (3.6.10)$$

The functions $\phi^{\pm}, \psi^{\pm}$ are analytic in $\mathbb{C} \setminus L$, and nonvanishing in spite of the points a_k, b_j respectively, having the orders $k_\phi := p - \chi_\lambda$, $k_\psi := n - \chi_\mu$ at infinity (where χ_λ, χ_μ are winding numbers of $\lambda(t), \mu(t)$ respectively). Our further aim is to construct *the canonical system* (see, e.g., [269, p. 70]) of the vector-matrix conjugation problem (3.6.4) using the given piece-wise meromorphic solution of problem (3.6.1). We begin with the case

$$\text{wind}_L(1 - b(t)c(t)) = 0.$$

Therefore the partial indices of problem (3.6.4) are opposite numbers, say $\chi \geq 0$ and $-\chi$. Let

$$(\phi_{-\chi}, \psi_{-\chi}), (\phi_\chi, \psi_\chi) \quad (3.6.11)$$

be the canonical system of (3.6.4), where $\phi_{-\chi}, \psi_{-\chi}$ have the order $-\chi$ at infinity. Any solution of problem (3.6.4) (in particular that from (3.6.10)) can be represented in the following form

$$\phi := p\phi_{-\chi} + q\phi_\chi, \quad \psi := p\psi_{-\chi} + q\psi_\chi, \quad (3.6.12)$$

with polynomial $p(z), q(z)$ in it. Using (3.6.12), and taking into account that (3.6.11) are solutions of problem (3.6.4), we get the following system of equations:

$$\begin{aligned} \phi^+_{-\chi} &= \frac{\phi^+}{\phi^-}\phi^-_{-\chi} + \frac{bq}{\chi+\phi^-}, & \psi^+_{-\chi} &= \frac{\psi^+}{\psi^-}\psi^-_{-\chi} - \frac{cq}{\chi+\psi^-}, \\ \phi^+_{-\chi} &= \frac{\phi^+}{\psi^-}\psi^-_{-\chi} - \frac{q}{\chi+\phi^-}, & \psi^+_{-\chi} &= \frac{\psi^+}{\phi^-}\phi^-_{-\chi} + \frac{q}{\chi+\phi^-}; \end{aligned} \quad (3.6.13)$$

$$\phi_x^+ = \tfrac{\phi^+}{\phi^-}\phi_x^- - \tfrac{bp}{X+\phi^-}, \quad \psi_x^+ = \tfrac{\psi^+}{\psi^-}\psi_x^- + \tfrac{cp}{X+\psi^-},$$
$$\phi_x^+ = \tfrac{\phi^+}{\psi^-}\psi_x^- + \tfrac{p}{X+\phi^-}, \quad \psi_x^+ = \tfrac{\psi^+}{\phi^-}\phi_x^- - \tfrac{p}{X+\phi^-}. \tag{3.6.14}$$

Besides, p, q in (3.6.12) do satisfy the boundary equations:
$$\begin{aligned} p &= X^+[\phi^+\psi_x^+ - \psi^+\phi_x^+] = X^-[\phi^-\psi_x^- - \psi^-\phi_x^-],\\ q &= X^+[\psi^+\phi_{-x}^+ - \phi^+\psi_{-x}^+] = X^-[\psi^-\phi_{-x}^- - \phi^-\psi_{-x}^-]. \end{aligned}$$

It shows immediately (due to Analytic Continuation Principle and the behaviour of the right hand-sides at infinity) that both functions p and q are polynomials. If q is already determined, then the solution of the first two problems of (3.6.13) can be represented in the form

$$\begin{aligned} \phi_{-x}^+ &= \phi^+ \left(\mathbf{P}\left[\tfrac{bq}{X+\phi+\phi^-} \right] + R_1 \right),\\ \phi_{-x}^- &= \phi^- \left(-\mathbf{Q}\left[\tfrac{bq}{X+\phi+\phi^-} \right] + R_1 \right), \end{aligned} \tag{3.6.15}$$

$$\begin{aligned} \psi_{-x}^+ &= \psi^+ \left(-\mathbf{P}\left[\tfrac{cq}{X+\psi+\psi^-} \right] + R_2 \right),\\ \psi_{-x}^- &= \psi^- \left(-\mathbf{Q}\left[\tfrac{cq}{X+\psi+\psi^-} \right] + R_2 \right), \end{aligned} \tag{3.6.16}$$

where $\mathbf{P}, \mathbf{Q}$ are mutually orthogonal projectors; $R_1 := \tfrac{P_m}{T_p}$, $R_2 := \tfrac{P_l}{T_n}$ are rational functions. Substituting (3.6.15) into (3.6.4) we obtain

$$\tfrac{bq}{X^+}\left(\tfrac{1}{\phi^-} - \tfrac{1}{\phi^+} + \tfrac{c}{\psi^-} \right) + (\phi^+ + \phi^-)\mathbf{S}\left[\tfrac{bq}{X+\phi+\phi^-} \right]$$
$$+ b\psi^+\mathbf{S}\left[\tfrac{cq}{X+\psi+\psi^-} \right] = 2\left[b\psi^+ R_2 - (\phi^+ + \phi^-) R_1 \right],$$

where $\mathbf{S} := \mathbf{P} - \mathbf{Q}$. Then for the function $\omega(z)$, introduced by the formula

$$\tfrac{q}{X^+} \tfrac{b\psi^+\psi^- + c\phi^+\phi^-}{\phi^+\phi^-\psi^+\psi^-} =: \omega^+ - \omega^-, \tag{3.6.17}$$

we get the following boundary value problem (ℂ-linear conjugation problem):

$$\omega^+ = \tfrac{X^-}{X+\lambda\mu}\omega^- - \tfrac{X^-}{X+\phi^+\psi^+}(b\psi^+\psi^- + c\phi^+\phi^-)(R_2 - R_1). \tag{3.6.18}$$

The solution of the latter in the class of analytic functions vanishing at infinity has the form

$$\begin{cases} \omega^+ = \tfrac{1}{X+\lambda^+\mu^+}\{-\mathbf{P}\left[\tfrac{X^-}{T_n T_p}(b\psi^+\psi^- + c\phi^+\phi^-)(R_2 - R_1) \right] + P_{j-1}\},\\ \omega^- = \tfrac{1}{X-\lambda^-\mu^-}\{\mathbf{Q}\left[\tfrac{X^-}{T_n T_p}(b\psi^+\psi^- + c\phi^+\phi^-)(R_2 - R_1) \right] + P_{j-1}\}, \end{cases} \tag{3.6.19}$$

3.6. LINEAR FRACTIONAL PROBLEM.

where $j := -(\chi_\lambda + \chi_\mu)$; $P_{j-1}(z) \equiv 0$ for $j \leq 0$. Now we are in position to determine the first component of the canonical system (3.6.11). To do it we choose the coefficients of the polynomials P_m, P_i, P_{j-1} so that the functions ϕ^-_χ in (3.6.15), ψ^-_χ in (3.6.16), and ω^- in (3.6.19) do vanish at infinity (if indices of corresponding boundary value problems are nonnegative). Then the number of remaining free constants is equal to the modulus of partial indices of problem (3.6.4). The functions (3.6.15), (3.6.16) of lowest order give the first component of canonical system (3.6.11). Substituting the polynomial q from (3.6.17) into (3.6.12), and determining then the polynomial p in terms of ϕ_χ, ψ_χ we reduce the first two equations of (3.6.14) to the pair of $\mathbb{C}$-linear conjugation problems for determining the second component of (3.6.11).

We have to note that any other canonical system (see, e.g., [269, p. 30]) has the form $(\alpha \phi_{-\chi}, \alpha \psi_{-\chi}), (P_{2\chi} \phi_{-\chi} + \beta \phi_\chi, (P_{2\chi} \psi_{-\chi} + \beta \psi_\chi)$, where α, β are constants ($P_{2\chi}$ is a polynomial of degree 2χ). Thus the first component of the canonical system is determined up to the constant factor.

Let us present the qualitative results concerning the determination of coefficients of the polynomials:

1) let $k_\phi \leq 0, k_\psi \leq 0$ for the solutions of (3.6.10), then for $\chi > 0$ we can take $(\phi_{-\chi}, \psi_{-\chi}) = (\phi, \psi)$ (with (ϕ, ψ) determined in (3.6.10)) and $-\chi = \max\{k_\phi, k_\psi\} = \max\{p - \chi_\lambda, n - \chi_\mu\}$. This choice can be used also for $\chi = 0$;

2) let $k_\phi \leq 0, k_\psi > 0$. Put $m := \chi_\lambda - 1, i := n - 1$ in (3.6.15), (3.6.16). If $j \leq 0$ in (3.6.19), then expressions (3.6.15), (3.6.16), and (3.6.19) depend on $\gamma := \chi_\lambda + n$ arbitrary constants. On the other hand the necessary behaviour of (3.6.19) at infinity leads to $-j = \chi_\lambda + \chi_\mu$ conditions, and that of (3.6.16) to $n - \chi_\mu$ conditions. If $j > 0$, then the number of conditions appearing for (3.6.16) coincides with the number of arbitrary constants $\gamma := n - \chi_\mu$.

3) let $k_\phi > 0, k_\psi \leq 0$. Then we put $m := p - 1, i := \chi_\lambda - 1$. If $j \leq 0$, then the number of arbitrary constants is equal to $\gamma := p + \chi_\mu$, but the number of conditions appearing from (3.6.15), (3.6.19) is equal to $p - \chi_\lambda$, and $-j = \chi_\lambda + \chi_\mu$ respectively. If $j > 0$, then the number of conditions appearing for (3.6.15) coincides with the number of arbitrary constants $\gamma := p - \chi_\lambda$.

4) let $k_\phi > 0, k_\psi > 0$. Then we put $m := p - 1, i := n - 1$. If $j \leq 0$, then we have $\gamma := p + n$ arbitrary constants, the same number as that of the conditions appearing from (3.6.15), (3.6.16), and (3.6.19). If $j > 0$, then the number of conditions coincides with the number of arbitrary constants $\gamma := p + n - \chi_\lambda - \chi_\mu$.

3.7 Cherepanov's nonlinear mixed boundary value problem.

3.7.1 Statement of the problem. Simple cases of solvability.

Let $L \subset \mathbb{R}$ be a collection of segments of real axes $\left(L := \bigcup_{k=1}^{n} L_k, L_k := (a_k, b_k)\right)$. Let $a, b, \alpha : L \to \mathbb{C}$ be functions, which are Hölder-continuous on L in spite of possibly a finite number of points $(a, b \in C(L \setminus \bigcup_i \{c_i\}), \alpha \in C(L \setminus \bigcup_s \{d_s\}))$. We consider the problem of finding a function ω analytic in the upper half-plane $\Pi_+ := \{z : \Im z > 0\}$, continuous in $cl(\Pi_+) \setminus (\bigcup_k \{a_k\} \bigcup_k \{b_k\} \bigcup_i \{c_i\} \bigcup_s \{d_s\}) =: cl(\Pi_+) \setminus N$, locally integrable in a neighbourhood of N, satisfying the boundary conditions

$$|\omega(t)| = \alpha(t), \ t \in L; \quad \Re[(a(t) - ib(t))\omega(t)] = 0, \ t \in M := \mathbb{R} \setminus cl\, L. \tag{3.7.1}$$

This problem is connected with the following $\mathbb{C}$-linear conjugation problem:

$$X^+(t) = G(t)X^-(t), t \in \mathbb{R} \setminus N, \tag{3.7.2}$$

where

$$G(t) := \begin{cases} -\frac{a(t)+ib(t)}{a(t)-ib(t)} &, \ t \in M, \\ 1 &, \ t \in L. \end{cases}$$

A special solution of problem (3.7.2)

$$X^\pm(z) := \prod_{k=1}^{n}(z-b_k)^{-\chi_k} e^{\Gamma(z)}, \ \Gamma(z) := \frac{1}{2\pi i}\int_{-\infty}^{\infty}\frac{\log G(\tau)}{\tau - z}d\tau, \ \pm\Im z > 0, \tag{3.7.3}$$

is called *the canonical function of problem* (3.7.1). Here $\chi := \sum_{k=1}^{n}\chi_k$ is the index of problem (3.7.2) determined in a standard way for $\mathbb{C}$-linear conjugation problems with a discontinuous coefficient (see, e.g., [91, pp. 440-441]). The canonical function has the same type of singularities as $a(t)$ and $b(t)$ at the points $t = c_i$, and is bounded in neighbourhoods of $t = a_k, t = b_k$. Using notation

$$\Phi(z) = \begin{cases} \omega(z)/X^+(z), & \Im z > 0, \\ \overline{\omega}(z)/X^-(z), & \Im z < 0, \end{cases} \tag{3.7.4}$$

one can reduce (3.7.1) to the following mixed problem

$$\Phi^+(t)\Phi^-(t) = \alpha^2 X^{-2}, t \in L; \quad \Phi^+(t) = \Phi^-(t), t \in M. \tag{3.7.5}$$

3.7. CHEREPANOV'S MIXED PROBLEM.

Let us suppose additionally that

$$\alpha^2(t) X^{-2}(t) \neq 0, \ t \in L, \tag{3.7.6}$$

i.e., $\alpha(t) \neq 0$ (hence, α is positive for all $t \in L$), and $X(t)$ is bounded on L. If the analytic solution Φ has m zeros at the points $z = z_j, j = 1, \ldots, m$, then it can be delivered by the formula

$$\Phi(z) = \prod_{j=1}^{m} \left(1 - \frac{z_j}{z}\right) e^{\Gamma_1(z)}, \tag{3.7.7}$$

where

$$\Gamma_1(z) := \frac{X_n(z)}{2\pi i} \int_L \log\left[\frac{\beta(\tau)}{\prod_{j=1}^{m}\left(1 - \frac{z_j}{\tau}\right)}\right] \frac{d\tau}{X_n^+(\tau)(\tau - z)}, \tag{3.7.8}$$

$$X_n(z) := \left\{\prod_{k=1}^{n}(z-a_k)(z-b_k)\right\}^{\frac{1}{2}}. \tag{3.7.9}$$

Besides, the following necessary conditions should be satisfied:

$$\int_L \log\left[\frac{\beta(\tau)}{\prod_{j=1}^{m}\left(1 - \frac{z_j}{\tau}\right)}\right] \frac{\tau^p}{X_n^+(\tau)} d\tau, p = 0, 1, \ldots, n-2. \tag{3.7.10}$$

The solution of (3.7.5) is bounded near $z = a_k, z = b_k$, and has power type singularities at the jump points of $\alpha(t)$ or $X(t)$. To prove the uniqueness of the solution of problem (3.7.5) (hence (3.7.1)) one needs to fix the points $z_j, j = 1, \ldots, m$. If not problem (3.7.1) has an infinite number of solutions

$$w(z) := X^+(z) \Phi^+(z),$$

with Φ^+ given in (3.7.7), and arbitrary numbers $z_1, \ldots, z_m, m$ in the last formula.

3.7.2 Meromorphic solutions.

Let us return to problem (3.7.1), and consider it in the class of meromorphic in Π_+ functions, continuous in $cl(\Pi_+)$ in spite of the beforehand given poles of $w(z)$. It is more suitable for the further calculations to interchange the boundary conditions, i.e., to consider problem

$$|w(t)| = \alpha(t), t \in M; \quad \Re[(a(t) - ib(t))w(t)] = 0, t \in L. \tag{3.7.11}$$

It is clear that (3.7.1) and (3.7.11) are equivalent. Besides, the usual transformation (see, e.g., [91, p. 280]) shows that one can consider only case $|a(t)+ib(t)| \equiv 1, t \in clL$. We note also that without loss of generality the argument of $\lambda(t) := a(t)+ib(t)$ is supposed to be fixed on the segments $L_k, k = 1, \ldots, n$ in the following way:

$$\arg \lambda(t) = \theta(t) - \theta_0(t) + \sigma_k(t)\pi, \qquad (3.7.12)$$

where $\theta_0(t)$ is a fixed Hölder-continuous function on L; $\sigma_k(t) \equiv \sigma_k(t \in L_k$ are some integers, subject of further determination). Our further aim is to obtain a representation of the solution $\omega(z)$ of problem (3.7.11) assuming its existence.

Lemma 3.14 *([215, p. 550]) The solution of problem (3.7.11) has at most finite number of zeros and poles in $cl\Pi_+$. Moreover its boundary zeros and poles belong to L and have integer orders.*

Proof. It suffices to get the result only with respect to zeros, because the function $\omega^{-1}(z)$ is the solution of the analogous problem $|\omega^{-1}(t)| = \alpha^{-1}(t), t \in M; \Re[(a(t) + ib(t))\omega^{-1}(t)] = 0, t \in L$. As $\omega(z)$ is continuous in a neighbourhood of its zeros, and $\alpha(t) > 0$ on M, then all boundary zeros of $\omega(z)$ belong to L. In order to describe the behaviour of $\omega(z)$ near a boundary zero we continue $\lambda(t)$ up to a function continuous on the whole real line $\mathbb{R}$, denoting it again $\lambda(t)$. Among all possible continuations one can choose those for which $|\lambda(t)| \equiv 1, t \in \mathbb{R}, wind_\mathbb{R}\lambda(t) = 0$. Let $r(t)$ be a *regularizing factor* (see, e.g., [91, p. 274]) for the function $\lambda(t)$ on $\mathbb{R}$, i.e., $r(t)\lambda(t) =: \Psi^+(t)$, where the latter is the boundary value of an analytic function $\Psi^+(z)$ in Π_+, such that $\Psi^+(z) \neq 0$ in $cl(\Pi_+)$. Hence, the function $\omega(z)[\Psi^+(z)]^{-1}$ has the same zeros as $\omega(z)$. Besides, $\Re\{\omega(z)[\Psi^+(z)]^{-1}\} = 0, t \in L$. Therefore by Schwarz's Symmetry Principle the function $\omega(z)[\Psi^+(z)]^{-1}$ can be analytically continued through L.

Let us suppose that $\omega(z)$ has an infinite number of zeros and x_0 is their limit point (which belongs necessarily to L due to uniqueness of the analytic function). Then by continuity $\omega(x_0) = 0$. Hence, $\omega(x_0)[\Psi^+(x_0)]^{-1} = 0$. It contradicts to the analyticity of $\omega(z)[\Psi^+(z)]^{-1}$ in a neighbourhood of x_0. ∎

Denote $z_j \in \Pi_+, j = 1, \ldots, \mu_1 (\bar{z}_j \in \Pi_+, j = \mu_1 + 1, \ldots, \mu)$ internal zeros (poles) of the solution $\omega(z)$, and $t_j \in L, j = 1, \ldots, \nu_1 (t_j \in L, j = \nu_1 + 1, \ldots, \nu)$ its boundary zeros (poles). It is convenient to use for further constructions the function $\alpha_j(z)$, which maps conformally the upper half-plane Π_+ onto the unit upper semi-disc $\mathbb{U} \cap \Pi_+$ such that $\alpha_j(L_k) = (-1, 1)$, and $\alpha_j(\tau_j) = 0$ for a given point $\tau_j \in L_k$. It has the form

$$\alpha_j(z) = \frac{2(1-s_j)s_j|L_k| + (1-2s_j)(z-\tau_j) + 2[(1-s_j)s_j(z-a_k)(b_k-z)]^{\frac{1}{2}}}{z - \tau_j},$$

(3.7.13)

3.7. CHEREPANOV'S MIXED PROBLEM.

where $|L_k|$ is the length of $L_k = (a_k, b_k)$, $s_j := \frac{\tau_j - a_k}{|L_k|}$, and the branch of the root in (3.7.13) is chosen in such a way that its values are negative on the upper side of the cut along L_k. The following properties of this function are of immediate verification:

$$|\alpha_j(t)| \equiv 1, \ t \in \mathbb{R} \setminus L_k; \ -1 \leq \alpha_j(t) \leq 1, \ t \in cl L_k, \qquad (3.7.14)$$

$$\alpha_j(z) = \frac{z - \tau_j}{z + i} \alpha_j^*(z), \qquad (3.7.15)$$

where $\alpha_j^*(z)$ is analytic in Π_+, continuous and nonvanishing in $cl\Pi_+$. Let us represent the solution $w(z)$ in the form

$$w(z) = \phi(z) \prod_{j=1}^{\mu} \frac{z - z_j}{z - \bar{z}_j} \prod_{j=1}^{\nu_1} \alpha_j(z) \prod_{j=\nu_1+1}^{\nu} \alpha_j^{-1}(z), \qquad (3.7.16)$$

where τ_j in formula (3.7.13) for $\alpha_j(z)$ coincides with boundary zeros or poles t_j of the function $w(z)$; $\phi(z)$ is analytic in Π_+, continuous and nonvanishing in $cl\Pi_+$. Substituting (3.7.16) into the boundary equation (3.7.11), and using (3.7.14) - (3.7.15) we get the following boundary value problem with respect to ϕ:

$$\Re\left\{\overline{\lambda_1(t)}\phi(t)\right\} = 0, t \in L; \ |\phi(t)| = \alpha(t), t \in M, \qquad (3.7.17)$$

where

$$\lambda_1(t) = \lambda(t) \overline{\prod_{j=1}^{\mu} \frac{t - z_j}{t - \bar{z}_j} \prod_{\substack{j=1 \\ t_j \notin L_k}}^{\nu_1} \alpha_j(t) \prod_{\substack{j=\nu_1+1 \\ t_j \notin L_k}}^{\nu} \alpha_j^{-1}(t)} =:$$

$$= \lambda(t) \overline{B(t)} \overline{A_k(t)}, t \in L_k, k = 1, \ldots, n.$$

It follows from (3.7.14) that $|\lambda_1| \equiv 1$ on L. Hence

$$\begin{cases} \lambda_1(t) = \exp\{i\theta_1(t)\}, \\ \theta_1(t) := \theta_0(t) + \sigma_k \pi - \arg \prod_{j=1}^{\mu} \frac{t - z_j}{t - \bar{z}_j} \\ \quad - \arg\left\{ \prod_{\substack{j=1 \\ t_j \notin L_k}}^{\nu_1} \alpha_j(t) \prod_{\substack{j=\nu_1+1 \\ t_j \notin L_k}}^{\nu} \alpha_j^{-1}(t) \right\}. \end{cases} \qquad (3.7.18)$$

Rewriting the boundary condition (3.7.17) in the form: $\overline{\lambda_1(t)}\phi(t) = -\lambda_1(t)\overline{\phi(t)}, t \in L$; $\phi(t)\overline{\phi(t)} = \alpha^2(t), t \in M$, and taking logarithms of both sides of these equations, we

obtain the following mixed boundary value problem (the problem of Keldysh–Sedov type):
$$\begin{cases} \Im\Phi(t) = \theta_1(t) + \frac{\pi}{2}, & t \in L, \\ \Re\Phi(t) = \log\alpha(t), & t \in M, \end{cases} \quad (3.7.19)$$

with respect to any branch of the function $\Phi(z) := \log\phi(z) \in \mathcal{C}_A(\Pi_+)$. The solution of this problem should be continuous at the points a_k, b_k as well. Such solution exist (see, e.g., [251]) iff the right-hand side of (3.7.19) satisfies some necessary and sufficient conditions, namely

$$i\int_L \frac{\theta_0(t)}{X_*(t)} t^j dt + \int_M \frac{\log\alpha(t)}{X_*^+(t)} t^j dt + \frac{i\pi}{2}\int_L \frac{t^j}{X_*(t)} dt$$
$$= i\int_L \frac{\arg B(t) + \arg \tilde{A}(t) - \sigma(t)}{X_*(t)} t^j dt, \quad j = 0, \ldots, n-2, \quad (3.7.20)$$

where $X_*(t)$ are the values on L, and $X_*^+(t)$ are the boundary values on the upper side of a cut along M of any branch of multi-valued function

$$X(z) = \left\{\prod_{k=1}^n (z-a_k)(z-b_k)\right\}^{\frac{1}{2}}, \quad z \in \mathbb{C}\setminus M; \quad (3.7.21)$$

$$\tilde{A}(t) := A_k(t), \; t \in L_k, k = 1, \ldots, n.$$

Under conditions (3.7.20) the solution of problem (3.7.19) has the form

$$\Phi(z) = \frac{X_*(z)}{\pi i}\int_M \frac{\log\alpha(t)}{X_*^+(t)}\frac{dt}{t-z} + \frac{X_*(z)}{\pi i}\int_L \frac{i[\theta_1 + \frac{\pi}{2}]}{X_*(t)}\frac{dt}{t-z}.$$

Now the question is whether it is possible to satisfy the conditions (3.7.20) by the choice of zeros and poles of solution, and numbers σ_k. Let us denote $X(z)$ the branch of analytic function (3.7.21) with the cut along L, which coincides in Π_+ with $X_*(z)$. Then
$$X(z) = \{X_*(z), z \in \Pi_+; \; -X_*(z), z \in \Pi_-\}.$$
Hence, $X^+(t) \equiv X_*(t), t \in L$. Therefore,

$$J(z) := \frac{1}{2\pi i}\int_L \frac{1}{X_*(t)}\frac{dt}{t-z} = \frac{1}{2\pi i}\int_L \frac{1}{X^+(t)}\frac{dt}{t-z}.$$

Besides,
$$J^*(z) := \frac{1}{2\pi i}\int_{L\cup L^-} \frac{1}{X(t)}\frac{dt}{t-z} = \frac{1}{X(z)}, \; z \in \mathbb{C}\setminus cl\,L$$

3.7. CHEREPANOV'S MIXED PROBLEM.

(here L^- is the lower side of the cut along L). From the other side

$$J^*(z) = \frac{1}{2\pi i}\left\{\int_L \frac{1}{X^+(t)}\frac{dt}{t-z} - \int_L \frac{1}{X^-(t)}\frac{dt}{t-z}\right\} = 2J(z).$$

Hence,

$$J(z) = \frac{1}{2X(z)}, \quad z \in \mathbb{C} \setminus cl L.$$

Therefore, $J(z)$ is analytic in a neighbourhood of $z = \infty$, having zero of order n at infinity. Expanding Cauchy integral into Taylor series at infinity we get

$$\int_L \frac{t^j}{X_*(t)} dt = 0, \quad j = 0, \ldots, n-2. \tag{3.7.22}$$

Consider now the integral

$$J_j := \int_L \frac{i \arg \tilde{A}(t)}{X_*(t)} t^j dt$$

$$= \sum_{k=1}^n \left\{ \sum_{\substack{r=1 \\ t_r \notin L_k}}^{\nu_1} \int_{L \setminus L_k} \frac{i \arg \alpha_r(t)}{X^+(t)} t^j dt - \sum_{\substack{r=\nu_1+1 \\ t_r \notin L_k}}^{\nu} \int_{L \setminus L_k} \frac{i \arg \alpha_r(t)}{X^+(t)} t^j dt \right\}. \tag{3.7.23}$$

The integrals in this formula can be rewritten as

$$J_{j,r} := \int_{L \setminus L_k} \frac{i \arg \alpha_r(t)}{X^+(t)} t^j dt \int_{L \setminus L_k} \frac{\log \alpha_r(t)}{X^+(t)} t^j dt = \int_{L \setminus L_k} \frac{\log \alpha_r^*(t)}{X^+(t)} t^j dt,$$

where the branch of the function $\log \alpha_r(z)$ is fixed in Π_+ by the conditions

$$\log \alpha_r(z) = \begin{cases} \pi, & t \in (a_k, t_r) =: L'_k \\ 0, & t \in (t_r, b_k) =: L''_k, \end{cases} \tag{3.7.24}$$

and $\alpha_r^*(z)$ is an analytic continuation of $\alpha_r(z)$ into Π_+ by the Symmetry Principle $\left(\text{i.e., } \alpha_r^*(z) = \begin{cases} \alpha_r(z), & z \in \Pi_+, \\ 1/\overline{\alpha_r(\bar{z})}, & z \in \Pi_-. \end{cases}\right)$ Then

$$\frac{\log \alpha_r^*(z)}{X(z)} = \frac{1}{2\pi i}\int_{L \cup L^-} \frac{\log \alpha_r^*(t)}{X(t)} \frac{dt}{t-z}$$

$$= \frac{1}{2\pi i}\int_L \frac{\log \alpha_r(t)}{X^+(t)} \frac{dt}{t-z} + \frac{1}{2\pi i}\int_L \frac{\log \alpha_r^{-1}(t)}{X^+(t)} \frac{dt}{t-z}$$

$$= \frac{1}{\pi i}\int_{L \setminus L_k} \frac{\log \alpha_r(t)}{X_*(t)} \frac{dt}{t-z} + \int_{L'_k} \frac{1}{X_*(t)} \frac{dt}{t-z}.$$

Hence,
$$\int_{L\setminus L_k} \frac{i\arg\alpha_r(t)}{X_*(t)}\frac{dt}{t-z} = \pi i\frac{\log\alpha_r^*(z)}{X(z)} - \pi i\int_{a_k}^{t_r}\frac{1}{X_*(t)}\frac{dt}{t-z}.$$

Expanding this integral in a neighbourhood of $z = \infty$, and taking into account that the first term in the right-hand side has the order n at infinity, we get

$$\int_{L\setminus L_k}\frac{i\arg\alpha_r(t)}{X_*(t)}t^j\,dt = -\pi i\int_{a_k}^{t_r}\frac{t^j}{X_*(t)}\,dt, \quad j = 0,\ldots,n-2.$$

So, the integral J_j has the form

$$J_j = -\pi i\sum_{k=1}^{n}\left\{\sum_{\substack{r=1\\t_r\notin L_k}}^{\nu_1}\int_{a_k}^{t_r}\frac{t^j}{X^+(t)}dt + \sum_{\substack{r=\nu_1+1\\t_r\notin L_k}}^{\nu}\int_{a_k}^{t_r}\frac{t^j}{X^-(t)}dt\right\}, j = 0,\ldots,n-2.$$

Theorem 3.10 *Problem (3.7.1) is unconditionally solvable in the class of meromorphic functions with n_1 internal and n_2 boundary zeros and poles if $2n_1 + n_2 = n-1$, and in general not solvable if $2n_1 + n_2 < n-1$.*

3.7.3 Analytic solutions for $n \le 2$.

Let us consider problem (3.7.1) in the class of analytic functions, i.e., in $\mathcal{C}_\mathcal{A}(\Pi_+)$. As it was already mentioned such solutions exist iff the generalized real analog of Jacobi inversion problem (3.7.20) has a solution in $\Pi_+ \cup L$. In the case of one segment L_k, i.e., for $n = 1$, the congruences in (3.7.20) disappeared, and problem (3.7.1) has analytic solutions. The formal statement of this result follows:

Lemma 3.15 *Problem (3.7.1) for $n = 1$ is unconditionally solvable in $\mathcal{C}_\mathcal{A}(\Pi_+)$. Its general solution has $\chi := \mu + \nu \ge 0$ zeros at the arbitrary points $z_k \in \Pi_+, k = 1,\ldots,\mu, t_k \in L, k = 1,\ldots,\nu$, and is given by the formula*

$$\omega(z) = \pm i\prod_{k=1}^{\mu}\frac{z-z_k}{z-\bar{z}_k}\prod_{k=1}^{\nu}\alpha_k(z)$$
$$\exp\left\{\frac{X(z)}{\pi i}\int_M\frac{\log\alpha(t)}{X(t)}\frac{dt}{t-z} + \frac{X(z)}{\pi i}\int_L\frac{i\theta_1(t)}{X^+(t)}\frac{dt}{t-z}\right\}, \quad (3.7.25)$$

where α_k are defined in (3.7.13),

$$\theta_1(t) = \theta_0(t) - \arg\prod_{k=1}^{\mu}\frac{t-z_k}{t-\bar{z}_k}.$$

3.7. CHEREPANOV'S MIXED PROBLEM.

Let us consider now the case $n = 2$. Using a linear fractional automorphism of upper half-plane Π_+ we can suppose that the segments L_k are symmetric with respect to zero: $L := (-\frac{1}{k}, -1) \cup (1, \frac{1}{k}), k > 1$. Introduce as before the branch of the multi-valued function

$$X(z) := \sqrt{(1-z^2)(1-k^2z^2)},$$

fixed in $\mathbb{C} \setminus L$ by the condition $X(t)|_{(-1,1)} > 0$. Denoting as usual

$$\hat{\omega}_0(z) - iB_{01}/2 := F(z,k) := \int\limits_0^z \frac{dt}{X(t)},$$

$$B_{01}/2 := -K' := \int\limits_{-\frac{1}{k}}^{-1} \frac{dt}{\sqrt{(1-t^2)(1-k^2t^2)}} < 0. \tag{3.7.26}$$

Inversion problem (3.7.20) consists now of one equation and has the form

$$\Im \left\{ \sum_{j=1}^{\nu} F(t_j, k) + 2 \sum_{j=1}^{\mu} F(z_j, k) \right\} = c_0 + (\sigma_1^* + \nu + 2\mu)K', \tag{3.7.27}$$

where $c_0 := \frac{1}{\pi} \int_L \frac{i\theta_0}{X^+(t)} dt + \frac{1}{\pi} \int_M \frac{\log \alpha(t)}{X(t)} dt$. The direct consequences of equation (3.7.27) are the following:

1^0. Problem (3.7.1) has a nowhere vanishing (i.e., $\nu = \mu = 0$) solution in $\mathcal{C}_A(\Pi_+)$ iff

$$c_0 \equiv 0 (mod K'). \tag{3.7.28}$$

This solution is given by the formula

$$\omega(z) = \pm i \exp\left\{ \frac{X(z)}{\pi i} \int_M \frac{\log \alpha(t)}{X(t)} \frac{dt}{t-z} + \frac{X(z)}{\pi i} \int_L \frac{i\theta_1(t)}{X^+(t)} \frac{dt}{t-z} \right\}, \tag{3.7.29}$$

where $\theta_1(t) = \theta_0(t) + \sigma_k \pi, t \in L_k, k = 1,2; \sigma_1 = -\frac{c_0}{K'}, \sigma_2 = 0$.

2^0. If (3.7.28) is valid, then problem (3.7.1) has no analytic solution with one simple boundary zero (i.e., $\nu = 1, \mu = 0$), since for any choice of $t_1 \in L$

$$\Im F(t_1, k) \not\equiv 0 (mod K').$$

In this case there exists an analytic solution of problem (3.7.1) with one simple internal zero (i.e., $\nu = 0, \mu = 1$), namely

$$z_1 = sn(u + iK'/2, k), |u| < K, \tag{3.7.30}$$

where
$$K := \int_0^1 \frac{dt}{X(t)} > 0,$$

and $sn(w, k)$ is the elliptic sine. It is known that the function $sn(w, k)$ maps conformally the rectangular D with wedge-points $\pm K, \pm K + iK'$ onto the upper half-plane Π_+, $(K, K + iK', -K + iK', -K) \mapsto (1, \frac{1}{k}, -\frac{1}{k}, -1)$. Hence, the unique solution with one simple zero is given by the formula

$$w(z) = \pm i \frac{z - z_1}{z - \bar{z}_1} \exp\left\{ \frac{X(z)}{\pi i} \int_M \frac{\log \alpha(t)}{X(t)} \frac{dt}{t - z} + \frac{X(z)}{\pi i} \int_L \frac{i\theta_1(t)}{X^+(t)} \frac{dt}{t - z} \right\}. \quad (3.7.31)$$

The above mentioned zero z_1 can be chosen arbitrarily on the image of median of the rectangular D under the mapping $sn(w, k)$. Besides $\sigma_1 = \sigma_1^* = -\frac{c_0}{K'} - 1, \sigma_2 = 0$.

3^0. If (3.7.28) is not valid, i.e.,

$$0 < \theta = \frac{c_0}{K'} - \left[\frac{c_0}{K'}\right] < 1, \quad (3.7.32)$$

then there exists a solution of problem (3.7.1) with one simple boundary zero (i.e., $\nu = 1, \mu = 0$). This zero is determined on L up to the sign:

$$t_1 := \pm sn(K + i\theta K', k). \quad (3.7.33)$$

The corresponding solution can be delivered by the formula

$$w(z) = \pm i\alpha_1(z) \exp\left\{ \frac{X(z)}{\pi i} \int_M \frac{\log \alpha(t)}{X(t)} \frac{dt}{t - z} + \frac{X(z)}{\pi i} \int_L \frac{i\theta_1(t)}{X^+(t)} \frac{dt}{t - z} \right\} \quad (3.7.34)$$

in which $\sigma_1 = -\left[\frac{c_0}{K'}\right]$, if $t_1 > 0$, $\sigma_1 = -\left[\frac{c_0}{K'}\right] - 1$, if $t_1 < 0$. The solution with one internal zero (i.e., $\nu = 0, \mu = 1$) exists in the case (3.7.32) if we choose

$$z_1 = sn(u + i\theta K'/2, k), |u| < K, \quad (3.7.35)$$

or

$$z_1 = sn(u + i(\theta + 1)K'/2, k), |u| < K, \quad (3.7.36)$$

and $\sigma_1 = -\left[\frac{c_0}{K'}\right] - 2$, or $\sigma_1 = -\left[\frac{c_0}{K'}\right] - 2$, respectively.

Theorem 3.11 *In the case $n = 2$ problem (3.7.1) has a solution in $C_A(\Pi_+)$ with at most $\chi = \nu + \mu$ zeros. $\chi - 1(\chi \geq 2)$ of these zeros can be chosen arbitrarily. The last* *zero is determined by formula (3.7.29), or by one of the formulae (3.7.31), (3.7.32), (3.7.33) in accordance with whether the constant*

$$c'_0 := c_0 - \Im\left\{\sum_{j=1}^{\nu'} F(t_j,k) + 2\sum_{j=1}^{\mu'} F(z_j,k)\right\}, \ \nu' + \mu' = \chi - 1,$$

satisfies condition (3.7.28), or (3.7.32).

Remark 3.4 *Condition (3.7.27) will be satisfied also in the limit case, when $u = \pm K$ in (3.7.30) (if $c_0 \equiv 0 (mod K')$), or in (3.7.35) or (3.7.36) (if $c_0 \not\equiv 0 (mod K')$). In this case the internal zero z_1 tends to the point on L in which the solution has a zero of order 2.*

3.8 Notes and Comments.

3.1.° Problem (3.1.1) is highly connected with the $\mathbb{C}$-linear conjugation (see, e.g., [205]) (or the Riemann boundary value) problem (see, e.g., [91]). Thus in our considerations we often refer to the results for the latter problem taking them from the classical book of F. D. Gakhov [91]. As the main purpose of this part is to construct the solution, we suppose the most simple conditions on G, g, L. We follow here the ideas and results obtained by G. V. Arzhanov ([23]–[25]), by F. D. Gakhov ([89]), by I. I. Komjak ([124]). It was Yu. V. Obnosov who first proposed to consider the nonlinear problems in the separate classes of the type $\mathcal{A}_{m,n}^{k,l}$ (see, e.g., [210], cf. also [15], [241]–[243]). The idea of considering the problem of power type with admissible zeros on contour belongs to F. D. Gakhov [90]; see also [259], [260], [239]. Some results on solvability of the nonlinear boundary value problem of power type (3.1.1) were obtained also in [256], [258].

3.2.° The nonlinear conjugation problem of the multiplication type on the closed curve was solved by G. P. Cherepanov [49] in connection with certain elasticity problems.

3.3.° It was N. V. Govorov who initiated an application of entire function methods in the study of multiplication type problems for an open arc (see [103]). We follow in Subsection 3.3.1 the results of this aricle as well as its continuation in [133], [134] (see also [18]). V. V. Kashevskii ([115]) has developed the ideas of [103] in the case of a compound contour enclosed 2-colourable chart on the plane. His results are described in Subsection 3.3.2. It should be noted that this approach allows obtaining some further results by using the technique of boundary value problems on a Riemann surface (cf., e.g., [294], [116], [47]).

3.4.° A number of cases in which the nonlinear Riemann–Hilbert problem is solvable in closed form were described in the survey by V. K. Natalevich [207] (see also [208], [209]). General results concerning problem (3.4.1) were obtained by Yu. V. Obnosov [210], [211], [213], [214] (cf. also [12]–[14], [16]). We describe in Section 3.4 the results in this direction quite similar to [217].

3.5.° Simple modulus problem (3.5.1) was studied among other nonlinear problems by V. K. Natalevich [207]. We use partly the description proposed by Yu. V. Obnosov in [217]. Other approaches to the study of the modulus and generalized modulus problem were described in [4], [5] (see also [233], [234][276]).

3.6.° The generalized modulus problem (3.5.14) is highly connected with the general linear fractional conjugation problem (3.6.1). It is closely related (see, e.g., [290]) to the problem of matrix factorization (see also [145]). The linear fractional conjugation problem was studied by S. N. Kiyasov [121]– [123]. We follow here his description of the results.

3.7.° The nonlinear mixed boundary value problem is a core tool for the study of the problems of plane elasticity (see [49], [50]), mechanics of brittle destruction [52], and other mechanical problems. The simple case of solvability of problem (3.7.20) was described by G. P. Cherepanov (e.g., [50]), the meromorphic solutions were studied by Yu.V.Obnosov [215]. Analytic solutions of the Cherepanov's problem were obtained by him only in the case $n \leq 2$ ([216]). We follow these results in Section 3.7.

Chapter 4

Method of functional equations and related topics

4.1 Dirichlet problem for a doubly connected domain and functional equations

In this section we study the simple functional equation to outline some of the fundamental ideas of the method of functional equations. Let us consider two discs $\mathbb{D}_k := \mathbb{D}(a_k, r_k) = \{z \in \mathbb{C} : |z - a_k| < r_k\}$ $(k = 1, 2)$, $cl\mathbb{D}_1 \cap cl\mathbb{D}_2 = \emptyset$, and the doubly connected domain $\mathbb{D} := \widehat{\mathbb{C}} \backslash (cl\mathbb{D}_1 \cup cl\mathbb{D}_2)$ on the extended complex plane $\widehat{\mathbb{C}}$ (see Figure 4.1). The orientation chosen on $\mathbb{T}_k := \{z \in \mathbb{C} : |z - a_k| = r_k\}$ leaves $\mathbb{D}$ to the left. The Dirichlet problem for the domain $\mathbb{D}$ consists in the following. Given Hölder continuous functions $f_k(t)$ on $\mathbb{T}_k$ ($f_k(t) \in \mathcal{H}(\mathbb{T}_k)$, $k = 1, 2$), find a function $u(z)$ harmonic in $\mathbb{D}$ and Hölder continuous on $cl\mathbb{D}$ satisfying the boundary condition

$$u(t) = f_k(t), \ t \in \mathbb{T}_k, \ k = 1, 2. \tag{4.1.1}$$

Using the Logarithmic Conjugation Theorem (Section 2.4) for a doubly connected domain we look for $u(z)$ in the following form

$$u(z) = \Re\left(\phi(z) + A\log\frac{z - a_1}{z - a_2}\right), \tag{4.1.2}$$

where the function $\phi(z)$ is analytic in $\mathbb{D}$ and Hölder continuous on $cl\mathbb{D}$, i.e., $\phi \in \mathcal{H}_A(\mathbb{D})$, A is an unknown real constant. Let us choose a curve L_1 connecting the points $z = a_1$ and infinity, which has no common points with $\mathbb{T}_1$. Suppose that

$\log(z-a_1)$ is a fixed analytic branch of the multi-valued logarithm in $\widehat{\mathbb{C}}\setminus L_1$. The function $\log(z-a_2)$ is defined in a similar way. Let us denote

$$z^*_{(k)} := z^*_{\mathbb{T}_k} = \frac{r_k^2}{z-a_k} + a_k$$

the inversion with respect to the circle $\mathbb{T}_k$ (see Section 2.1). We have $t^*_{(k)} = t$ on $|t-a_k| = r_k$. A function $\varphi(z)$ is analytic in $|z-a_k| < r_k$ if and only if $\varphi\left(z^*_{(k)}\right)$ is analytic in $|z-a_k| > r_k$ (see Section 2.1). Using the Decomposition Theorem (see Section 2.4) we find the function $\phi(z)$ in the form

$$\phi(z) = \overline{\phi_1\left(z^*_{(1)}\right)} + \overline{\phi_2\left(z^*_{(2)}\right)}, \tag{4.1.3}$$

where the function $\phi_k(z)$ is analytic in $|z-a_k| < r_k$ and Hölder continuous on $|z-a_k| \leq r_k$ $(k=1,2)$, i.e., $\phi_k \in \mathcal{H}_A(\mathbb{D}_k)$. Substituting (4.1.3) in (4.1.2) and using (4.1.1) we arrive at the following boundary conditions

$$\overline{\Re\phi_1\left(t^*_{(1)}\right)} + \overline{\Re\phi_2\left(t^*_{(2)}\right)} + A\log\left|\frac{t-a_1}{t-a_2}\right| = f_k(t),\ t\in\mathbb{T}_k,\ k=1,2. \tag{4.1.4}$$

On the circle on $\mathbb{T}_1$ we have $\overline{\Re\phi_1\left(t^*_{(1)}\right)} = \Re\phi_1(t)$, $A\log\left|\frac{t-a_1}{t-a_2}\right| = A[\log r_1 - \log|t-a_2|]$. Then the first condition (4.1.4) becomes

$$\Re\left[\phi_1(t) + \overline{\phi_2\left(t^*_{(2)}\right)}\right] = \Re[g_1(t) + A\log(t-a_2)] - A\log r_1,\ t\in\mathbb{T}_1, \tag{4.1.5}$$

where

$$g_1(z) = -\frac{1}{\pi i}\int_{\mathbb{T}_1}\frac{f_1(\tau)}{\tau-z}d\tau + \frac{1}{2\pi i}\int_{\mathbb{T}_1}\frac{f_1(\tau)}{\tau-a_1}d\tau + i\gamma_1 \tag{4.1.6}$$

is a solution of the Schwarz problem $\Re g_1(t) = f_1(t)$, $t\in\mathbb{T}_1$, with respect to $g_1 \in \mathcal{H}_A(\mathbb{D}_1)$ (see Section 2.7.1 and take into account the orientation of $\mathbb{T}_1$). One can consider (4.1.5) as the Schwarz problem with respect to the function $\phi_1(z) + \overline{\phi_2\left(z^*_{(2)}\right)}$ analytic in the disc $|z-a_1| < r_1$ with the prescribed real part of boundary values $\Re[g_1(t) + A\log(t-a_2)] - A\log r_1$. It is easily seen that

$$\phi_1(z) + \overline{\phi_2\left(z^*_{(2)}\right)} = g_1(z) + A[\log(z-a_2) - \log r_1],\ z\in cl\mathbb{D}_1. \tag{4.1.7}$$

4.1. DIRICHLET PROBLEM FOR A DOUBLY CONNECTED DOMAIN.

Similar arguments for the second relation (4.1.4) yield

$$\phi_2(z) + \overline{\phi_1\left(z^*_{(1)}\right)} = g_2(z) - A\left[\log(z - a_1) - \log r_2\right], \quad z \in cl\mathbb{D}_2, \tag{4.1.8}$$

where the known function g_2 belongs to $\mathcal{H}_A(\mathbb{D}_2)$. The equalities (4.1.7) and (4.1.8) constitute a system of two functional equations with respect to the functions $\phi_1(z)$ and $\phi_2(z)$. Excluding $\overline{\phi_2\left(z^*_{(2)}\right)}$ from equality (4.1.7) we obtain the simple functional equation

$$\phi_1(z) = \phi_1[\alpha(z)] + g(z), \quad z \in cl\mathbb{D}_1, \tag{4.1.9}$$

where

$$g(z) := g_1(z) - \overline{g_2\left(z^*_{(2)}\right)} + A\left[\log(z - a_2) + \log\left(z^*_{(2)} - a_1\right) - \log r_1 r_2\right] + i\gamma \tag{4.1.10}$$

belongs to $\mathcal{H}_A(\mathbb{D}_1)$. The pure imaginary constant $i\gamma$ contains the additive constants $i\gamma_1$ and $i\gamma_2$ appearing in the definition of $g_1(z)$ and $g_2(z)$. The conformal mapping

$$\alpha(z) := \left(z^*_{(2)}\right)^*_{(1)} = \frac{r_1^2(z - a_2)}{r_2^2 + (a_2 - a_1)(z - a_2)} - a_1$$

transfers the closed disc $|z - a_1| \leq r_1$ into the open one $|z - a_1| < r_1$.

Now we outline some notations and facts of the general theory of the functional equation which are connected with those equations appearing at the study of boundary value problems. In particular we need the following

Theorem 4.1 *(Denjoy–Wolff) Let f be an analytic function in $\mathbb{U}$ mapping $\mathbb{U}$ into itself (but not a Möbius transformation of $\mathbb{U}$ onto $\mathbb{U}$). Then there exists a point z_0, $|z_0| \leq 1$, such that the sequence of iterations $f^n(z)$ converges to z_0 uniformly on each compact subset of $\mathbb{U}$. Moreover, if $|z_0| = 1$, then $\lim_{r \to 1-0} f(rz_0) = z_0$ and $s := \lim_{r \to 1-0} f'(rz_0)$ exists and $0 < |s| \leq 1$.*

Theorem 4.2 *Let a function $f \in \mathcal{C}_A(cl\mathbb{U})$ be mapping the closed unit disc $cl\mathbb{U}$ into the open unit disc $\mathbb{U}$. Then f has a unique fixed point z_0 in $\mathbb{U}$ and $|f'(z_0)| < 1$. The sequence $f^n(z)$ converges uniformly in $|z| \leq 1$ to the point z_0.*

The point z_0 is called *the attractive point* of $f(z)$.

Theorem 4.3 *The operator $\varphi \mapsto \varphi[f(z)]$ is compact in the spaces $\mathcal{H}_A(\mathbb{U})$, $\mathcal{C}_A(\mathbb{U})$ and in the Hardy space $\mathcal{H}_p(\mathbb{U})$, $1 < p < +\infty$.*

Theorem 4.4 *Let $G, g \in \mathcal{C}_A(\mathbb{U})$. If $G(z_0)[f'(z_0)]^k \neq 1$ for all $k = 0, 1, 2, \ldots$, then the functional equation*

$$\varphi(z) = G(z)\varphi[f(z)] + g(z), \quad |z| \leq 1 \tag{4.1.11}$$

has a unique solution $\varphi \in \mathcal{C}_A(\mathbb{U})$. If for some k we have $G(z_0)[f'(z_0)]^k = 1$, then (4.1.11) has a solution in $\mathcal{C}_A(\mathbb{U})$ if and only if a solvability condition on $G(z)$ and $g(z)$ is fulfilled. If so then the general solution of (4.1.11) depends on an arbitrary complex constant.

Remark 4.1 *The solvability condition of (4.1.11) involves Taylor's coefficients of the known functions $f, G,$ and g. See for details [132].*

Remark 4.2 *The index k for which $G(z_0)[f'(z_0)]^k = 1$ can be only unique.*

Our functional equation (4.1.9) has the same form as (4.1.11). We only consider the disc $|z - a_1| < r_1$ instead of $|z| < 1$. The function $\alpha(z)$ has a unique attractive fixed point satisfying the quadric equation $\alpha(z) = z$ in $|z - a_1| < r_1$. Let z_0 and z^0 be the roots of the quadric equation $\alpha(z) = z$. It follows from the simple geometric observations that the points z_0 and z^0 belong to the segment connecting a_1 and a_2, (for the definiteness $z_0 \in \mathbb{D}_1$ and $z^0 \in \mathbb{D}_2$). Moreover, $(z_0)^*_{(1)} = (z_0)^*_{(2)} = z^0$, and $(z^0)^*_{(1)} = (z^0)^*_{(2)} = z_0$ (see Figure 4.1).

The necessary and sufficient solvability condition for (4.1.9) has the form $g(z_0) = 0$. By virtue of (4.1.9) the relation $\Re g(z_0) = 0$ yields

$$A = \Re\left[g_1(z_0) - \overline{g_2\left((z_0)^*_{(2)}\right)}\right] \log \frac{\left|(z_0 - a_2)\left(\overline{(z_0)^*_{(2)}} - \overline{a_2}\right)\right|}{r_1 r_2}. \tag{4.1.12}$$

The relation $\Im g(z_0) = 0$ determines the constant $i\gamma$ in (4.1.10). We apply now the method of successive approximations to equation (4.1.9).

Theorem 4.5 *Let the given function g belongs to $\mathcal{C}_A(\mathbb{D}_1)$ and $g(z_0) = 0$, where z_0 is the fixed point of $\alpha(z)$ in $|z - a_1| < r_1$. Then the general solution of functional equation (4.1.9) has the form*

$$\phi_1(z) = \sum_{k=0}^{\infty} g[\alpha^k(z)] + C, \tag{4.1.13}$$

where C is an arbitrary constant.. The series (4.1.13) converges absolutely and uniformly in $|z - a_1| \leq r_1$.

4.1. DIRICHLET PROBLEM FOR A DOUBLY CONNECTED DOMAIN. 133

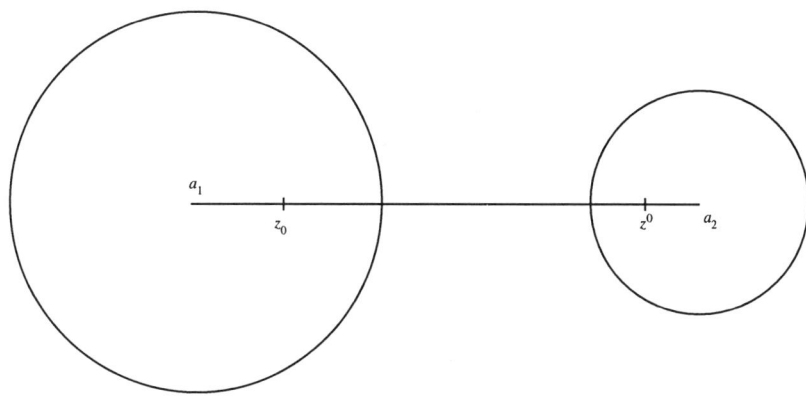

Figure 4.1: Doubly connected circular domain

Proof. First we prove that the following inequality is valid in the condition of theorem:
$$|\alpha(z) - z_0| \leq q|z - z_0|, \ q < 1, \ z \in cl\mathbb{D}_1. \tag{4.1.14}$$
The function $(z - z_0)^{-1}(\alpha(z) - z_0)$ belongs to $\mathcal{C}_\mathcal{A}(\mathbb{D}_1)$. The Maximum Modulus Principle applied to this function yields the inequality

$$\max_{cl\mathbb{D}_1} \frac{|\alpha(z)-z_0|}{|z-z_0|} = \max_{cl\mathbb{D}_1} \frac{r_1^2 r_2^2}{\left|(r_2^2+(a_2-a_1)(z-a_2))(r_2^2+(a_2-a_1)(z_0-a_2))\right|} \leq$$

$$\frac{r_1^2 r_2^2}{\left|(r_1+r_2+\delta)r_2-r_2^2\right|^2} = \left(\frac{r_1^2 r_2^2}{r_1 r_2 + \delta r_2}\right)^2 =: q,$$

where δ is the distance between the circles $\mathbb{T}_1$ and $\mathbb{T}_2$. Hence, inequality (4.1.14) is proved. Let $g(z_0) = 0$, then $h(z) := (z-z_0)^{-1}g(z)$ belongs to $\mathcal{C}_\mathcal{A}(\mathbb{D}_1)$. The series

$$\sum_{k=0}^{\infty} g[\alpha^k(z)] = \sum_{k=0}^{\infty} (\alpha^k(z) - \alpha^k(z_0)) h[\alpha^k(z)] \tag{4.1.15}$$

is majorized by a convergent nonnegative series, since $\left|h[\alpha^k(z)]\right| \leq \max_{cl\mathbb{D}_1} |h(z)|$,

$$\left|\alpha^k(z) - \alpha^k(z_0)\right| \leq q\left|\alpha^{k-1}(z) - \alpha^{k-1}(z_0)\right| \leq \ldots \leq q^{k-1} \max_{cl\mathbb{D}_1} |z - z_0|.$$

Therefore, (4.1.15) converges absolutely and uniformly on $c l \mathbb{D}_1$ to an analytic function φ_1 satisfying (4.1.9). Let φ_2 be another solution of (4.1.9). Then $\varphi_3 := \varphi_2 - \varphi_1$ is a solution of the equation $\phi(z) = \phi[\alpha(z)]$. We have $\varphi_3(z) = \varphi_3[\alpha^N(z)]$ for each natural N. Hence, $\varphi_3(z) = \lim_{N \to \infty} \varphi_3[\alpha^N(z)] = \varphi_3(z_0)$ is a constant C. This implies that the general solution of (4.1.9) has the form (4.1.13).

The theorem is proved. ∎

We apply here the Pumping Principle (see Section 2.3) to (4.1.9) in the following way. The properties of the Cauchy integral (Section 2.6) imply that g from (4.1.10) belongs to $\mathcal{H}_A(\mathbb{D}_1)$, since the data $f_k \in \mathcal{H}(\mathbb{T}_k)$ for $k = 1, 2$ (cf. (4.1.6)). We solved equation (4.1.9) in the space $\mathcal{C}_A(\mathbb{D}_1)$. Then the function $\phi_1[\alpha(z)]$ is analytic in $|z - a_1| \leq r_1$; hence the right-hand part of (4.1.9) belongs to $\mathcal{H}_A(\mathbb{D}_1)$. Therefore, (4.1.15) is the general solution of (4.1.9) in the space $\mathcal{H}_A(\mathbb{D}_1)$.

4.2 A nonlinear boundary value problem.

4.2.1 General approach.

We consider the following nonlinear boundary value problem. Find two functions $\phi^+(z)$ and $\phi^-(z)$ which are analytic in the connected domains D^+ and D^- separated by a curve L (see Section 2.1), Hölder continuous in the closures of the domains ($\phi^+ \in \mathcal{H}_+(L)$, $\phi^- \in \mathcal{H}_-(L)$), satisfying the following conjugation condition:

$$\phi^+(t) = a(t)\phi^-(t) + \left(\mathbf{W}\overline{\phi^-}\right)(t) + c(t), \ t \in L, \ \phi^-(\infty) = 0, \quad (4.2.1)$$

where $a(t)$ and $c(t)$ are given Hölder continuous functions on L, $a(t) \neq 0$, $\mathbf{W}$ is a nonlinear operator acting on the space of Hölder continuous functions $\mathcal{H}(L)$.

Theorem 4.6 Let $\|\mathbf{S}\|_\alpha$ be the norm of the singular integral $\mathbf{S}$ in $\mathcal{H}^\alpha(L)$ (see Section 2.6). Let

$$\|\mathbf{W}f_1 - \mathbf{W}f_2\| \leq W \|f_1 - f_2\| \quad (4.2.2)$$

for all $f_1, f_2 \in \mathcal{H}^\alpha(L)$, where W is a positive constant, and

$$W \frac{1 + \|\mathbf{S}\|_\alpha}{2} \leq q < 1. \quad (4.2.3)$$

If $a(t) = 1$, then problem (4.2.1) has a unique solution.

Proof. Applying the Sokhotsky–Plemelj formulae (2.6.6) to the relation

$$\phi^+(t) - \phi^-(t) = \left(\mathbf{W}\overline{\phi^-}\right)(t) + c(t), \ t \in L,$$

4.2. A NONLINEAR BOUNDARY VALUE PROBLEM.

we arrive at the following singular integral equation

$$\phi^-(t) = (\mathbf{A}\phi^-)(t), \ t \in L, \tag{4.2.4}$$

to be solved in $\mathcal{H}(L)$. Here the nonlinear operator $\mathbf{A}$ is defined by the formula

$$(\mathbf{A}f)(t) := \frac{1}{2}\left[-(\mathbf{W}\overline{f})(t) + (\mathbf{S}\mathbf{W}\overline{f})(t)\right] + d(t), \ t \in L,$$

The known function $d(t) := (\mathbf{Q}c)(t)$, where $\mathbf{Q}$ is the standard projector (see Section 2.6). It follows from (4.2.2), (4.2.3) that the operator $\mathbf{A}$ satisfies the inequality

$$\|\mathbf{A}f_1 - \mathbf{A}f_2\| \leq \frac{1+\|\mathbf{S}\|_\alpha}{2}\|\mathbf{W}\overline{f_1} - \mathbf{W}\overline{f_2}\| \leq W\frac{1+\|\mathbf{S}\|_\alpha}{2}\|f_1 - f_2\| \leq q\|f_1 - f_2\|,$$

where $0 < q < 1$. Using the Banach Theorem we obtain that equation (4.2.4) has the unique solution

$$\phi^-(t) = \lim_{k \to \infty}(\mathbf{A}^k d)(t), \ t \in L. \tag{4.2.5}$$

The limit (4.2.5) is uniform on L, since the topology of $\mathcal{H}^\alpha(L)$ is stronger than that of $C(L)$. Moreover due to the Maximum Modulus Principle, (4.2.5) implies the uniform convergence of $(\mathbf{A}^k d)(t)$ on D^-, and $\phi^- \in \mathcal{H}^\alpha_A(D^-)$. Finally, $(\mathbf{A}f)(\infty) = 0$ for each $f \in \mathcal{H}(L)$. Hence, $\phi^-(\infty) = 0$.

The theorem is proved. ∎

Theorem 4.7 *Let $a(t) = t^\chi X^+(t)[X^-(t)]^{-1}$ be the factorization of $a(t)$ on L (see Subsection 2.8.1), and X^- satisfies the inequality*

$$\max_{t \in L}|t^{-\chi}X^-(t)|W\frac{1+\|\mathbf{S}\|_\alpha}{2\min_{t\in L}|X^-(t)|} \leq q < 1, \tag{4.2.6}$$

in notations of Theorem 4.6. If $\chi > 0$, then the general solution of problem (4.2.1) depends on χ arbitrary complex constants. If $\chi < 0$, then problem (4.2.1) has a solution if and only if $|\chi|$ solvability conditions are fulfilled. In the latter case the solution of (4.2.1) is unique.

Proof. Rewrite (4.2.1) in the form

$$\frac{\phi^+(t)}{X^+(t)} = t^\chi \frac{\phi^-(t)}{X^-(t)} + \frac{1}{X^-(t)}(\mathbf{W}\overline{\phi^-})(t) + \frac{c(t)}{X^-(t)}, \ t \in L. \tag{4.2.7}$$

Let $\chi > 0$. Put
$$\psi^+(z) := \phi^+(z)\,[X^+(t)]^{-1},$$
$$\psi^-(z) + P_{\chi-1}(z) := z^\chi \phi^-(z)\,[X^-(t)]^{-1},$$
$$c_1(t) := c(t)\,[X^-(t)]^{-1} + P_{\chi-1}(t),$$
$$(\mathbf{W}_1 f)(t) := [X^-(t)]^{-1}\,\mathbf{W}\left[(f(t)+P_{\chi-1}(t))\,t^{-\chi}X^-(t)\right](t),$$

where a polynomial $P_{\chi-1}(z) = p_0 + p_1 z + \ldots + p_{\chi-1}z^{\chi-1}$ with complex coefficients p_j ($j=0,1,\ldots,\chi-1$) coincides with the principal part of $z^\chi \phi^-(z)\,[X^-(t)]^{-1}$ at infinity. One can see that $\psi^+ \in \mathcal{H}_+(L)$, $\psi^- \in \mathcal{H}_-(L)$ and $\psi^-(\infty) = 0$, since the functions $X^\pm(t)$, and $[X^\pm(t)]^{-1}$ are bounded in $cl D^\pm$, respectively. The operator $\mathbf{W}_1$ satisfies the inequality

$$\|\mathbf{W}_1 f_1 - \mathbf{W}_1 f_2\| \leq W \frac{\max_{t\in L} |t^{-\chi} X^-(t)|}{\min_{t\in L} |X^-(t)|} \|f_1 - f_2\|, \quad \forall f_1, f_2 \in \mathcal{H}^\alpha(L). \tag{4.2.8}$$

It follows from (4.2.7) that

$$\psi^+(t) = \psi^-(t) + \left(\mathbf{W}_1\overline{\psi^-}\right)(t) + c_1(t), \quad t \in L. \tag{4.2.9}$$

Using inequalities (4.2.6), (4.2.8) and applying Theorem 4.6 to (4.2.9) we conclude that problem (4.2.9) has solutions depending on χ arbitrary constants p_j ($j=0,1,\ldots,\chi-1$). Then the general solution of (4.2.1) has the form

$$\phi^+(z) = \psi^+(z) X^+(t), \quad \phi^-(z) = z^{-\chi}\left[\psi^-(z) + P_{\chi-1}(z)\right] X^-(t),$$

i.e., also depends on χ arbitrary constants p_j. Let $\chi < 0$. Put

$$\psi^+(z) := \phi^+(z)\left[X^+(t)\right]^{-1}, \quad \psi^-(z) := z^\chi \phi^-(z)\left[X^-(t)\right]^{-1}$$

Then (4.2.7) is reduced to (4.2.9) with $P_{\chi-1}(z) \equiv 0$. In this case problem (4.2.9) has a unique analytic solution. Using (4.2.5) we get

$$\psi^-(z) = \lim_{k\to\infty}(\mathbf{A}^k d)(z), \quad z \in cl D^-, \tag{4.2.10}$$

where $\mathbf{A} := \tfrac{1}{2}[-\mathbf{W}_1 \mathbf{C} + S\mathbf{W}_1 \mathbf{C}] + d$, $\mathbf{C}$ is the operator of complex conjugation, $d := \mathbf{P} c_1$ (see Section 2.6). The function $\psi^-(z)$ must have zero-point of the order $|\chi|+1$ at infinity, since $\phi^-(\infty) = 0$ and $X^-(\infty) = 1$. This condition yields the necessary and sufficient solvability conditions for problem (4.2.1):

$$\lim_{z\to 0} z^{-l}\psi^-(z) = 0, \quad l = 1, 2, \ldots, |\chi|,$$

4.2. A NONLINEAR BOUNDARY VALUE PROBLEM.

where $\psi^-(z)$ has the form (4.2.10). The last condition can be rewritten in the form

$$\int_L t^l (\mathbf{B}c)(t)dt = 0, \ l = 1, 2, ..., |\chi|,$$

where $\mathbf{B}$ is a nonlinear operator.
The theorem is proved. ∎

4.2.2 A problem for a doubly connected domain.

Let us consider the problem (4.2.1) in the case when the contour L consists of the two circles $\mathbb{T}_j := \{t \in \mathbb{C} : |t - a_j| = r_j\}$, $j = 1, 2$. We assume that $a(t) = 1$, $c(t) = 0$, $(\mathbf{W}\overline{\phi^-})(t) := F_j(t, \overline{\phi^-(t)})$ on $\mathbb{T}_j$, where $F_j(z, \xi)$ is the entire function with respect to ξ, F belongs to $\mathcal{C}_A(\mathbb{D})$ with respect to z, and $F_j(\infty, \xi) = 0$. Thus (4.2.1) has the form

$$\phi(t) = \phi_j(t) + F_j\left(t, \overline{\phi_j(t)}\right), \ t \in \mathbb{T}_j, \ j = 1, 2, \ (\phi(\infty) = 0). \tag{4.2.11}$$

The unknown functions $\phi(z)$ and $\phi_j(z)$ belong to $\mathcal{C}_A(\mathbb{D})$ and $\mathcal{C}_A(\mathbb{D}_j)$ respectively.

Remark 4.3 *If $F_j(t, \xi) = -\xi + f_j(t)$, then (4.2.11) becomes the $\mathbb{R}$-linear conjugation problem*

$$\phi(t) = \phi_j(t) - \overline{\phi_j(t)} + f_j(t), \ |t - a_j| = r_j, \ j = 1, 2, \ (\phi(\infty) = 0).$$

The real part of the last relation coincides with the boundary condition of the Schwarz problem.

Let us introduce the function

$$\Phi(z) := \begin{cases} \phi_1(z) + F_2\left(z, \overline{\phi_2\left(z^*_{(2)}\right)}\right), & |z - a_1| \le r_1, \\ \phi_2(z) + F_1\left(z, \overline{\phi_1\left(z^*_{(1)}\right)}\right), & |z - a_2| \le r_2, \\ \phi(z) + \sum_{j=1,2} F_j\left(z, \overline{\phi_j\left(z^*_{(j)}\right)}\right), & z \in \mathbb{D}, \end{cases}$$

sectionally analytic in $\widehat{\mathbb{C}} \backslash L$. Put

$$\Phi^+(t) := \lim_{z \to t, \ z \in \mathbb{D}} \Phi(z), \quad \Phi^-(t) := \lim_{z \to t, \ z \in \mathbb{D}_1} \Phi(z).$$

Let us calculate the jump of $\Phi(z)$ along $\mathbb{T}_1$

$$\Delta_1 := \Phi^+(t) - \Phi^-(t) = \phi(t) - \phi_1(t) - F_1\left(t, \overline{\phi_1\left(t^*_{(1)}\right)}\right).$$

Using (4.2.11) we obtain $\Delta_1 = 0$. Similar arguments for the jump Δ_2 along $\mathbb{T}_2$ yield the relation $\Delta_2 = 0$. The Analytic Continuation Principle and Liouville's theorem imply that $\Phi(z) = $ constant. We have $\Phi(z) \equiv 0$, since $\Phi(\infty) = 0$. Therefore

$$\begin{cases} \phi_1(z) = -F_2\left(z, \overline{\phi_2\left(z^*_{(2)}\right)}\right), & |z - a_1| \leq r_1, \\ \phi_2(z) = -F_1\left(z, \overline{\phi_1\left(z^*_{(1)}\right)}\right), & |z - a_2| \leq r_2. \end{cases} \quad (4.2.12)$$

Substituting $\phi_2(z)$ from the second equation into the first one we arrive at the simple nonlinear functional equation

$$\phi_1(z) = h(z, \phi_1[\alpha(z)]), \quad |z - a_1| \leq r_1, \quad (4.2.13)$$

where $h(t, \xi) := -F_2\left(z, \overline{-F_1\left(z^*_{(1)}, \overline{\xi}\right)}\right)$. The properties of the transformation $\alpha(z) = \left(z^*_{(2)}\right)^*_{(1)}$ are discussed in the previous section. The function $h(t, \xi)$ belongs to $\mathcal{C}_A(\mathbb{D}_1)$ in z, and is entire in ξ. A detailed study of (4.2.13) is given in the next section.

4.2.3 A nonlinear boundary value problem and functional equation.

In the present section we consider a nonlinear problem which is reduced to a functional equation. This problem is similar to problem (4.2.11). The main reason why we discuss it is to pose a functional equation which can have a singularity at fixed point.

The problem is in the following: to find functions $\phi(z)$, $\phi_1(z)$, $\phi_2(z)$ analytic in $r < |z| < 1$, $|z| < r$, $|z| > 1$ respectively and continuous in the closures of the domains considered with the boundary conditions

$$\phi(t) = a_1(t)\phi_1(t) + F_1\left(t, \overline{\phi_1(t)}\right), \quad |t| = r, \quad (4.2.14)$$

$$\phi(t) = a_2(t)\phi_2(t) + F_2\left(t, \overline{\phi_2(t)}\right), \quad |t| = 1, \ \phi(\infty) = 0,$$

4.2. A NONLINEAR BOUNDARY VALUE PROBLEM.

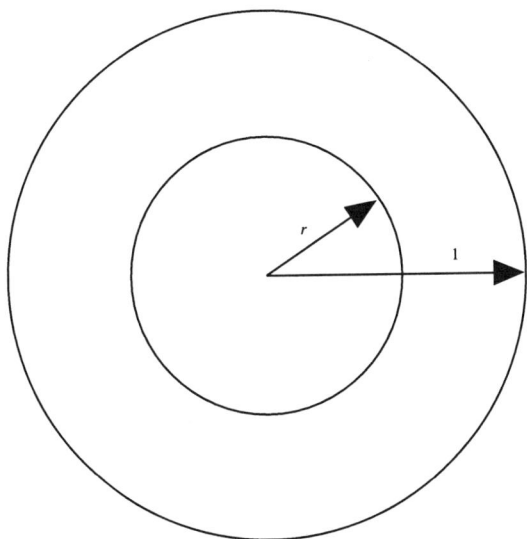

Figure 4.2: Annulus $r < |z| < 1$

where given Hölder continuous functions $a_1(t)$ and $a_2(t)$ are not vanishing on $|t| = r$ and $|t| = 1$, respectively. The known functions $F_j(t,\xi)$ ($j = 1, 2$) are entire on ξ and satisfy some conditions of the analytic continuation on t written below. Let

$$a_1(t) = t^{\chi_1} X_1^+(t) \left[X_1^-(t)\right]^{-1}, \quad |t| = r,$$

be the factorization of $a_1(t)$ (see Section 2.8.1), i.e., $X_1^+(z)$ and $X_1^-(z)$ are analytic in $|z| < r$ and $|z| > r$, respectively. Let

$$a_2(t) = t^{\chi_2} X_2^+(t) \left[X_2^-(t)\right]^{-1}, \quad |t| = 1,$$

be the factorization of $a_2(t)$, i.e., $X_2^+(z)$ and $X_2^-(z)$ are analytic in $|z| < 1$ and $|z| > 1$, respectively. Introduce new functions

$$G_1(t,\xi) := \frac{X_1^-(t)}{X_2^+(t)} F_1\left(t, \frac{\overline{X_2^+(t)}}{X_1^+(t)}\xi\right), \quad G_2(t,\xi) := \frac{X_1^-(t)}{X_2^+(t)} F_1\left(t, \frac{\overline{X_2^-(t)}}{X_1^-(t)}\xi\right).$$

We additionally assume that $G_1(z,\xi) \in \mathcal{H}_-(|z|=r)$ and $G_2(z,\xi) \in \mathcal{H}_+(|z|=1)$ on the variable z. Introduce unknown functions

$$\psi(z) := \frac{X_1^-(z)}{X_2^+(z)}\phi(z) \in \mathcal{H}_A(r < |z| < 1), \quad \psi_1(z) := \frac{X_1^+(z)}{X_2^+(z)}\phi_1(z) \in \mathcal{H}_A(|z| < r),$$

$$\psi_2(z) := \frac{X_1^-(z)}{X_2^-(z)}\phi_2(z) \in \mathcal{H}_A(|z| > 1).$$

Then (4.2.14) becomes

$$\psi(t) = t^{\chi_1}\psi_1(t) + G_1\left(t,\overline{\psi_1(t)}\right), \quad |t| = r, \tag{4.2.15}$$

$$\psi(t) = t^{\chi_2}\psi_2(t) + G_2\left(t,\overline{\psi_2(t)}\right), \quad |t| = 1, \ \psi(\infty) = 0.$$

The index of problem (4.2.15) is $\chi := \chi_2 - \chi_1$.

Following the previous section we introduce the function

$$\Phi(z) := \begin{cases} z^{\chi_1}\psi_1(z) + G_2\left(z,\overline{\psi_1\left(\frac{1}{z}\right)}\right), & |z| \leq r, \\ z^{\chi_2}\psi_2(z) + G_1\left(z,\overline{\psi_2\left(\frac{r^2}{z}\right)}\right), & |z| \geq 1, \\ \psi(z) + G_1\left(z,\overline{\psi_2\left(\frac{r^2}{z}\right)}\right) + G_2\left(z,\overline{\psi_1\left(\frac{1}{z}\right)}\right), & r < |z| < 1, \end{cases}$$

analytic in $\widehat{\mathbb{C}}$ except eventually the points $z = 0$ and $z = \infty$. If $\chi_1 < 0$, then $\Phi(z)$ can have a pole of the order $|\chi_1|$ at $z = 0$. If $\chi_2 > 0$, then a pole of the order $\chi_2 - 1$ is possible at $z = \infty$. Applying the generalized Liouville's theorem we conclude that

$$\Phi(z) = z^{\chi_1} P_{|\chi_1|-1}(z) + Q_{\chi_2-1}(z),$$

where $P_{|\chi_1|-1}(z) = p_0 + p_1 z + ... + p_{|\chi_1|-1} z^{|\chi_1|-1}$, $Q_{|\chi_2|-1}(z) = q_0 + q_1 z + ... + q_{|\chi_2|-1} z^{|\chi_2|-1}$ are polynomials with complex coefficients. If $\chi_1 > 0$, then $P_{|\chi_1|-1}(z) \equiv 0$; if $\chi_2 \leq 0$, then $Q_{|\chi_2|-1}(z) \equiv 0$. The definition of $\Phi(z)$ yields the set of functional equations

$$z^{\chi_1}\psi_1(z) = -G_2\left(z,\overline{\psi_1\left(\frac{1}{z}\right)}\right) + \Phi(z), \quad |z| \leq r, \tag{4.2.16}$$

$$z^{\chi_2}\psi_2(z) = -G_1\left(z,\overline{\psi_2\left(\frac{r^2}{z}\right)}\right) + \Phi(z), \quad |z| \geq 1.$$

4.2. A NONLINEAR BOUNDARY VALUE PROBLEM.

Let us consider the case when $\chi_1 \leq 0$ and $\chi_2 \geq 0$. Then $\chi \geq 0$, $\Phi(z)$ contains χ constants. Eliminating ψ_2 from (4.2.16) we obtain an equation of type (4.2.13). In the opposite case (4.2.16) is reduced to an equation with singularity at the fixed point $z = 0$. We consider the case $\chi_2 = 0$ and $\chi = -\chi_1 \leq 0$. Then (4.2.16) becomes

$$\psi_1(z) = z^\chi h(z, \psi_1(r^2 z)), \quad |z| \leq r, \qquad (4.2.17)$$

where $h(z, \xi) := -G_2\left(z, \overline{G_1\left(\frac{1}{\bar z}, \xi\right)}\right)$. First we solve equation (4.2.17) with $\chi = 0$ following J. Matkowski and W. Smajdor [132]. The same method is applied to equation (4.2.13). At the beginning we are looking for local solutions of the functional equation

$$\psi_1(z) = h(z, \psi_1(r^2 z)), \quad |z| \leq r, \qquad (4.2.18)$$

i.e., solutions analytic at $z = 0$. Represent the unknown function in the form of its Taylor expansion

$$\psi_1(z) = \sum_{m=0}^{\infty} \alpha_m z^m.$$

Substituting $z = 0$ into (4.2.17) we arrive at the equation

$$\alpha_0 = h(0, \alpha_0) \qquad (4.2.19)$$

with respect to α_0. Without loss of generality we assume that $\frac{\partial h}{\partial \xi}(0, \alpha_0) \neq 0$. If number equation (4.2.19) has no solution, then functional equation (4.2.17) is unsolvable. Assume now that (4.2.19) has solutions. Let us fix one of them. The next coefficients $\alpha_1, \alpha_2, \ldots$ are dependent of α_0. The algorithm to calculate $\alpha_1, \alpha_2, \ldots$ is the following. Differentiate (4.2.18) and substitute $z = 0$. We have

$$\alpha_1 = \frac{\partial h}{\partial \xi}(0, \alpha_0) r^2 \alpha_1 + \frac{\partial h}{\partial z}(0, \alpha_0), \qquad (4.2.20)$$

$$\alpha_2 = \frac{\partial h}{\partial \xi}(0, \alpha_0) r^4 \alpha_2 + \frac{\partial^2 h}{\partial \xi^2}(0, \alpha_0) \frac{r^4}{2} \alpha_1^2 + \frac{\partial^2 h}{\partial z \partial \xi}(0, \alpha_0) r^2 \alpha_1 + \frac{1}{2} \frac{\partial^2 h}{\partial z^2}(0, \alpha_0),$$

$$\ldots.$$

One can see that if $\frac{\partial h}{\partial \xi}(0, \alpha_0) r^{2K} \neq 1$ for all natural k, then system (4.2.20) has a unique formal solution $(\alpha_1, \alpha_2, \ldots)$. In the other case, since $r < 1$, there may exist only one natural K such that $\frac{\partial h}{\partial \xi}(0, \alpha_0) r^{2K} = 1$ and a condition on $\alpha_1, \alpha_2, \ldots, \alpha_{K-1}$ has to be fulfilled. If it is fulfilled, then α_K is arbitrary and system (4.2.20) has a family of solutions depending on α_K. Otherwise system (4.2.20) has no solution.

Remark 4.4 *A similar situation takes place for the linear equation (4.1.11) (cf. Remark 4.1).*

Theorem 4.8 *(J. Matkowski and W. Smajdor [132]). Each formal solution of (4.2.20) $(\alpha_0, \alpha_1, \alpha_2, ...)$ generates a series $\sum_{m=0}^{\infty} \alpha_m z^m$ with a positive radius of convergence.*

Proof. Let us know a formal solution $(\alpha_0, \alpha_1, \alpha_2, ...)$. The function $h(z,\xi)$ satisfies the inequality
$$|h(z,\xi_1) - h(z,\xi_2)| \leq C |\xi_1 - \xi_2| \qquad (4.2.21)$$
in some neighborhood of the point $(0, \alpha_0)$. Here C is a positive constant. We now show that equation (4.2.18) can be always reduced to the same equation with $0 < C < 1$. Let us represent the unknown function in the form
$$\psi_1(z) = P(z) + z^p \omega(z), \qquad (4.2.22)$$
where $P(z) = \sum_{k=0}^{p-1} \alpha_k z^k$. The natural number p is chosen in such a way that
$$\left|\frac{\partial h}{\partial \xi}(0, \alpha_0)\right| r^{2p} < \delta$$
for arbitrary fixed positive δ. Substituting (4.2.22) into (4.2.18) we obtain the functional equation
$$\omega(z) = h_1(z, \omega(r^2 z)), \qquad (4.2.23)$$
where $h_1(z,\xi) := z^{-p} [h(z, P(r^2 z) + (r^2 z)^p \xi) - P(z)]$. The function $h_1(z,\xi)$ is analytic at $z = 0$, since
$$h(z, P(r^2 z) + (r^2 z)^p \xi) - P(z) = \sum_{k=0}^{\infty} \beta_k(z) \left[P(r^2 z) + (r^2 z)^p \xi - \alpha_0\right]^k - P(z) =$$
$$\sum_{k=0}^{\infty} \gamma_k(z) z^{pk} \xi^k.$$

Here $h(z,\xi) = \sum_{k=0}^{\infty} \beta_k(z)(\xi - \alpha_0)^k$; the functions $\gamma_k(z)$ are represented by terms of $\beta_k(z)$ and $P(z)$. It follows from the relation
$$z^p \omega(z) = \gamma_0(z) + z^p \sum_{k=1}^{\infty} \gamma_k(z) z^{p(k-1)} \left[\omega(r^2 z)\right]^k$$

4.2. A NONLINEAR BOUNDARY VALUE PROBLEM.

that $\gamma_0(z)$ has the order p at $z = 0$, i.e., $h_1(z, w)$ is analytic at $z = 0$. We have

$$\frac{\partial h_1}{\partial \xi}(z, \xi) = r^{2p} \frac{\partial h}{\partial \xi}(z, P(r^2 z) + (r^2 z)^p \xi).$$

Then $\left|\frac{\partial h_1}{\partial \xi}(0, \alpha_0)\right| = \left|r^{2p} \frac{\partial h}{\partial \xi}(0, \alpha_0)\right| < \delta$ and hence $|h_1(z, \xi_1) - h_1(z, \xi_2)| \leq C |\xi_1 - \xi_2|$ in some neighbourhood of $(0, \alpha_0)$ with $0 < C < 1$.

It proves the theorem. ∎

We now proceed to study functional equation (4.2.17) with singularity at $z = 0$ ($\chi < 0$). We have seen that the general case is reduced to the case when

$$|h(z, \xi_1) - h(z, \xi_2)| \leq C |\xi_1 - \xi_2| \tag{4.2.24}$$

with $0 < C < 1$ for $|z| \leq \varepsilon$, $|\xi| \leq E$ for sufficiently small ε and fixed E. Let us consider the boundary value problem

$$\psi^+(t) - \psi^-(t) = b\overline{\psi^-(t)}, \quad |t| = \varepsilon r, \tag{4.2.25}$$

$$\psi^+(t) - \psi^-(t) = H(t, \overline{\psi^-(t)}), \quad |t| = \varepsilon, \quad \psi^-(\infty) = 0,$$

with respect to $\psi^- \in C_A(\{|z| < \varepsilon r\} \cup \{|z| > \varepsilon\})$, $\psi^+ \in C_A(\varepsilon r < |z| < \epsilon)$. The function $H(z, \xi)$ has the form

$$H(z, \xi) = h(z, b^{-1}\xi + \alpha_0) + b\overline{\alpha_0}, \tag{4.2.26}$$

with b a sufficiently small given positive constant and α_0 is root of the equation

$$h(0, \alpha_0) = 0. \tag{4.2.27}$$

If equation (4.2.27) has no solution, then (4.2.17) is unsolvable too. We fix a root α_0 and further only that root is meant. According to Theorem 4.6 problem (4.2.25) has a unique solution

$$\psi_0(z) = \begin{cases} \psi^+(z), & \varepsilon r \leq |z| \leq r \\ \psi^-(z), & |z| \leq \varepsilon r \text{ or } |z| \geq \varepsilon. \end{cases} \tag{4.2.28}$$

Theorem 4.9 *Let equation (4.2.17) have the following properties:*
i) (4.2.17) has a unique formal solution $(\alpha_0, \alpha_1, \alpha_2, ...)$,
ii) the positive value C from (4.2.24) is sufficiently small in the disc $|z| \leq \varepsilon$.
Then functional equation (4.2.17) has a local solution if and only if $|\chi|$ conditions on $h(z, \xi)$ are fulfilled. These conditions can be written in the form

$$\lim_{z \to 0} z^{-l} \psi_0(z) = 0, \ l = 1, 2, ..., |\chi| - 1, \tag{4.2.29}$$

where $\psi_0(z)$ is a solution of (4.2.25). If conditions (4.2.29) hold, then functional equation (4.2.17) has a unique local solution.

Proof. 1. Let conditions (4.2.29) be fulfilled. Then problem (4.2.25) has a unique solution $\psi_0(z)$ written as (4.2.28). It is possible to introduce a function $\varphi(z) = z^{-x}\psi_0(z) \in \mathcal{C}_A(\{|z| < \varepsilon r\}$, since (4.2.29) holds. Relations (4.2.25) imply

$$\psi^+(t) - t^x \varphi(t) = b\overline{\varphi(t)}, \quad |t| = \varepsilon r, \qquad (4.2.30)$$

$$\psi^+(t) - \psi^-(t) = H(t, \overline{\psi^-(t)}), \quad |t| = \varepsilon, \quad \psi^-(\infty) = 0.$$

Define the function $\Psi(z)$ by

$$\Psi(z) := \begin{cases} z^x \varphi(z) - H\left(z, \overline{\psi^-\left(\frac{\varepsilon^2}{\bar{z}}\right)}\right), & |z| \leq \varepsilon r, \\ \psi^-(z) - b\overline{\varphi\left(\frac{\varepsilon^2 r^2}{\bar{z}}\right)}, & |z| \geq \varepsilon, \\ \psi^+(z) - b\overline{\varphi\left(\frac{\varepsilon^2 r^2}{\bar{z}}\right)} - H\left(z, \overline{\psi^-\left(\frac{\varepsilon^2}{\bar{z}}\right)}\right), & \varepsilon r < |z| < \varepsilon. \end{cases}$$

The jump Δ of $\Psi(z)$ along $|t| = \varepsilon r$ is calculated as follows

$$\Delta = \psi^+(t) - b\overline{\varphi\left(\frac{\varepsilon^2 r^2}{\bar{t}}\right)} - t^x \varphi(t).$$

Applying the first relation (4.2.30) we obtain $\Delta = 0$. Similar arguments for $|t| = \varepsilon$ yield that the jump of $\Psi(z)$ along $|t| = \varepsilon$ is equal to zero too. By the Analytic Continuation Principle $\Psi(z)$ is analytic in $\widehat{\mathbb{C}}$. Using Liouville's theorem we conclude that $\Psi(z)$ is a constant. Substituting $z = 0$ in $\Psi(z)$ we obtain this constant:

$$\Psi(z) = \Psi(0) = -H(0,0) = -b\overline{\alpha_0}. \qquad (4.2.31)$$

It follows from the definition of $\Psi(z)$ that

$$z^x \varphi(z) - H\left(z, \overline{\psi^-\left(\frac{\varepsilon^2}{\bar{z}}\right)}\right) = -H(0,0), \quad |z| \leq \varepsilon r, \qquad (4.2.32)$$

$$\psi^-(z) - b\overline{\varphi\left(\frac{\varepsilon^2 r^2}{\bar{z}}\right)} = -H(0,0), \quad |z| \geq \varepsilon.$$

Eliminating $\psi^-(z)$ from (4.2.31) we arrive at the functional equation

$$z^x \varphi(z) = H(z, b\varphi(r^2 z) - \overline{H(0,0)}) - H(0,0), \quad |z| \leq \varepsilon r.$$

4.3. LINEAR FUNCTIONAL EQUATIONS.

By virtue of (4.2.26) one can see that this is equation (4.2.13). Since equation (4.2.13) has a unique formal solution, $\varphi(z)$ is the unique local solution of (4.2.13).

2. Let the function $\varphi(z)$ be a local solution of (4.2.13), i.e., $\varphi(z)$ satisfies (4.2.13) in some disc $|z| \leq \varepsilon$. Then the limit values of the functions

$$\psi^+(z) := z^X \varphi(z) + \overline{b\varphi\left(\frac{\varepsilon^2 r^2}{\bar{z}}\right)}, \quad \varepsilon r \leq |z| \leq \varepsilon,$$

$$\psi^-(z) := \begin{cases} z^X \varphi(z), & |z| \leq \varepsilon r, \\ \overline{b\varphi\left(\frac{\varepsilon^2 r^2}{\bar{z}}\right)} - H(0,0), & |z| \geq \varepsilon, \end{cases} \quad (4.2.33)$$

satisfy (4.2.30). Let us note that $\psi^-(\infty) = 0$ follows from the relations (4.2.26), $\psi^-(\infty) = \overline{b\varphi(0)} - H(0,0)$ and $h(0,\alpha_0)$, where $\alpha_0 = \varphi(0)$. The functions $\psi^+(z)$ and $\psi^-(z)$ satisfy the boundary value problem (4.2.30), which has the unique solution $\psi^-(z) = \psi_0(z)$. In particular, $z^X \varphi(z) = \psi_0(z)$ which implies condition (4.2.29).

The theorem is proved. ∎

We have constructed local solutions of the functional equation (4.2.13). Global solutions, i.e., solutions in $|z| \leq r$, can be obtained by successive application of the formula

$$\psi_1(r^{-2}z) = (r^{-2}z)^X h(r^{-2}z, \psi_1(z)).$$

4.3 Linear functional equations.

Let us consider mutually disjointed discs $\mathbb{D}_k := \mathbb{D}(a_k, r_k) = \{z \in \mathbb{C} : |z - a_k| < r_k\}$ ($k = 1, 2, ..., n$) in the complex plane $\mathbb{C}$. Let $\mathbb{D} := \widehat{\mathbb{C}} \setminus \cup_{k=0}^{n} cl\mathbb{D}_k$. The chosen ("positive") orientation on $\mathbb{T}_k := \mathbb{T}(a_k, r_k) = \{t \in \mathbb{C} : |t - a_k| = r_k\}$ leaves $\mathbb{D}$ to the left; $\mathbb{T}_k \cap \mathbb{T}_m = \emptyset$ for $k \neq m$. Let us first consider some auxiliary notations and assertions. Introduce the following mappings:

$$z^*_{(k_m k_{m-1} ... k_1)} := \left(z^*_{(k_{m-1} ... k_1)}\right)^*_{(k_m)},$$

where $z^*_{(k)} = \frac{r_k^2}{\bar{z} - \bar{a}_k} + a_k$ is the symmetry with respect to the circle $\mathbb{T}_k$ (see Section 2.1). Hence, $z^*_{(k_m k_{m-1} ... k_1)}$ is the composition of successive symmetries with respect to circles $\mathbb{T}_{k_1}, \mathbb{T}_{k_2}, ..., \mathbb{T}_{k_m}$. In the sequence $k_1, k_2, ..., k_m$ no two neighbouring numbers are equal. The number m is called the *level* of the mapping. When m is even, these

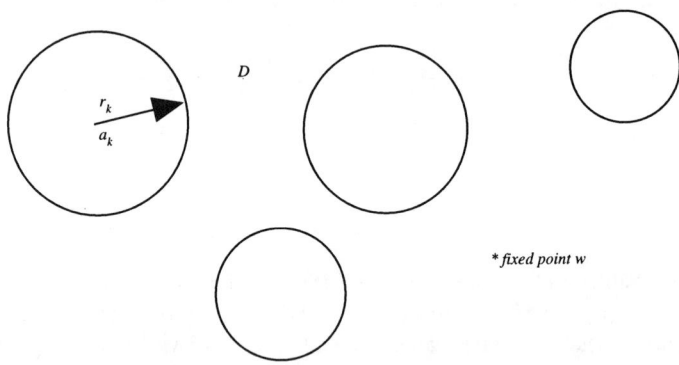

Figure 4.3: Multiply connected circular domain $\mathbb{D}$ and discs $\mathbb{D}_k$

are Möbius transformations. If m is odd, we have anti-Möbius transformations, i.e., Möbius transformations in $\bar{z}$. Thus these mappings can be written in the form

$$\gamma_j(z) = \left(\widehat{a}_j z + \widehat{b}_j\right) / \left(\widehat{c}_j z + \widehat{d}_j\right), \quad m \text{ is even}$$
$$\gamma_j(\bar{z}) = \left(\widehat{a}_j \bar{z} + \widehat{b}_j\right) / \left(\widehat{c}_j \bar{z} + \widehat{d}_j\right), \quad m \text{ is odd},$$

where $\widehat{a}_j \widehat{d}_j - \widehat{b}_j \widehat{c}_j = 1$. Here

$$\gamma_0(z) = z, \ \gamma_1(\bar{z}) = z^*_{(1)}, \ \gamma_2(\bar{z}) = z^*_{(2)}, \ldots, \gamma_n(\bar{z}) = z^*_{(n)}, \ \gamma_{n+1}(z) = z^*_{(21)}, \ \gamma_{n+2}(z) = z^*_{(31)}. \tag{4.3.1}$$

The functions γ_j generate a Schottky group $\mathcal{K}$ [84]. Let $\mathcal{G} \subset \mathcal{K}$ be the subgroup consisting of γ_j of the even level, $\mathcal{F}$ be the set of γ_j of the odd level. Let us fix a point $w \in cl\mathbb{D}$.

Lemma 4.1 *Let the given numbers ν_k have the form $\nu_k := \exp(-i\mu_k)$, $\mu_k \in \mathbb{R}$. The system of functional equations*

$$\phi_k(z) = -\nu_k \sum_{m \neq k} \overline{\nu_m} \left[\overline{\phi_m\left(z^*_{(m)}\right)} - \overline{\phi_m\left(w^*_{(m)}\right)} \right], \ |z - a_k| \leq r_k; (k = 1, 2, ..., n), \tag{4.3.2}$$

4.3. LINEAR FUNCTIONAL EQUATIONS.

with respect to the functions $\phi_k(z)$ analytic in $\mathbb{D}_k$ has only the trivial solution.

Proof. First we note that if $\phi_m(z)$ is a non-trivial solution of (4.3.2), then the functions $\phi_k(z)$ are analytic in $|z - a_k| \leq r_k$ respectively. Let us introduce the function

$$\psi(z) := -\sum_{m=1}^{n} \overline{\nu_m} \left[\overline{\phi_m\left(z_{(m)}^*\right)} - \overline{\phi_m\left(w_{(m)}^*\right)} \right],$$

analytic in $cl\mathbb{D}$. Then the functions ψ, ϕ_k satisfy the $\mathbb{R}$-linear boundary conditions (see Section 2.12)

$$\nu_k \psi(t) = \phi_k(t) - \overline{\phi_k(t)} + \overline{\phi_k\left(w_{(k)}^*\right)}, \quad t \in \mathbb{T}_k, \quad k = 1, \ldots, n.$$

One can write the latter relations in the following form

$$\Re \nu_k \psi(t) = c_k, \quad t \in \mathbb{T}_k, \quad k = 1, \ldots, n, \tag{4.3.3}$$

$$2\Im \phi_k(t) = \Im \nu_k \psi(t) + d_k, \quad t \in \mathbb{T}_k, \quad k = 1, \ldots, n. \tag{4.3.4}$$

Here $\phi_k\left(w_{(k)}^*\right) = c_k + id_k$. We may consider (4.3.3) as a boundary value problem with respect to the function $\psi(z) \in \mathcal{C}_A(\mathbb{D})$. The real constants c_k have to be determined later (cf. Subsection 2.7.2). Problem (4.3.3) has only constant solutions: $\psi(z) \equiv c$, $c_k = \Re \nu_k c$. So, let $\psi(\mathbb{D}) := \{\varsigma \in \widehat{\mathbb{C}} : z \in \mathbb{D}, \varsigma = \psi(z)\}$ be the image of $\mathbb{D}$ under mapping ψ. From the Boundary Correspondence Principle for conformal mapping we conclude that the boundary of $\psi(\mathbb{D})$ consists of the segments $\Re \nu_k \varsigma = c_k$ ($k = 1, 2, \ldots, n$). But in this case the point $\varsigma = \infty \in \psi(\mathbb{D})$ corresponds to a inner point of $\mathbb{D}$. It contradicts to boundedness of the function $\psi(z)$ in $cl\mathbb{D}$. Then the equalities (4.3.4) imply that $\phi_k(t) = $ constant [91]. Using (4.3.2) we have $\phi_k(z) \equiv 0$.
The lemma is proved. ∎

Let us consider the Banach space $\mathcal{C}_A := \mathcal{C}_A\left(\cup_{k=0}^{n} \mathbb{D}_k\right)$ introduced in Section 2.2.

Lemma 4.2 *Let $h \in \mathcal{C}_A$, $|\nu_k| = 1$. Then the system of functional equations*

$$\phi_k(z) = -\nu_k \sum_{m \neq k} \overline{\nu_m} \left[\overline{\phi_m\left(z_{(m)}^*\right)} - \overline{\phi_m\left(w_{(m)}^*\right)} \right] + h_k(z), \tag{4.3.5}$$

$$|z - a_k| \leq r_k \quad (k = 1, 2, \ldots, n),$$

has a unique solution $\Phi \in \mathcal{C}_A$. Here $\Phi(z) := \phi_k(z)$ in $|z - a_k| \leq r_k$, $k = 1, 2, \ldots, n$. This solution can be found by the method of successive approximations. The approximations are converging in $\mathcal{C}_A$.

Proof. Let us write system (4.3.5) on $\mathbb{T}_k$ in the form of a system of integral equations:

$$\phi_k(t) = -\nu_k \sum_{m \neq k} \overline{\nu_m} \overline{\frac{1}{2\pi i} \int_{\mathbb{T}_m^-} \phi_m(\tau) \left(\frac{1}{\tau - t_{(m)}^*} - \frac{1}{\tau - w_{(m)}^*} \right) d\tau} + h_k(t), \quad (4.3.6)$$

$$|t - a_k| = r_k \ (k = 1, 2, ..., n).$$

The orientation on $\mathbb{T}_m^-$ leaves $\mathbb{D}_m$ to the left. System (4.3.6) can be written as an equation in the space $\mathcal{C} := \mathcal{C}\left(\cup_{k=1}^n \mathbb{T}_k\right)$:

$$\Phi = \mathbf{A}\Phi + h. \quad (4.3.7)$$

Since the integral operators from (4.3.6) are compact in $\mathcal{C}(\mathbb{T}_k)$ and multiplication by $\overline{\nu_m}$ and complex conjugation are bounded operators in $\mathcal{C}$, it follows that $\mathbf{A}$ is a compact operator in $\mathcal{C}$. If Φ is a solution of (4.3.7) in $\mathcal{C}$, then $\Phi \in \mathcal{C}_A$ (see the Pumping Principle from Section 2.3). This follows from the properties of the Cauchy integral and the condition $h \in \mathcal{C}_A$. Therefore, equation (4.3.7) in $\mathcal{C}$ and equation (4.3.5) in $\mathcal{C}_A$ are equivalent when $h \in \mathcal{C}_A$. It follows from Lemma 4.1 that the homogeneous equation $\Phi = \mathbf{A}\Phi$ has only a trivial solution. Then the Fredholm theorem implies that equation (4.3.7) or the system (4.3.5) has a unique solution.

Let us show the convergence of the method of successive approximations. By virtue of the Successive Approximation Theorem (see Section 2.3) it is sufficient to prove the inequality $\rho(\mathbf{A}) < 1$, where $\rho(\mathbf{A})$ is the spectral radius of the operator $\mathbf{A}$. The inequality $\rho(\mathbf{A}) < 1$ is satisfied if for all complex numbers λ such that $|\lambda| \leq 1$ the equation

$$\Phi = \lambda \mathbf{A} \Phi$$

has only a trivial solution. This equation can be rewritten in the form

$$\phi_k(z) = -\lambda \nu_k \sum_{m \neq k} \overline{\nu_m} \left[\overline{\phi_m\left(z_{(m)}^*\right)} - \overline{\phi_m\left(w_{(m)}^*\right)} \right], \ |z - a_k| \leq r_k. \quad (4.3.8)$$

Let $|\lambda| < 1$. Let us introduce the function

$$\psi(z) = -\lambda \sum_{m=0}^n \overline{\nu_m} \left(\overline{\phi_m\left(z_{(m)}^*\right)} - \overline{\phi_m\left(w_{(m)}^*\right)} \right),$$

which is analytic in $cl\mathbb{D}$. Then $\psi(z)$ and $\phi_k(z)$ satisfy the $\mathbb{R}$-linear boundary value problem

$$\nu_k \psi(t) = \phi_k(t) - \lambda \overline{\phi_k(t)} + \gamma_k, \ |t - a_k| = r_k,$$

4.4. HARMONIC MEASURES AND SCHWARZ OPERATOR.

where $\gamma_k := \overline{\lambda \phi_k \left(w^*_{(k)} \right)}$. For $|\lambda| < 1$ (see Section 2.12) this problem has a solution depending on an arbitrary additive constant. In our case it gives immediately that $\psi(z) = constant$. Applying the uniqueness assertion from the Decomposition Theorem to the boundary condition we have $\phi_k(z) = constant$. Hence by (4.3.8) $\phi_k(z) \equiv 0$.

Let $|\lambda| = 1$. Then by substituting $\omega_k(z) = \phi_k(z)/\sqrt{\lambda}$ the system (4.3.8) is reduced to the same system with $\lambda = 1$. It follows from Lemma 4.1 that $\omega_k(z) = \phi_k(z) = 0$. Hence, $\rho(\mathbf{A}) < 1$.

This inequality proves the lemma. ∎

4.4 Harmonic measures and Schwarz operator for circular multiply connected domains

4.4.1 Harmonic measures.

According to Subsection 2.7.2 the harmonic measure $\alpha_s(z)$ ($s = 1, 2, ..., n$) is a function harmonic in the multiply circular domain $\mathbb{D}$, continuous in $cl\mathbb{D}$, satisfying the boundary conditions

$$\alpha_s(t) = \delta_{sk}, \ |t - a_k| = r_k, \ k = 1, 2, ..., n, \quad (4.4.1)$$

where δ_{sk} is the Kronecker symbol. It follows from Subsection 2.7.2 that $\alpha_s \in C^\infty_{\mathcal{A}}(\mathbb{D})$. Using the Logarithmic Conjugation Theorem we look for $\alpha_s(z)$ in the form

$$\alpha_s(z) = \Re \phi(z) + \sum_{m=1}^{n} A_m \log |z - a_m| + A, \quad (4.4.2)$$

where the function $\phi(z)$ belongs to $C^\infty_{\mathcal{A}}(\mathbb{D})$, $\phi(w) = 0$, w is a fixed point from $cl\mathbb{D}$, and A_m and A are real constants,

$$\sum_{m=1}^{n} A_m = 0. \quad (4.4.3)$$

Using the boundary condition (4.4.1) and the representation (4.4.2) we arrive at the following boundary value problem

$$\Re \phi(t) + \sum_{m=1}^{n} A_m \log |t - a_k| + A = \delta_{sk}, \ |t - a_k| = r_k. \quad (4.4.4)$$

150   CHAPTER 4. METHOD OF FUNCTIONAL EQUATIONS

The boundary value problem (4.4.4) is equivalent to the $\mathbb{R}$-linear problem (see Section 2.12)
$$\phi(t) = \phi_k(t) - \overline{\phi_k(t)} + f_k(t), \ t \in \mathbb{T}_k, \ k = 1, 2, ..., n, \tag{4.4.5}$$
where the unknown functions $\phi \in \mathcal{C}_\mathcal{A}(\mathbb{D})$, $\phi_k \in \mathcal{C}_\mathcal{A}(\mathbb{D}_k)$,
$$\phi(w) = 0, \tag{4.4.6}$$
$$f_k(z) := \delta_{sk} - A - A_k \log r_k - \sum_{m \neq k} A_m \log(z - a_m), z \in \mathbb{D}_k. \tag{4.4.7}$$

The branch of the function $\log(z - a_m)$ is fixed in such a way that the cut connecting the points $z - a_m$ and $z = \infty$ does not intersect the circles $|t - a_k| = r_k$ for $k \neq m$ and does not pass through the point $z = w$. The function $f_k(z)$ satisfies the boundary condition
$$\Re f_k(t) := \delta_{sk} - A - \sum_{m=1}^{n} A_m \log |t - a_m|, \ |t - a_k| = r_k$$
and belongs to $\mathcal{C}_\mathcal{A}(\mathbb{D}_k)$.

Remark 4.5 *More precisely the functions ϕ, ϕ_k and f_k are infinitely differentiable in the closures of the domains considered.*

Let us introduce the function
$$\Phi(z) := \begin{cases} \phi_k(z) + \sum_{m \neq k} \left[\overline{\phi_m\left(z_{(m)}^*\right)} - \overline{\phi_m\left(w_{(m)}^*\right)} \right] - \overline{\phi_k\left(w_{(k)}^*\right)} + f_k(z), \ z \in cl\mathbb{D}_k, \\ \phi(z) + \sum_{m=1}^{n} \left[\overline{\phi_m\left(z_{(m)}^*\right)} - \overline{\phi_m\left(w_{(m)}^*\right)} \right], \ z \in \mathbb{D}. \end{cases}$$

Calculating the jump along the circle $\mathbb{T}_k$
$$\Delta_k := \Phi^+(t) - \Phi^-(t), \ t \in \mathbb{T}_k,$$
where $\Phi^+(t) := \lim_{z \to t, z \in \mathbb{D}} \Phi(z)$, $\Phi^-(t) := \lim_{z \to t, z \in \mathbb{D}_k} \Phi(z)$ and using (4.4.5), (4.4.7) we get $\Delta_k = 0$. It follows from the Analytic Continuation Principle and the Liouville's theorem that $\Phi(z)$ is a constant. Then condition (4.4.6) and the definition of $\Phi(z)$ give us immediately that $\Phi(z) \equiv 0$. Thus the definition of $\Phi(z)$ yields the following system of functional equations
$$\phi_k(z) = -\sum_{m \neq k} \left[\overline{\phi_m\left(z_{(m)}^*\right)} - \overline{\phi_m\left(w_{(m)}^*\right)} \right] - \delta_{sk} + A + A_k \log r_k + \tag{4.4.8}$$

4.4. HARMONIC MEASURES AND SCHWARZ OPERATOR.

$$\sum_{m \neq k} A_m \log(z - a_m) + \overline{\phi_k\left(w_{(k)}^*\right)}, \quad |z - a_k| \leq r_k,$$

with respect to the functions $\phi_k(z) \in \mathcal{C_A}(\mathbb{D}_k)$. The branches of logarithmic functions are chosen as in the definition of f_k. The initial function $\phi(z)$ is related to the auxiliary functions by the formula

$$\phi(z) = -\sum_{m=1}^{n}\left[\overline{\phi_m\left(z_{(m)}^*\right)} - \overline{\phi_m\left(w_{(m)}^*\right)}\right], \quad z \in cl\mathbb{D}, \tag{4.4.9}$$

where the function $\phi_m(z)$ belongs to $\mathcal{C}_{\mathcal{A}}^{\infty}(\mathbb{D}_m)$. It is convenient to rewrite the system of functional equations (4.4.8) as two systems

$$\varphi_k(z) = -\sum_{m \neq k}\left[\overline{\varphi_m\left(z_{(m)}^*\right)} - \overline{\varphi_m\left(w_{(m)}^*\right)}\right] + \sum_{m \neq k} A_m \log\frac{z - a_m}{w - a_m}, \tag{4.4.10}$$

$$|z - a_k| \leq r_k, \quad k = 1, \ldots, n,$$

$$\phi_k^0(z) = -\sum_{m \neq k}\left[\overline{\phi_m^0\left(z_{(m)}^*\right)} - \overline{\phi_m^0\left(w_{(m)}^*\right)}\right] - \delta_{sk} + A + A_k \log r_k$$
$$-\sum_{m \neq k} A_m \log(w - a_m) + \overline{\phi_k^0\left(w_{(k)}^*\right)} + \overline{\varphi_k\left(w_{(k)}^*\right)}, \tag{4.4.11}$$
$$|z - a_k| \leq r_k, \quad k = 1, \ldots, n.$$

The functions in the left-hand parts are related to initial functions ϕ_k by the equality

$$\phi_k(z) = \varphi_k(z) + \phi_k^0(z), \quad k = 1, \ldots, n.$$

According to Lemma 4.2 from Section 4.3 (4.4.10) and (4.4.11) can be solved by the method of successive approximations. It follows from (4.4.11) that $\phi_k^0(z)$ ($k = 1, 2, \ldots, n$) are constants, since the zero-th approximation is a constant and the right-hand part of (4.4.11) transfers constants to constants. Using the method of successive approximations to (4.4.10) we obtain

$$\varphi_k(z) = \sum_{m \neq k} A_m \log\frac{z - a_m}{w - a_m} - \sum_{k_1 \neq k}\sum_{m \neq k_1} A_m \log\frac{\overline{z_{(k_1)}^*} - a_m}{\overline{w_{(k_1)}^*} - a_m} + \tag{4.4.12}$$

$$\sum_{k_1 \neq k}\sum_{k_2 \neq k_1}\sum_{m \neq k_2} A_m \log\frac{z_{(k_2 k_1)}^* - a_m}{w_{(k_2 k_1)}^* - a_m} - \cdots,$$

where the sum $\sum\limits_{k_j \neq k_{j-1}}$ contains the terms with $k_j = 1, 2, \ldots, n;\ k_j \neq k_{j-1}$. By virtue of Lemma 4.2 with $\nu_m = 1$ the series (4.4.12) converges uniformly in $|z - a_k| \leq r_k$. One can see from (4.4.9) that the constant $\phi_k^0(z)$ does not impact on $\phi(z)$, hence

$$\phi(z) = -\sum_{k=1}^{m}\left[\overline{\varphi_k\left(z_{(k)}^*\right)} - \overline{\varphi_k\left(w_{(k)}^*\right)}\right]. \tag{4.4.13}$$

Therefore

$$\phi(z) = -\sum_{k=1}^{n}\sum_{m\neq k} A_m \log \frac{\overline{z_{(k)}^* - a_m}}{\overline{w_{(k)}^* - a_m}} + \sum_{k=1}^{n}\sum_{k_1\neq k}\sum_{m\neq k_1} A_m \log \frac{\overline{z_{(k_1 k)}^* - a_m}}{\overline{w_{(k_1 k)}^* - a_m}} \tag{4.4.14}$$

$$-\sum_{k=1}^{n}\sum_{k_1\neq k}\sum_{k_2\neq k_1}\sum_{m\neq k_2} A_m \log \frac{\overline{z_{(k_2 k_1 k)}^* - a_m}}{\overline{w_{(k_2 k_1 k)}^* - a_m}} + \ldots,\ z \in c l\mathbb{D}.$$

In order to transform (4.4.14) we need the following

Lemma 4.3 *There holds the equality*

$$\sum_{k=1}^{n}\sum_{k_1\neq k}\ldots\sum_{k_s\neq k_{s-1}}\sum_{m\neq k_s} A_m R_{m k_s k_{s-1}\ldots k_1 k} = \sum_{m=1}^{n}\sum_{k_1\neq m}\ldots\sum_{k_s\neq k_{s-1}}\sum_{k\neq k_s} A_m R_{m k_1 k_2 \ldots k_s k}. \tag{4.4.15}$$

Proof. It is sufficient to note that both parts of equality (4.4.15) contain the same terms.
This proves the lemma. ■

Applying the result of Lemma 4.3 to (4.4.14) we get

$$\phi(z) = -\sum_{m=1}^{n} A_m \sum_{k\neq m} \log \frac{\overline{z_{(k)}^* - a_m}}{\overline{w_{(k)}^* - a_m}} + \sum_{m=1}^{n} A_m \sum_{k_1\neq m} \sum_{k\neq k_1} \log \frac{\overline{z_{(k_1 k)}^* - a_m}}{\overline{w_{(k_1 k)}^* - a_m}}$$

$$-\sum_{m=1}^{n} A_m \sum_{k_1\neq m} \sum_{k_2\neq k_1} \sum_{k\neq k_2} \log \frac{\overline{z_{(k_1 k_2 k)}^* - a_m}}{\overline{w_{(k_1 k_2 k)}^* - a_m}} + \ldots,\ z \in c l\mathbb{D}. \tag{4.4.16}$$

Let us rewrite formulae (4.4.16) in the form

$$\phi(z) = \sum_{m=1}^{n} A_m \psi_m(z),$$

4.4. HARMONIC MEASURES AND SCHWARZ OPERATOR.

where

$$\psi_m(z) = \log \prod_{k \neq m} \overline{\frac{w^*_{(k)} - a_m}{z^*_{(k)} - a_m}} + \log \prod_{k_1 \neq m} \prod_{k \neq k_1} \frac{z^*_{(k_1 k)} - a_m}{w^*_{(k_1 k)} - a_m} \qquad (4.4.17)$$
$$+ \log \prod_{k_1 \neq m} \prod_{k_2 \neq k_1} \prod_{k \neq k_2} \overline{\frac{w^*_{(k_1 k_2 k)} - a_m}{z^*_{(k_1 k_2 k)} - a_m}} + ..., \ z \in c \mathbb{D}.$$

Let us rewrite (4.4.17) in terms of the group $\mathcal{K}$ (see Section 4.3)

$$\psi_m(z) = \log \left[\prod_{j=1,\ j \neq m}^{\infty} \psi_m^j(z) \right], \qquad (4.4.18)$$

where

$$\psi_m^j(z) = \begin{cases} \frac{\gamma_j(z) - a_m}{\gamma_j(w) - a_m}, & \text{if } \gamma_j \in \mathcal{G}, \\[2mm] \overline{\frac{\gamma_j(\overline{w}) - a_m}{\gamma_j(\overline{z}) - a_m}}, & \text{if } \gamma_j \in \mathcal{F}. \end{cases}$$

The numeration on j in (4.4.18) is in accordance with (4.3.1), i.e., $\gamma_0(z) = z$, $\gamma_j(\overline{z}) = z_j^*$ $(j = 1, 2, ..., n)$ and so on.

Let us find the constants A and A_m. Substituting $z = w^*_{(k)}$ into the real parts of (4.4.8) we obtain

$$0 = -\sum_{m \neq k} \Re \left[\varphi_m \left(\left(w^*_{(k)} \right)^*_{(m)} \right) - \varphi_m \left(w^*_{(m)} \right) \right] \qquad (4.4.19)$$
$$-\delta_{sk} + A + A_k \log r_k + \sum_{m \neq k} A_m \log \left| w^*_{(k)} - a_m \right|, \ k = 1, 2, ..., n,$$

where φ_m has the form (4.4.12) and linearly contains the unknown constants A_m. The equalities (4.4.3), (4.4.19) generate a system of $n+1$ linear algebraic equations with respect to $n+1$ unknowns $A, A_1, ..., A_n$. It is easy to prove that this system has a unique solution (if not, it contradicts to uniqueness of the solution of the Dirichlet problem).

Theorem 4.10 *The harmonic measures have the form*

$$\alpha_s(z) = \sum_{m=1}^{n} A_m \left[\Re \psi_m(z) + \log |z - a_m| \right] + A, \qquad (4.4.20)$$

where $\psi_m(z)$ is given in (4.4.18). The infinite product (4.4.18) converges uniformly on each compact subset of $\mathbb{D} \setminus \{\infty\}$. The real constants A and A_m are uniquely defined from the system (4.4.3), (4.4.19).

Remark 4.6 *The constants A and A_m depend on the choice of w.*

Remark 4.7 *The logarithmic terms in (4.4.20) can be included into infinite product (4.4.17). Then (4.4.20) becomes*

$$\alpha_s(z) = \sum_{m=1}^{n} A_m \log \prod_{j=0,\, j\neq m}^{\infty} |\psi_m^j(z)| + A^0,$$

where $A^0 := A - \sum_{m=1}^{n} A_m \log|w - a_m|$.

4.4.2 Schwarz operator.

Following the previous section we construct the complex Green function and the Schwarz operator (cf. Subsection 2.7.2) for the circular multiply connected domain $\mathbb{D}$. We use the representation for the Green function (see Subsection 2.7.2)

$$M(z,\zeta) = M_0(z,\zeta) + \sum_{k=1}^{n} \alpha_k(\zeta) \log(z - a_k) - \log(\zeta - z) + A(\zeta),$$

where α_k is a harmonic measure of $\mathbb{D}$, $A(\zeta)$ is a real function in ζ. The point w and the branches of $\log(z - a_k)$ are fixed as in the previous section. The function $M_0(z,\zeta)$ belongs to $\mathcal{C}_{\mathcal{A}}^{\infty}(\mathbb{D})$ in z and satisfies the boundary value problem (see Subsection 2.7.2):

$$\Re\left[M_0(t,\zeta) + \sum_{k=1}^{n} \alpha_k(\zeta)\log(t - a_k) - \log(\zeta - t) + A(\zeta)\right] = 0, \quad (4.4.21)$$

$$t \in \mathbb{T}_k, \ k = 1, 2, \ldots, n; \ M_0(w, \zeta) = 0.$$

Problem (4.4.21) has a unique solution. It is reduced to the following system of functional equations

$$\phi_k(z) = -\sum_{m\neq k}\left[\phi_m\left(z^*_{(m)}\right) - \overline{\phi_m\left(w^*_{(m)}\right)}\right] - \log(\zeta - z) + A$$

$$+ \alpha_k(\zeta) \log r_k + \sum_{m\neq k} \alpha_m(\zeta)\log(z - a_m) + \overline{\phi_k\left(w^*_{(k)}\right)}, \ |z - a_k| \leq r_k,\ k = 1,\ldots,n, \quad (4.4.22)$$

where $\phi_k(z) \in \mathcal{C}_{\mathcal{A}}^{\infty}(\mathbb{D}_k)$. The initial function M_0 is related to the auxiliary functions $\phi_k(z)$ by the equality

$$M_0(z,\zeta) = -\sum_{k=1}^{n}\left[\phi_k\left(z^*_{(k)}\right) - \overline{\phi_k\left(w^*_{(k)}\right)}\right],\ z \in cl\mathbb{D} \setminus \{\zeta\}. \quad (4.4.23)$$

4.4. HARMONIC MEASURES AND SCHWARZ OPERATOR.

Let us consider two auxiliary systems of functional equations

$$\Psi_k(z) = -\sum_{m \neq k} \left[\overline{\Psi_m\left(z^*_{(m)}\right)} - \overline{\Psi_m\left(w^*_{(m)}\right)} \right] + A + \alpha_k(\zeta) \log r_k$$

$$+ \sum_{m \neq k} \alpha_m(\zeta) \log(z - a_m) + \overline{\phi_m\left(w^*_{(k)}\right)}, \quad |z - a_k| \leq r_k,$$

$$\Omega_k(z) = -\sum_{m \neq k} \left[\overline{\Omega_m\left(z^*_{(m)}\right)} - \overline{\Omega_m\left(w^*_{(m)}\right)} \right] - \log(\zeta - z), \quad |z - a_k| \leq r_k, \; k = 1, \ldots, n.$$

The first system coincides with the system (4.3.6) ($\nu_k = 1$), and thus can be solved by the method of successive approximations (cf. Lemma 4.2). Let us consider the second system. If $|\zeta - z| \leq r_s$ for some s, then the right-hand part $-\log(\zeta-z)$ does not belong to $\mathcal{C}_A(\mathbb{D})$. But by introducing a new unknown function $\Omega_s^0(z) := \Omega_s(z) - \log(\zeta - z)$ we get a system in the space $\mathcal{C}_A(\mathbb{D})$. Therefore, the method of successive approximations can be applied to the second system too, and

$$\overline{\Psi_k\left(z^*_{(k)}\right)} - \overline{\Psi_k\left(w^*_{(k)}\right)} = \sum_{m \neq k} \alpha_m(\zeta) \log \frac{z^*_{(k)} - a_m}{w^*_{(k)} - a_m} \qquad (4.4.24)$$

$$- \sum_{k_1 \neq k} \sum_{m \neq k_1} \alpha_m(\zeta) \log \frac{z^*_{(k_1 k)} - a_m}{w^*_{(k_1 k)} - a_m} + \sum_{k_1 \neq k} \sum_{k_2 \neq k_1} \sum_{m \neq k_2} \alpha_m(\zeta) \log \frac{z^*_{(k_2 k_1 k)} - a_m}{w^*_{(k_2 k_1 k)} - a_m} - \ldots,$$

$$\overline{\Omega_k\left(z^*_{(k)}\right)} - \overline{\Omega_k\left(w^*_{(k)}\right)} = \log \frac{\zeta - w^*_{(k)}}{\zeta - z^*_{(k)}} + \sum_{k_1 \neq k} \log \frac{\zeta - z^*_{(k_1 k)}}{\zeta - w^*_{(k_1 k)}} \qquad (4.4.25)$$

$$+ \sum_{k_1 \neq k} \sum_{k_2 \neq k_1} \log \frac{\zeta - z^*_{(k_2 k_1 k)}}{\zeta - w^*_{(k_2 k_1 k)}} + \ldots, \quad |z - a_k| \geq r_k.$$

The series (4.4.24), (4.4.25) converge uniformly in every compact subset of $c\mathbb{D} \setminus \{\zeta\}$. We have $\phi_k(z) = \Psi_k(z) + \Omega_k(z)$; hence, the values

$$\overline{\phi_k\left(z^*_{(k)}\right)} - \overline{\phi_k\left(w^*_{(k)}\right)} = \overline{\Psi_k\left(z^*_{(k)}\right)} - \overline{\Psi_k\left(w^*_{(k)}\right)} + \overline{\Omega_k\left(z^*_{(k)}\right)} - \overline{\Omega_k\left(w^*_{(k)}\right)} \qquad (4.4.26)$$

are completely determined. It follows from (4.4.23) that

$$M_0(z, \zeta) = \sum_{m=1}^{n} \alpha_m(\zeta) \psi_m(z) + \omega(z, \zeta),$$

where the functions $\psi_m(z)$ are presented in (4.4.17) or (4.4.18), $\alpha_m(\zeta)$ are given in Theorem 4.10,

$$\omega(z,\zeta) = \log\left(\prod_{k=1}^{n} \overline{\frac{\zeta - z^*_{(k)}}{\zeta - w^*_{(k)}}}\right)\left(\prod_{k=1}^{n}\prod_{k_1\neq k} \overline{\frac{\zeta - w^*_{(k_1k)}}{\zeta - z^*_{(k_1k)}}}\right)\left(\prod_{k=1}^{n}\prod_{k_1\neq k}\prod_{k_2\neq k_1} \overline{\frac{\zeta - z^*_{(k_2k_1k)}}{\zeta - w^*_{(k_2k_1k)}}}\right)\cdots \quad (4.4.27)$$

This infinite product can be represented in the form

$$\omega(z,\zeta) = \log \prod_{j=1}^{\infty} \omega_j(z,\zeta), \qquad (4.4.28)$$

where

$$\omega_j(z,\zeta) = \begin{cases} \dfrac{\zeta - \gamma_j(z)}{\zeta - \gamma_j(w)}, & \text{if } \gamma_j \in \mathcal{G}, \\[6pt] \dfrac{\zeta - \gamma_j(\overline{w})}{\zeta - \gamma_j(\overline{z})}, & \text{if } \gamma_j \in \mathcal{F}. \end{cases}$$

In order to find $A(\zeta)$ we substitute $w^*_{(k)}$ in the real part of (4.4.22). We have

$$0 = -\sum_{m\neq k} \Re\left[\overline{\phi_m\left(w^*_{(mk)}\right)} - \overline{\phi_m\left(w^*_{(m)}\right)}\right] - \log\left|\zeta - w^*_{(k)}\right| + A(\zeta) \qquad (4.4.29)$$

$$+\alpha_k(\zeta)\log r_k + \sum_{m\neq k}\alpha_m(\zeta)\log\left|w^*_{(k)} - a_m\right|, \quad k = 1,\ldots,n.$$

The harmonic measures satisfy the equality

$$\sum_{m=1}^{n} \alpha_m(\zeta) = 1. \qquad (4.4.30)$$

One can consider (4.4.29), (4.4.30) as a system of $n+1$ real linear algebraic equations with respect to $n+1$ real unknowns $\alpha_1(\zeta), \alpha_2(\zeta), \ldots, \alpha_n(\zeta), A(\zeta)$. The systems (4.4.29), (4.4.30) and (4.4.3), (4.4.19) have the same homogeneous part. Therefore, the system (4.4.29), (4.4.30) has a unique solution. We may at the beginning look for the complex Green function $M(z,\zeta)$ with undetermined periods $\alpha_k(\zeta)/2\pi$, find $\alpha_k(\zeta)$ from (4.4.29), (4.4.30) and after assert that $\alpha_k(\zeta)$ is a harmonic measure. Thus to determine $A(\zeta)$ we fix, for instance $k = n$, in (4.4.29) and find

$$A(\zeta) = \sum_{m=1}^{n-1} \Re\left[\overline{\phi_m\left(\left(w^*_{(n)}\right)^*_{(m)}\right)} - \overline{\phi_m\left(w^*_{(m)}\right)}\right] + \log\left|\zeta - w^*_{(n)}\right|$$

4.4. HARMONIC MEASURES AND SCHWARZ OPERATOR.

$$-\alpha_k(\zeta)\log r_k - \sum_{m=1}^{n-1}\alpha_m(\zeta)\log\left|w^*_{(k)} - a_m\right|, \qquad (4.4.31)$$

where $\left[\overline{\phi_m\left(z^*_{(m)}\right)} - \overline{\phi_m\left(w^*_{(m)}\right)}\right]$ has the form (4.4.24), (4.4.25), (4.4.26).

It follows from (2.7.14) that

$$T(z,\zeta) = \sum_{m=1}^{n}\frac{\partial\alpha_m}{\partial\nu}(\zeta)[\psi_m(z) + \log(z - a_m)] + \frac{\partial\omega}{\partial\nu}(z,\zeta) - \frac{1}{\zeta-z}\frac{\partial\zeta}{\partial\nu}d\sigma + \frac{\partial A}{\partial\nu}(\zeta), \qquad (4.4.32)$$

where ν is the outward (in sense of orientation) normal vector at the point $\zeta \in \partial\mathbb{D}$.
The function

$$\phi(z) = \frac{1}{2\pi}\sum_{k=1}^{n}\int_{\mathbb{T}_k} f(\zeta)T(z,\zeta)\,d\sigma \qquad (4.4.33)$$

is single-valued in $\mathbb{D}$ if and only if

$$\sum_{k=1}^{n}\int_{\mathbb{T}_k} f(\zeta)\frac{\partial\alpha_m}{\partial\nu}(\zeta)d\sigma = 0, \quad m = 1, 2, ..., n. \qquad (4.4.34)$$

Let us note that one of relation (4.4.34) follows from the other ones. For instance, let (4.4.34) be valid for $m = 1, 2, ..., n-1$. Then (4.4.34) for $m = n$ is fulfilled, since

$$\sum_{k=1}^{n}\int_{\mathbb{T}_k} f(\zeta)\frac{\partial\alpha_n}{\partial\nu}(\zeta)d\sigma = -\sum_{m=1}^{n-1}\sum_{k=1}^{n}\int_{\mathbb{T}_k} f(\zeta)\frac{\partial\alpha_m}{\partial\nu}(\zeta)d\sigma = 0.$$

Here we use the identity (4.4.30). Using (4.4.32) we separate single- and multi-valued components of the Schwarz operator

$$T(z,\zeta) = T_s(z,\zeta) + T_m(z,\zeta),$$

$$T_s(z,\zeta) = \sum_{m=1}^{n}\frac{\partial\alpha_m}{\partial\nu}(\zeta)[\psi_m(z) + \log(z - a_m)],$$

$$T_m(z,\zeta) = \frac{\partial\omega}{\partial\nu}(z,\zeta) - \frac{1}{\zeta-z}\frac{\partial\zeta}{\partial\nu}d\sigma + \frac{\partial A}{\partial\nu}(\zeta).$$

We now proceed to calculate the normal derivatives in the later formulae. We have

$$\frac{\partial f}{\partial\nu}d\sigma = -\frac{1}{i}\left[\frac{\partial f}{\partial\zeta} + \left(\frac{r_k}{\zeta-a_k}\right)^2\frac{\partial f}{\partial\bar\zeta}\right]d\tau, \quad |\zeta-a_k| = r_k, \qquad (4.4.35)$$

for any $f \in C^1(\partial \mathbb{D})$. Recall that we deal with the outward normal to $\mathbb{D}$. In order to apply (4.4.35) to $\omega(z,\zeta)$ we find from (4.4.27)

$$\frac{\partial \omega}{\partial \bar\zeta}(z,\zeta) = \sum_{k=1}^{n}\sum_{k_1 \neq k}\left(\frac{1}{\zeta - w^*_{(k_1 k)}} - \frac{1}{\zeta - z^*_{(k_1 k)}}\right)$$

$$+ \sum_{k=1}^{n}\sum_{k_1 \neq k}\sum_{k_2 \neq k_1}\sum_{k_3 \neq k_2}\left(\frac{1}{\zeta - w^*_{(k_3 k_2 k_1 k)}} - \frac{1}{\zeta - z^*_{(k_3 k_2 k_1 k)}}\right) + \ldots \quad (4.4.36)$$

$$= \sum_{j=1}^{\infty}{}''\left(\frac{1}{\zeta - \gamma_j(w)} - \frac{1}{\zeta - \gamma_j(z)}\right),$$

where the terms in the later sum are ordered due to an increasing even level. Analogously

$$\frac{\partial \omega}{\partial \bar\zeta}(z,\tau) = \sum_{j=1}^{\infty}{}'\left(\frac{1}{\zeta - \gamma_j(\bar z)} - \frac{1}{\zeta - \gamma_j(\overline{w})}\right), \quad (4.4.37)$$

where elements γ_j have the odd level. Substituting (4.4.36), (4.4.37) into (4.4.32), (4.4.33) we obtain the following

Theorem 4.11 *The Schwarz operator of $\mathbb{D}$ has the form*

$$\phi(z) = \frac{1}{2\pi i}\sum_{k=1}^{n}\int_{\mathbb{T}_k} f(\zeta)\left\{\sum_{j=2}^{\infty}{}''\left[\frac{1}{\zeta - \gamma_j(w)} - \frac{1}{\zeta - \gamma_j(z)}\right]\right.$$

$$\left. + \left(\frac{r_k}{\zeta - a_k}\right)^2 \sum_{j=1}^{\infty}{}'\left[\frac{1}{\zeta - \gamma_j(\bar z)} - \frac{1}{\zeta - \gamma_j(\overline{w})}\right] - \frac{1}{\zeta - z}\right\}d\zeta \quad (4.4.38)$$

$$+ \frac{1}{2\pi i}\sum_{k=1}^{n}\int_{\mathbb{T}_k} f(\zeta)\frac{\partial A}{\partial \nu}(\zeta)d\sigma + \sum_{m=1}^{n} A_m\left[\log(z - a_m) + \psi_m(z)\right] + i\varsigma,$$

where

$$A_m := \frac{1}{2\pi i}\sum_{k=1}^{n}\int_{\mathbb{T}_k} f(\zeta)\frac{\partial \alpha_m}{\partial \nu}(\zeta)d\sigma, \quad m = 1, 2, \ldots, n,$$

$A(\zeta)$ has the form (4.4.31), the functions $\alpha_m(\zeta)$ and $\psi_m(z)$ are derived in Theorem 4.10, ς is an arbitrary real constant, $\sum'$ contains γ_j of odd level ($\gamma_j \in \mathcal{F}$), and $\sum''$ of even level ($\gamma_j \in \mathcal{G}$). The series converges uniformly in each compact subset of $cl\mathbb{D}\setminus\{\infty\}$.

4.5. LINEAR RIEMANN–HILBERT PROBLEM.

The single-valued part of the Schwarz operator appears in the modified Dirichlet problem (see Subsection 2.7.2):

$$\Re\phi(t) = f(t) + c_k, \; t \in \mathbb{T}_k, \; k = 1, 2, ..., n, \qquad (4.4.39)$$

where a given function $f \in C(\partial \mathbb{D})$, c_k are undetermined real constants. If one of the constants c_k is fixed arbitrary, then the remaining ones are determined uniquely and $\phi(z)$ is determined up to an arbitrary additive purely imaginary constant (see Subsection 2.7.2). Thus we have

Theorem 4.12 *The single-valued part of the Schwarz operator of $\mathbb{D}$ corresponding to the modified Dirichlet problem (4.4.39) has the form*

$$\phi(z) = \frac{1}{2\pi i} \sum_{k=1}^{n} \int_{\mathbb{T}_k} (f(\zeta) + c_k) \left\{ \sum_{j=2}^{\infty} {}'' \left[\frac{1}{\zeta - \gamma_j(w)} - \frac{1}{\zeta - \gamma_j(z)} \right] \right.$$

$$\left. + \left(\frac{r_k}{\zeta - a_k} \right)^2 \sum_{j=1}^{\infty} {}' \left[\frac{1}{\zeta - \gamma_j(\overline{z})} - \frac{1}{\zeta - \gamma_j(\overline{w})} \right] - \frac{1}{\zeta - z} \right\} d\zeta \qquad (4.4.40)$$

$$+ \frac{1}{2\pi i} \sum_{k=1}^{n} \int_{\mathbb{T}_k} f(\zeta) \frac{\partial A}{\partial \nu}(\zeta) d\sigma + i\varsigma.$$

One of the real constants c_k can be fixed arbitrarily; the remaining ones are determined uniquely from the linear algebraic system

$$\sum_{k=1}^{n} \int_{\mathbb{T}_k} (f(\zeta) + c_k) \frac{\partial \alpha_m}{\partial \nu}(\zeta) d\sigma = 0, \; m = 1, 2, ..., n-1. \qquad (4.4.41)$$

Further we write (4.4.40) in the following form $\phi = \mathbf{T}f + i\varsigma$. For the definiteness we assume that the operator $\mathbf{T}$ contains the constants c_k as parameters, i.e., the conditions (4.4.41) on c_k are satisfied and $\mathbf{T}f$ is single-valued.

4.5 Linear Riemann–Hilbert problem for multiply connected domains

In the present section we consider $n+1$ discs $\mathbb{D}_k$ ($k = 0, 1, ..., n$) beginning from the zero one. This changing of the numeration is related to the traditional study [91], where n-connected bounded domains are considered. Let $\lambda(t)$, $f(t)$, $t \in \partial\mathbb{D}$ be given

Hölder continuous functions ($\lambda, f \in \mathcal{H}(\partial\mathbb{D})$), $\lambda(t) \neq 0$. Find a function $\phi(z)$ analytic in $\mathbb{D}$, continuous in $cl\mathbb{D}$ ($\phi \in \mathcal{C}_A(\mathbb{D})$) satisfying the following boundary condition

$$\Re\overline{\lambda(t)}\phi(t) = f(t), \quad t \in \partial\mathbb{D}. \tag{4.5.1}$$

Let $\chi_k = wind_{\mathbb{T}_k} \lambda(t) := \frac{1}{2\pi i} \int_{\mathbb{T}_k} d\log \lambda(t)$. According to Subsection 2.8.2 the value $\chi := \sum_{k=0}^{n} \chi_k$ is called the index of the problem. Let us consider first problem (4.5.1) with $\chi \geq 0$. Introduce a new unknown function $\psi(z) := \phi(z)/R(z) - \sum_{s=1}^{\chi} \delta_s z^s$ belonging to $\mathcal{C}_A(\mathbb{D})$. Here

$$R(z) := \prod_{m=0}^{n} (z - a_m)^{\chi_m},$$

$\sum_{s=1}^{\chi} \delta_s z^s$ is the principal part of the function $\phi(z)/R(z)$ at infinity. If $\chi = 0$ then the sum $\sum_{s=1}^{\chi}$ is identically equal to zero and $R(z) \equiv 1$. The boundary value problem (4.5.1) transforms to the following problem

$$\Re R(t)\overline{\lambda(t)}\psi(t) = f(t) - \Re\left(\overline{\lambda(t)} \sum_{s=1}^{\chi} \delta_s t^s R(t)\right), \quad t \in \partial\mathbb{D}, \tag{4.5.2}$$

with respect to $\psi(z) \in \mathcal{C}_A(\mathbb{D})$. The index of this problem is equal to zero but the right-hand side contains χ undetermined constants δ_s. Let us apply the factorization method (see Section 2.8) to problem (4.5.2). Introduce the function $p(t) := \overline{R(t)\lambda(t)}\left(R(t)\overline{\lambda(t)}\right)^{-1}$ on $\partial\mathbb{D}$. First we note that $\log p(t)$ is a continuous function on every $\mathbb{T}_k$, since $wind_{\mathbb{T}_k} p(t) = 0$, $k = 0, 1, ..., n$. Hence, we can consider the modified Dirichlet problem (see Subsection 2.7.2)

$$X_1(t) - \overline{X_1(t)} = \log p(t) - 2i\mu_k, \quad |t - a_k| = r_k, \; k = 0, 1, ..., n, \tag{4.5.3}$$

with respect to the function $X_1(z) \in \mathcal{H}_A(\mathbb{D})$. Here μ_k are unknown real constants to be determined. The solution of problem (4.5.3) is given in Theorem 4.12. So the function $X(z) := \exp X_1(z)$ is known. It is analytic in $\mathbb{D}$ and Hölder continuous in $cl\mathbb{D}$. Since

$$X(t)\left[\overline{X(t)}\right]^{-1} \exp 2i\mu_k = p(t), \quad |t - a_k| = r_k,$$

hence the boundary condition (4.5.2) can be rewritten in the form

$$\Re \nu_k \omega(t) = h(t) + \sum_{s=1}^{2\chi} p_s \beta_s(t), \quad |t - a_k| = r_k, \tag{4.5.4}$$

4.5. LINEAR RIEMANN–HILBERT PROBLEM.

where $\nu_k := \exp(-i\mu_k)$, the function $\omega(z) := \psi(z)[X(z)]^{-1}$ belongs to $\mathcal{C}_A((\mathbb{D}))$, $h(t) := f(t)\nu_k \left[R(t)\overline{\lambda(t)} X(t) \right]^{-1}$ is a known real function, and the real constants p_s and functions $\beta_s(t)$ are defined by the relations

$$-\Re\left(\nu_k [X(t)]^{-1} \sum_{s=1}^{\chi} \delta_s t^s\right) = \sum_{s=1}^{2\chi} p_s \beta_s(t),$$

where

$$\delta_s = p_s - ip_{s+\chi}, \; \nu_k [X(t)]^{-1} t^s = -(\beta_s(t) + i\beta_{s+\chi}(t)), \; s = 1, 2, ..., \chi; \; k = 0, 1, ..., n. \tag{4.5.5}$$

Following [91] we consider two cases for (4.5.4): a) all numbers ν_k are equal (i.e., the conditions of single-valuedness hold); b) $\nu_k \neq \nu_m$ for some $k \neq m$ (i.e., the conditions of single-valuedness do not hold). In the first case problem (4.5.4) is solved using the Schwarz operator $\mathbf{T}$ constructed in Theorem 4.12, namely

$$\omega(z) = \rho^{-1}\left[(\mathbf{T}h)(z) + \sum_{s=1}^{2\chi} p_s (\mathbf{T}\beta_s)(z) + i\varsigma\right], \tag{4.5.6}$$

where ς is an arbitrary real constant. Conditions (4.4.34) imply

$$\sum_{m=1}^{n} \int_{\mathbb{T}_m} h(\zeta) \frac{\partial \alpha_k}{\partial n_\zeta}(\zeta) ds + \sum_{m=1}^{n} \sum_{s=1}^{2\chi} p_s \int_{\mathbb{T}_m} \beta_s(\zeta) \frac{\partial \alpha_k}{\partial n_\zeta}(\zeta) ds = 0, \; k = 1, 2, ..., n. \tag{4.5.7}$$

The relation (4.5.7) for $k = 0$ is satisfied automatically (see Subsection 4.4.2, where another numeration on k is used). Therefore (4.5.7) is a system of n real linear algebraic equations with respect to 2χ unknowns $p_1, p_2, ..., p_{2\chi}$. This is Bojarski's system [43] which corresponds to solvability of the Riemann–Hilbert problem when the conditions of single-valuedness hold. If system (4.5.7) is solvable then the solution of (4.5.1) has the form

$$\phi(z) = R(z)\left[X(z)\omega(z) + \sum_{s=1}^{\chi} \delta_s z^s\right], \; z \in cl\mathbb{D}, \tag{4.5.8}$$

where δ_s are defined by (4.5.5).

We now consider the case b). Let us reduce problem (4.5.4) to a system of functional equations. The problem (4.5.4) can be rewritten as a $\mathbb{R}$-linear problem

$$\nu_k \omega(t) = \phi_k(t) - \overline{\phi_k(t)} + f_k(t), \; |t - a_k| = r_k.$$

Here the unknown functions $\phi_k(z)$ belong to $\mathcal{C}_A(\mathbb{D}_k)$, $f_k(z) := -h_k(z) - \sum_{s=1}^{2\chi} p_s \beta_{sk}(z)$; the function $h_k(z)$ is a solution of the Schwarz problem $\Re h_k(t) = h(t)$, $|t - a_k| = r_k$ (see Subsection 2.7.1), with respect to $h_k(z) \in \mathcal{C}_A(\mathbb{D}_k)$. According to (2.7.2)

$$h_k(z) = -\frac{1}{\pi i} \int_{\mathbb{T}_k} \frac{h(\tau)}{\tau - z} d\tau + \frac{1}{2\pi i} \int_{\mathbb{T}_k} \frac{h(\tau)}{\tau - a_k} d\tau.$$

The functions $\beta_{sk}(z)$ are defined by $\beta_s(t)$ by the same formulae. Let us introduce the function

$$\Phi(z) := \begin{cases} \overline{\nu_k} \phi_k(z) + \sum_{m \neq k} \overline{\nu_m} \left[\overline{\phi_m\left(z_{(m)}^*\right)} - \overline{\phi_m\left(w_{(m)}^*\right)} \right] - \overline{\nu_k} \left(\overline{\phi_k(w_{(k)}^*)} - f_k(z) \right), \\ \qquad\qquad\qquad\qquad\qquad\qquad z \in cl\mathbb{D}_k, \\ \omega(z) + \sum_{m=1}^n \overline{\nu_m} \left[\overline{\phi_m\left(z_{(m)}^*\right)} - \overline{\phi_m\left(w_{(m)}^*\right)} \right], \ z \in \mathbb{D}. \end{cases}$$

Taking into account the Analytic Continuation Principle and Liouville's theorem we conclude that $\Phi(z)$ is a constant: $\Phi(z) = Q + \sum_{m=1}^n \overline{\nu_m} \overline{\phi_m\left(w_{(m)}^*\right)}$, where $Q = \omega(w)$.
The definition of $\Phi(z)$ implies the system of functional equations

$$\phi_k(z) = -\nu_k \sum_{m \neq k} \overline{\nu_m} \left[\overline{\phi_m\left(z_{(m)}^*\right)} - \overline{\phi_m\left(w_{(m)}^*\right)} \right] + \overline{\phi_k\left(w_{(k)}^*\right)} - f_k(z) + \nu_k Q, \quad (4.5.9)$$

$$|z - a_k| \leq r_k, \ k = 0, 1, ..., n,$$

with respect to the functions $\phi_k(z) \in \mathcal{C}_A(\mathbb{D}_k)$. The function $\omega(z)$ is related to the auxiliary functions by the formula

$$\omega(z) = Q - \sum_{m=1}^n \overline{\nu_m} \left[\overline{\phi_m\left(z_{(m)}^*\right)} - \overline{\phi_m\left(w_{(m)}^*\right)} \right]. \tag{4.5.10}$$

System (4.5.9) is solved in Lemma 4.2:

$$\phi_k(z) = \left(\mathbf{A}_k h^+\right)(z) + \sum_{s=1}^{2\chi} p_s (\mathbf{A}_k \beta_s^+)(z) + \nu_k Q + \overline{\phi_k\left(w_{(k)}^*\right)}, \ z \in cl(\mathbb{D}_k), \ k = 0, 1, \ldots, n.$$
$$\tag{4.5.11}$$

The operators $\mathbf{A}_k$ are defined by the following formula

$$(\mathbf{A}_k F)(z) = \nu_k \sum_{m=0}^\infty (-1)^m \sum_{k_1 \neq k} \sum_{k_2 \neq k_1} \cdots \sum_{k_m \neq k_{m-1}} C^m \nu_{km} [F(z_{(k_m k_{m-1} \ldots k_1)}^*)$$
$$- F(w_{(k_m k_{m-1} \ldots k_1)}^*)] + F(z), \ |z - a_k| \leq r_k, \ k = 0, 1, \ldots, n.$$
$$\tag{4.5.12}$$

4.5. LINEAR RIEMANN–HILBERT PROBLEM.

Here $F(z)$ is a function of $\mathcal{C}_A(\cup_{k=0}^n \mathbb{D}_k)$; $\mathbf{C}$ is the operator of complex conjugation.

Let us calculate the values

$$R_{km} := \overline{\phi_m\left(w^*_{(mk)}\right)} - \overline{\phi_m\left(w^*_{(m)}\right)} = \overline{(\mathbf{A}_k f_k)\left(w^*_{(mk)}\right)} - \overline{(\mathbf{A}_k f_k)\left(w^*_{(m)}\right)},$$

$$k = 0, 1, ..., n; \quad m = 0, 1, ..., n, \; m \neq k.$$

Let us substitute $z = w^*_{(k)}$ into (4.5.9) and take the real parts of these equalities. The result is

$$\Re\left(-\nu_k \sum_{m \neq k} \overline{\nu_m} R_{km} - h_k(w^*_{(k)}) - \sum_{s=1}^{2\chi} p_s \beta^+_{sk}(w^*_{(k)}) + \nu_k Q\right) = 0, \quad k = 0, 1, ..., n.$$

(4.5.13)

Relation (4.5.13) is a system of $(n+1)$ real linear algebraic equations with respect to $(2\chi + 2)$ real unknowns $p_1, p_2, ..., p_{2\chi}$, $\Re Q$, $\Im Q$. System (4.5.13) gives the necessary and sufficient solvability condition of the Riemann–Hilbert problem (4.5.1) in the case b). If (4.5.13) is satisfied then the solution $\phi(z)$ has the form (4.5.8), (4.5.10), (4.5.11). Let us note that $\overline{\phi_k\left(w^*_{(k)}\right)}$ remain undetermined. It follows from (4.5.10) that they have no impact on the function $w(z)$.

Theorem 4.13 ($\chi \geq 0$) *If all numbers ν_k are equal (case a)) then the linear Riemann–Hilbert boundary value problem (4.5.1) is solvable if and only if system (4.5.7) is satisfied. If so, then the general solution to problem (4.5.1) has the form (4.5.8), (4.5.6). If at least two numbers ν_k are non-equal (case b)), then the linear Riemann–Hilbert boundary value problem (4.5.1) is solvable if and only if the system (4.5.13) is satisfied. If so then the general solution to problem (4.5.1) has the form (4.5.8), (4.5.10), (4.5.11).*

The case of the negative index χ is studied in the same way. In this case we introduce the auxiliary unknown function $w(z) := \phi(z)[R(z)X(z)]^{-1}$ from the space $\mathcal{C}_A(\mathbb{D}_k)$ with the following additional condition:

$$w(z) \text{ has zero at } z = \infty \text{ of the order } |\chi|. \tag{4.5.14}$$

Thus the function $w(z)$ has to satisfy the following $\mathbb{R}$-linear problem

$$\nu_k w(t) = \phi_k(t) - \overline{\phi_k(t)} + f_k(t), \; |t - a_k| = r_k$$

with the additional condition (4.5.14). We solve first this problem without (4.5.14) and then rewrite condition (4.5.14) in terms of the obtained solutions. Applying the standard consideration we have the following

Theorem 4.14 ($\chi < 0$) *If all numbers ν_k are equal (case a)) then the linear Riemann–Hilbert boundary value problem (4.5.1) with respect to the function $\phi(z)$ having the order $|\chi|$ at infinity has a solution if and only if*

$$\sum_{m=1}^{n} \int_{\mathbb{T}_m} h(\zeta)\frac{\partial \alpha_k}{\partial n_\zeta}(\zeta)ds = 0, \quad k = 1,2,...,n.$$

If these conditions are valid then the general solution to problem (4.5.1)

$$\phi(z) := \omega(z)R(z)X(z) = \rho^{-1}\left[(\mathbf{T}h)(z) + i\varsigma\right]R(z)X(z)$$

contains one arbitrary real constant ς. If at least two numbers ν_k are non-equal (case b)), then the linear Riemann–Hilbert boundary value problem (4.5.1) with respect to the function $\phi(z)$ having the order $|\chi|$ at infinity has a solution if and only if $(n-1)$ conditions

$$\Re\left(-\nu_k \sum_{m \neq k} \overline{\nu_m} R_{km} - h_k(w_k^*) + \nu_k Q\right) = 0, \quad k = 0,1,...,n.$$

If these conditions are valid then a unique solution to problem (4.5.1) has the form $\phi(z) := \omega(z)R(z)X(z)$.

In all cases the solution $\phi(z)$ is regular at infinity if and only if the function $R(z)X(z)$ satisfies condition (4.5.14).

4.6 Poincaré series.

Let us recall the main idea of Section 4.4 to construct the harmonic measures and the Schwarz operator of the multiply connected circular domain $\mathbb{D}$. First we state a boundary value problem for a harmonic measure. Then we reduce this problem to a system of functional equations. Finally we solve the system of functional equations by the method of successive approximations and ultimately obtain the harmonic measure in the form of a series. This series is closely related to the remarkable Poincaré series

$$\theta_{2q}(z) := \sum_{j=0}^{\infty} H(\gamma_j(z))(c_j z + d_j)^{-2q}, \quad (q \in \mathbb{Z}/2) \tag{4.6.1}$$

for $q = 1$ associated with the subgroup $\mathcal{G} := \left\{\gamma_j = z^*_{(k_m k_{m-1}...k_1)} : \text{the level of } \gamma_j \text{ is even}\right\}$ of the group $\mathcal{K}$ from Section 4.3. It is assumed that $H(z)$ is a rational function. Let $S^*_{(k)} := \left\{z \in \widehat{\mathbb{C}} : z^*_{(k)} \in S\right\}$, $S^*_{(k_m k_{m-1}...k_1)} := \left(S^*_{(k_{m-1}...k_1)}\right)^*_{(k_m)}$, where S is a set of $\widehat{\mathbb{C}}$.

4.6. POINCARÉ SERIES.

Definition 4.1 *A point z_0 is called a limit point of the group $\mathcal{G}$ if z_0 is a point of accumulation of the sequence $\gamma_j(z)$ for some $z \in \widehat{\mathbb{C}}$. A point which is not a limit point is called an ordinary point.*

In other words, if z runs over the extended complex plane, then the accumulation points of the sequence $\gamma_j(z)$ generate the limit set $\Lambda(\mathcal{G})$. We assume that in the formula (4.6.1) $z \in B := \widehat{\mathbb{C}} \backslash (B_1 \cup \Lambda(\mathcal{G}))$, B_1 is the set of poles of all $H(\gamma_j(z))$ and $\gamma_j(z)$. Detailed study of the limit sets is given by L. Ford [84] and J. Lehner [138]. We only expose properties of $\Lambda(\mathcal{G})$. Ordinary points are characterized by the following property.

Lemma 4.4 *A point z is a regular point of $\mathcal{G}$ if there exist a sequence of numbers $k_1, k_2, ..., k_m$ such that $z^*_{(k_m k_{m-1}...k_1)}$ belongs to $cl\mathbb{D}$.*

The points z_1 and z_2 are called congruent if there exists such $\gamma_j \in \mathcal{G}$ that $\gamma_j(z_1) = z_2$. All limit points of the Schottky group $\mathcal{G}$ lie within the discs $cl\mathbb{D}_k$. In the neighborhood of a limit point ς there is an infinite number of distinct points congruent to any point of $\widehat{\mathbb{C}}$, at most, exception of ς itself and of one other point. The limit set $\Lambda(\mathcal{G})$ is transformed into itself by any $\gamma_j \in \mathcal{G}$; $\Lambda(\mathcal{G})$ is closed and dense itself.

Let us use for brevity the notation $\mathcal{C}_A$ instead of $\mathcal{C}_A(\cup_{k=1}^n cl\mathbb{D}_k)$. It is customary to introduce some Banach spaces of functions meromorphic in $\cup_{k=1}^n cl\mathbb{D}_k$. Let $\{z_{kj}\}_{j=1}^m$ be a finite collection of points in $cl\mathbb{D}_k := \{z : |z - a_k| \leq r_k\}$, and let $p_k = \sum_{j=1}^{m} \sum_{s=1}^{\alpha_j} \frac{a_{kjs}}{(z - z_{kj})^s}$ be a given rational function (so-called sum of partial fractions), $\mathbf{p} := (p_1, \ldots, p_n)$. We denote

$$\mathcal{C}_A(\cup_{k=1}^n cl\mathbb{D}_k; \mathbf{p}) := \left\{ f : \cup_{k=1}^n cl\mathbb{D}_k \to \widehat{\mathbb{C}} : f - p_k \in \mathcal{C}_A(cl\mathbb{D}_k), \forall k = 1, \ldots, n \right\}$$

the space of meromorphic functions in $\cup_{k=1}^n cl\mathbb{D}_k$ with prescribed principle parts at their poles. This is a Banach space endowed with the norm

$$\|f\|_{\mathcal{C}_A(\cup_{k=1}^n cl\mathbb{D}_k; \mathbf{p})} := \max_{1 \leq k \leq n} \sup_{z \in cl\mathbb{D}_k} |f - p_k|.$$

Let us now prove some auxiliary facts about the functional equation

$$\Phi_k(z) = -\sum_{m \neq k} \left(\overline{z^*_{(m)}} \right)' \overline{\Phi_m \left(z^*_{(m)} \right)} + g_k(z), \ |z - a_k| \leq r_k, \ k = 1, 2, ..., n. \quad (4.6.2)$$

Following Section 4.3 we write (4.6.2) as the equation $\Phi = \mathbf{A}\Phi + g$ in the space $\mathcal{C}_A$, with $g(z) := g_k(z)$, $z \in \mathbb{D}_k$, $k = 1, \ldots, n$.

CHAPTER 4. METHOD OF FUNCTIONAL EQUATIONS

Lemma 4.5 *The homogeneous equation* $\Phi = \mathbf{A}\Phi$ *has in* $\mathcal{C}_A$ *only trivial solution.*

Proof. If $\Phi_m(z)$ is a solution of the system

$$\Phi_k(z) = -\sum_{m \neq k} \left(\overline{z^*_{(m)}}\right)' \overline{\Phi_m\left(z^*_{(m)}\right)}, \quad |z - a_k| \leq r_k, \tag{4.6.3}$$

then the $\Phi_k(z)$ are analytic in $|z - a_k| \leq r_k$. Let $\phi'_k(z) = \Phi_k(z)$. Then integrating (4.6.3), we have

$$\phi_k(z) = -\sum_{m \neq k} \overline{\phi_m\left(z^*_{(m)}\right)} + c_k, \quad |z - a_k| \leq r_k, \tag{4.6.4}$$

where c_k are arbitrary constants of integration. Lemma 4.2 implies that (4.6.4) has only constant solutions. Therefore, $\Phi_k(z) = \phi'_k(z) \equiv 0$.

The lemma is proved. ∎

Lemma 4.6 *The equation* $\Phi = \mathbf{A}\Phi + g$ *is a Fredholm equation in* $\mathcal{C}(\partial \mathbb{D})$ *and has a unique solution.*

Proof. The shift operator $\Phi_m \mapsto \Phi_m\left(z^*_{(m)}\right)$ is compact in $\mathcal{C}(\partial \mathbb{D}_m)$ (see (4.3.6)). The operator of complex conjugation is bounded in $\mathcal{C}(\partial \mathbb{D})$. Therefore the operator $\mathbf{A}$ is compact in $\mathcal{C}(\partial \mathbb{D})$, and equation $\Phi = \mathbf{A}\Phi + g$ is a Fredholm equation in $\mathcal{C}(\partial \mathbb{D})$. If $g \in \mathcal{C}_A$, then $\Phi \in \mathcal{C}_A$ (Pumping Principle from Section 2.3). In particular the identically zero function belongs to $\mathcal{C}_A$. Hence, from Lemma 4.5 the homogeneous equation in $\mathcal{C}$ has only the trivial solution. Then, according to the Fredholm alternative, the inhomogeneous equation has a unique solution.

The lemma is proved. ∎

Lemma 4.7 *The system of functional equations (4.6.2) has a unique solution in* $\mathcal{C}_A$ *for* $g(z) \in \mathcal{C}_A$, $k = 1, \ldots, n$. *This solution can be found by the method of successive approximations.*

Proof. According to Theorem 2.3 and Corollary 2.2 from Section 2.3 it is sufficient to show that for $|\lambda| \leq 1$ the equation $\Phi = \lambda \mathbf{A}\Phi$ has only trivial solution. Let us note that the case $\lambda = 1$ has been investigated in Lemma 4.6. Integrating the relations

$$\Phi_k(z) = -\lambda \sum_{m \neq k} \left(\overline{z^*_{(m)}}\right)' \overline{\Phi_m\left(z^*_{(m)}\right)}, \quad |z - a_k| \leq r_k,$$

4.6. POINCARÉ SERIES.

we obtain

$$\phi_k(z) = -\lambda \sum_{m \neq k} \overline{\phi_m \left(z^*_{(m)}\right)} + c_k, \quad |z - a_k| \leq r_k. \tag{4.6.5}$$

According to Section 4.3 the system (4.6.5) has only constant solutions. Hence, $\Phi_k(z) = \phi'_k(z) \equiv 0$.

The lemma is proved. ∎

Applying Lemma 4.7 to g_k/i instead of g_k, and setting $\Omega = i\Phi$, we get the following result

Lemma 4.8 *The system of functional equations*

$$\Omega_k(z) = \sum_{m \neq k} \left(\overline{z^*_{(m)}}\right)' \overline{\Omega_m \left(z^*_{(m)}\right)} + g_k(z), \quad |z - a_k| \leq r_k, \quad k = 1, 2, \ldots, n \tag{4.6.6}$$

has a unique solution in $\mathcal{C}_\mathcal{A}(\mathbb{D})$ *for* $g_k(z) \in \mathcal{C}_\mathcal{A}(\mathbb{D}_k)$, $k = 1, \ldots, n$. *This solution can be found by the method of successive approximations.*

Theorem 4.15 *Let a rational function $H(z)$ have poles only at regular points of $\mathcal{G}$. Then the Poincaré θ_2-series converges uniformly on every compact subset of each domain* $\mathbb{D}^*_{(k_m k_{m-1} \ldots k_1)} \cap B$. *The order of summation depends on the domain* $\mathbb{D}^*_{(k_m k_{m-1} \ldots k_1)}$.

Proof. Let us consider the systems (4.6.2) and (4.6.6) when $g_k(z) = -H(z)$. At first we assume that $H(z)$ has poles only in the domain $\mathbb{D}$. Let $\Phi_k(z)$ be a sum of (4.6.2) successive approximations for the equation (4.6.2):

$$\Phi_k(z) = -H(z) + \sum_{k_1 \neq k} \left(\overline{z^*_{(k_1)}}\right)' \overline{H\left(z^*_{(k_1)}\right)}$$

$$- \sum_{k_1 \neq k} \sum_{k_2 \neq k_1} \left(z^*_{(k_2 k_1)}\right)' H\left(z^*_{(k_2 k_1)}\right) + \ldots \quad z \in cl(\mathbb{D}_k).$$

The solution of (4.6.6) can be formally written as

$$\Omega_k(z) = -H(z) - \sum_{k_1=1}^{n} \sum_{k_1 \neq k} \left(\overline{z^*_{(k_1)}}\right)' \overline{H\left(z^*_{(k_1)}\right)}$$

$$- \sum_{k_1=1}^{n} \sum_{k_1 \neq k} \sum_{k_2=1}^{n} \sum_{k_2 \neq k_1} \left(z^*_{(k_2 k_1)}\right)' H\left(z^*_{(k_2 k_1)}\right) - \ldots, \quad z \in cl(\mathbb{D}_k).$$

These series converge in $\mathcal{C}_\mathcal{A}(\mathbb{D}_k)$, i.e., uniformly in $|z - a_k| \le r_k$. Introduce the functions

$$\phi(z) := \sum_{k=1}^{n} \overline{\left(z^*_{(k)}\right)'} \Phi_k\left(z^*_{(k)}\right), \quad \omega(z) := \sum_{k=1}^{n} \overline{\left(z^*_{(k)}\right)'} \Omega_k\left(z^*_{(k)}\right). \qquad (4.6.7)$$

Let $z^*_{(km)}$ ($k, m = 1, 2, \ldots, n$; $k \neq m$) be generating elements of $\mathcal{G}$ [84, 228]. Then the equalities $\gamma'_j(z) = (c_j z + d_j)^{-2}$, $\gamma_j \in \mathcal{G}$ hold [84, 228]. Hence, we have from (4.6.7)

$$\frac{1}{2}[\phi(z) - \omega(z)] = \sum_{j=1}^{\infty} H[\gamma_j(z)](c_j z + d_j)^{-2}.$$

Remember that the order of summation is fixed in accordance with the level of $\gamma_j \in \mathcal{G}$. Thus the Poincaré θ_2-series (4.6.1) converges uniformly in $cl\mathbb{D}$, since

$$\theta_2(z) = -\frac{1}{2}[(\phi(z) + H(z)) - (\omega(z) - H(z))]. \qquad (4.6.8)$$

Let us study system (4.6.2) for the functions $g_k(z) = -H(z)$ meromorphic in $cl\mathbb{D}_k$. Let $H(z)$ have a pole at a regular point w belonging to some disc $cl\mathbb{D}_p$. Let us describe the process of removing this pole. Let $H(z) = H_0(z) + h(z)$, where $H_0(z)$ is analytic near $\mathbb{D}_p$ and $h(z)$ is the principal part of $H(z)$ at w. Let us introduce the new p-th function $\Phi_p(z) = \Phi_p^0(z) - h(z)$. If $w^*_{(p)} \notin cl\mathbb{D}_k$ for each $k \neq p$, then the process stops. If $w^*_{(p)} \in cl\mathbb{D}_q$ for certain $q \neq p$, then we make the change of the q-th function $\Phi_q(z) = \Phi_q^0(z) + \overline{\left(z^*_{(p)}\right)'} \overline{h(z^*_{(p)})}$, etc. After a finite number of steps that process stops because w is a regular point of $\mathcal{G}$. The same can be done for another poles of $H(z)$. Therefore we reduce the equation $\Phi = \mathbf{A}\Phi - H$ for $H \in \mathcal{C}_\mathcal{A}(\cup_{k=1}^n cl\mathbb{D}_k, \mathbf{p})$ with respect to $\Phi \in \mathcal{C}_\mathcal{A}(\cup_{k=1}^n cl\mathbb{D}_k, \mathbf{p})$ to the equation $\Psi = \mathbf{A}\Psi - G$ in $\mathcal{C}_\mathcal{A}$ with $\mathbf{p}(z)$ being the principle part of $-H(z)$ in $\cup_{k=1}^n cl\mathbb{D}_k$, the elements $\Psi = \Phi - q$, $G = H + q - \mathbf{A}q$ belong to $\mathcal{C}_\mathcal{A}$, q is a meromorphic function in $\mathbb{C}$. By virtue of Lemma 4.7 the series $\Psi = -\sum_{k=0}^{\infty} \mathbf{A}^k G$ converges in $\mathcal{C}_\mathcal{A}$. This means that the series $\Phi = q + \Psi = q - \sum_{k=0}^{\infty} \mathbf{A}^k(H + q - \mathbf{A}q)$ converges uniformly in every compact subset of $\cup_{k=1}^n \overline{\mathbb{D}_k} \setminus \{poles\ of\ q\}$. Let us prove that the last series coincides with $-\sum_{k=0}^{\infty} \mathbf{A}^k H$. The finite sum $q - \sum_{k=0}^{N} \mathbf{A}^k(H + q - \mathbf{A}q) = -\sum_{k=0}^{N} \mathbf{A}^k H + \mathbf{A}^{N+1} q$. Since q has poles only at regular points, there exists a number M such that $\mathbf{A}^{N+1} q \in \mathcal{C}_\mathcal{A}$. Therefore $\mathbf{A}^{N+1} q \to 0$, as $N \to \infty$. The same argument is valid for the system (4.6.6) when the function $g_k(z) = -H(z)$ is meromorphic in $cl\mathbb{D}_k$. Using relations (4.6.7) and (4.6.8) let us construct the function

$$\theta_2(z) = \sum_{j=1}^{\infty} H[\gamma_j(z)](c_j z + d_j)^{-2}$$

4.6. POINCARÉ SERIES.

meromorphic in $cl\mathbb{D}$. This series converges uniformly on every compact subset of $cl\mathbb{D} \cap B$.

The theorem is proved. ∎

Theorem 4.16 *The Poincaré θ_2-series (4.6.1) is an automorphic function, i.e.,*

$$\theta_2(z) = \theta_2(\gamma_j(z))(c_j z + d_j)^{-2} \text{ for each } \gamma_j \in \mathcal{G}. \tag{4.6.9}$$

Proof. It follows from (4.6.7) that

$$\phi(t) = -\overline{\frac{t-a_k}{t-a_k}} \Phi_k(t) + \sum_{m \neq k} \overline{\left(t^*_{(m)}\right)'} \overline{\Phi_m\left(t^*_{(m)}\right)}, \quad |t - a_k| = r_k, \tag{4.6.10}$$

since $\overline{\left(t^*_{(k)}\right)'} = -\overline{\frac{t-a_k}{t-a_k}}$, $|t-a_k| = r_k$. Using (4.6.10) and (4.6.2) with $g_k(z) = -H(z)$, we calculate

$$\Im(t-a_k)\phi(t) = \tag{4.6.11}$$

$$\Im(t-a_k)\left[\Phi_k(t) + \sum_{m \neq k} \overline{\left(t^*_{(m)}\right)'} \overline{\Phi_m\left(t^*_{(m)}\right)}\right] = -\Im(t-a_k)H(t), \quad |t-a_k| = r_k.$$

The domain $\mathbb{D}^*_{(k)}$ is symmetric to the domain $\mathbb{D}$ with respect to $|t-a_k| = r_k$. The circle $(\mathbb{T}_m)^*_{(k)}$ is symmetric to the circle $|t-a_m| = r_m$ with respect to $|t-a_k| = r_k$ $(m \neq k)$; the domain $\mathbb{D}^*_{(km)}$ is symmetric to the domain $\mathbb{D}^*_{(k)}$ with respect to $(\mathbb{T}_m)^*_{(k)}$. Let us note that the numbers k and m are fixed in the definitions of these domains; $cl(\mathbb{D}_k \cup \mathbb{D}^*_{(k)})$ and $cl(\mathbb{D}_k \cup \mathbb{D}^*_{(km)})$ are the fundamental regions of $\mathcal{G}$ [84, 228]. The relation (4.6.11) implies meromorphic continuation of $(\phi(z) + H(z))$ to $\mathbb{D}_k$ and $\mathbb{D}^*_{(km)}$. Using the Reflection Principle from Section 2.1 we have

$$\phi(z) + H(z) = -\overline{\left(z^*_{(k)}\right)'}\left[\overline{\phi\left(z^*_{(k)}\right)} + \overline{H\left(z^*_{(k)}\right)}\right], \quad z \in (\mathbb{D})^*_{(k)},$$

$$\phi(z) + H(z) = \gamma'_p(z)\left[\phi\left(\gamma'_p(z)\right) + H\left(\gamma'_p(z)\right)\right], \quad z \in (\mathbb{D})^*_{(km)},$$

where $\gamma'_p(z)$ is the composition of symmetries with respect to $(\mathbb{T}_m)^*_{(k)}$ and $\mathbb{T}_k$. The transformation $\gamma'_p(z)$ is an element of the group $\mathcal{G}$.

Similar arguments for the function $\omega(z)$ from (4.6.7) yield

$$\Re(t-a_k)[\omega(t) - H(t)] = 0, |t-a_k| = r_k, \quad k = 1, 2, ..., n.$$

Hence, the function $[\omega(z) - H(z)]$ can be meromorphically continued into $\mathbb{D}^*_{(k)}$ and $\mathbb{D}^*_{(km)}$, by

$$\omega(z) - H(z) = \overline{\left(t^*_{(k)}\right)}' \left[\overline{\omega\left(t^*_{(k)}\right)} - \overline{H\left(t^*_{(k)}\right)}\right], \ z \in \mathbb{D}^*_{(k)},$$

$$\omega(z) - H(z) = \gamma'_p(z) \left[\omega\left(\gamma'_p(z)\right) - H\left(\gamma'_p(z)\right)\right], \ z \in \mathbb{D}^*_{(km)}.$$

It follows from (4.6.8) that the function $\theta_2(z)$ can be meromorphically continued into $\mathbb{D}^*_{(k)}$ and $\mathbb{D}^*_{(km)}$. Moreover the values of $\theta_2(z)$ in $\mathbb{D}^*_{(k)}$ and $\mathbb{D}^*_{(km)}$ are related by the equality

$$\theta_2(z) = \theta_2(\gamma'_p(z))\gamma'_p(z), \ z \in \mathbb{D}^*_{(km)} \cap B = \gamma'_p(\mathbb{D} \cap B).$$

Using the Reflection Principle we can continue the function $\theta_2(z)$ through $(\mathbb{T}_m)^*_{(k)}$ to the next symmetric domain and so on up to the domain of discontinuity of $\mathcal{G}$. Moreover the values of $\theta_2(z)$ in $\widehat{\mathbb{C}}\backslash\{poles\}$ are related by the equality (4.6.9).

The theorem is proved. ∎

It follows from the proof of Theorem 4.16 the functions $\phi(z)$ and $\omega(z)$ are represented as a series in each $\mathbb{D}^*_{(k_m,...,k_1)}$. Substituting these series into (4.6.9) we obtain a θ_2-series with another order of summation. Thus $\theta_2(z)$ can be represented as the series

$$\sum_{j=1}^{\infty} H\left[\gamma_{\sigma(j)}(z)\right]\gamma'_{\sigma(j)}(z)$$

in each image $\mathbb{D}^*_{(k_m,...,k_1)}$ of $\mathbb{D}$ under the mapping $z \mapsto z^*_{(k_m,...,k_1)}$, where σ is the bijection of the set of all non-negative integers that corresponds to $\mathbb{D}^*_{(k_m,...,k_1)}$.

4.7 Mixed problem for multiply connected domains

Let as in the previous sections $\mathbb{D}$ denote a circular multiply connected domain. Let $\partial \mathbb{D}$ consist of two disjoint sets of the circles $L' := \cup_{k=1}^N \mathbb{T}_k$ and $L'' := \cup_{k=N+1}^n \mathbb{T}_k$. In the present section we consider the mixed nonlinear problem

$$\Re \overline{\lambda(t)}\phi(t) = 0, \ t \in L', \qquad (4.7.1)$$

$$|\phi(t)| = g(t), \ t \in L''. \qquad (4.7.2)$$

We introduce the function

$$R(z) := (z - a_{N+1})^\chi \prod_{k=1}^N (z - a_k)^{-\chi_k} \qquad (4.7.3)$$

4.7. MIXED PROBLEM FOR MULTIPLY CONNECTED DOMAINS. 171

analytic in the domain $G := \widehat{\mathbb{C}} \setminus \cup_{n=1}^{N+1} cl\mathbb{D}_k$, and the function $p(t) := \overline{R(t)}\lambda(t)\left(R(t)\overline{\lambda(t)}\right)^{-1}$ on ∂G. The function $\log p(t)$ can be considered continuous on each $\mathbb{T}_k$, $k = 1, \ldots, N+1$, since $wind_{\mathbb{T}_k} p(t) = 0$ for each $k = 1, 2, \ldots, N+1$. Following Subsection 4.4.2 we consider the modified Dirichlet problem (4.5.3) and introduce the function $X(z) \in \mathcal{H}_A(\mathbb{D})$ non-zero in clG and satisfying the equalities

$$X(t)\left[\overline{X(t)}\right]^{-1} \exp 2i\eta_k = p(t), \quad |t - a_k| = r_k,$$

where $\pi/2 \leq \eta_k < 3\pi/2$ and $\exp 2i\eta_k = \exp 2i\mu_k$. Then (4.7.1) becomes

$$\Re \exp(-i\eta_k) \omega(t) = 0, \quad t \in L', \tag{4.7.4}$$

where $\omega(z) := \phi(z)[R(z)X(z)]^{-1}$ belongs to $\mathcal{C}_A(\mathbb{D})$. The boundary condition (4.7.2) becomes

$$|\omega(t)| = g_1(t), \quad t \in L'', \tag{4.7.5}$$

where $g_1(t) := g(t) |R(z)X(z)|^{-1}$. We have problem (4.7.4), (4.7.5) with respect to $\omega(z)$. We now reduce the latter nonlinear problem to a linear one. Let us consider an arbitrary fixed branch of the $\Phi(z) := \log \omega(z)$. Then the boundary condition (4.7.4) can be rewritten as

$$\Im \phi(t) = \eta_k + \pi/2 + \pi m_k, \quad t \in L', \; m_k \in \mathbb{Z}, \tag{4.7.6}$$

with arbitrary chosen $m_k \in \mathbb{Z}$ for each $\mathbb{T}_k \subset L', k = 1, \ldots, N$. Condition (4.7.5) is equivalent to

$$\Re \overline{\lambda(t)} \Phi(t) = h(t), \quad t \in L'', \tag{4.7.7}$$

where

$$\lambda(t) = \begin{cases} i, & t \in L', \\ 1 & t \in L'', \end{cases}, \quad h(t) = \begin{cases} \eta_k + \pi/2 + \pi m_k, & t \in L', \\ g_1(t) & t \in L''. \end{cases}$$

For any fixed set $\{m_k \in \mathbb{Z}, k = 1, \ldots, N\}$ (4.7.6), (4.7.7) is a linear Riemann–Hilbert problem (4.5.1) which was already solved (see Section 4.5) by reducing to the system of functional equations

$$\phi_k(z) = -\nu_k \sum_{m \neq k} \overline{\nu_m} \left[\phi_m\left(z_{(m)}^*\right) - \overline{\phi_m\left(w_{(m)}^*\right)}\right] + \overline{\phi_k\left(w_{(k)}^*\right)} - f_k(z) + \nu_k Q,$$

$$|z - a_k| \leq r_k, \quad k = 1, 2, \ldots, n. \tag{4.7.8}$$

Here
$$\nu_k := -i, \quad f_k(z) := \eta_k + \pi/2 + \pi m_k \text{ for } k = 1, 2, ..., N;$$
$$\nu_k := 1, \quad f_k(z) = -\frac{1}{\pi i}\int_{\mathbb{T}_k}\frac{h(\tau)}{\tau - z}d\tau + \frac{1}{2\pi i}\int_{\mathbb{T}_k}\frac{h(\tau)}{\tau - w}d\tau \text{ for } k = N+1, ..., n,$$
$Q := \omega(w)$. If ϕ_k are known then ω is calculated by the formula
$$\Phi(z) := Q - \sum_{k=1}^{n} \overline{\nu_k}\left[\overline{\phi_k\left(z^*_{(k)}\right)} - \overline{\phi_k\left(w^*_{(k)}\right)}\right].$$

Applying the method of successive approximations to (4.7.8) (see Section 4.3) we have
$$\phi_k(z) = (\mathbf{A}_k f_k)(z) + \nu_k Q + \overline{\phi_k\left(w^*_{(k)}\right)}, \quad |z - a_k| \leq r_k, \ k = 1, 2, ..., n, \quad (4.7.9)$$
where the operator $\mathbf{A}_k$ has the form (4.5.12), Q and $\overline{\phi_k\left(w^*_{(k)}\right)}$ are undetermined constants. Using (4.7.9) we can find
$$R_{km} := \overline{\phi_m\left(w^*_{(mk)}\right)} - \overline{\phi_m\left(w^*_{(m)}\right)} = \overline{(\mathbf{A}_k f_k)\left(w^*_{(mk)}\right)} - \overline{(\mathbf{A}_k f_k)\left(w^*_{(m)}\right)},$$
$$k = 1, ..., n; \ m = 1, ..., n, \ m \neq k.$$
Let us substitute $z = w^*_{(k)}$ into the real part of (4.7.8)
$$\Re\left[-\nu_k \sum_{m \neq k}\overline{\nu_m}R_{km} - f_k(w^*_{(k)}) + \nu_k Q\right] = 0, \ k = 1, ..., N. \quad (4.7.10)$$

Rewrite (4.7.10) in the form
$$\Im Q = \Im \sum_{m \neq k}\overline{\nu_m}R_{km} + \eta_k + \pi/2 + \pi m_k, \ k = 1, 2, ..., N, \quad (4.7.11)$$
$$\Re Q = \Re \sum_{m \neq k}\overline{\nu_m}R_{km} + \Re f_k(w^*_{(k)}), \ k = N+1, ..., n. \quad (4.7.12)$$

One can consider (4.7.11) and (4.7.12) as a linear algebraic system with respect to real variables $\Re Q$, $\Im Q$ and integer variables m_k, $k = 1, 2, ..., N$. This system corresponds to solvability of problem (4.7.1) and (4.7.2). Thus we have proved the following

4.8. CIRCULAR POLYGONS WITH ZERO ANGLES.

Theorem 4.17 *Problem (4.7.1) and (4.7.2) are solvable if and only if the following conditions are valid:*

i) the values $\Re \sum_{m \neq k} \overline{\nu_m} R_{km} + \Re f_k(w_k^*)$ are independent of $k = N+1, ..., n$,

ii) there exists real number $\Im Q$ such that the numbers

$$m_k = \frac{1}{\pi} \Im Q - \Im \sum_{m \neq k} \overline{\nu_m} R_{km} - \eta_k - \pi/2, \ k = 1, 2, ..., N,$$

are integers. If these conditions are fulfilled then the solution to (4.7.1) and (4.7.2) has the form

$$\phi(z) = \omega(z) R(z) X(z),$$

where $\omega(z) = Q - \sum_{k=1}^n \overline{\nu_k} \left[\overline{(\mathbf{A}_k f_k)(z_{(k)}^*)} - \overline{(\mathbf{A}_k f_k)(w_{(k)}^*)} \right]$, the operators $\mathbf{A}_k$ are defined by the formulae (4.5.12), the function $R(z)$ has the form (4.7.3), and $X(z)$ is a solution of the modified Dirichlet problem (4.5.3).

4.8 Modulus problem for circular polygons with zero angles

The connection between the modulus problem and conformal mapping were described in Section 2.11. Here we use the method of functional equations to solve the modulus problem for a circular polygon with zero angles and apply this solution to the constructing of the conformal mapping of the above domain onto the unit disc.

Let us consider mutually disjoint discs $\mathbb{D}_k$, $k = 1, ..., n$ (see Fig.4.8). Let $\mathbb{T}_k$ be tangent to $\mathbb{T}_{k+1}$ at the point w_{k+1}, $k = 1, ..., n$, and $w_1 := w_{n+1}$. The union $\cup_{k=1}^n cl\mathbb{D}_k$ separates two disjoint domains, bounded (denoted G_1), and unbounded (denoted G_2), $G_2 = \widehat{\mathbb{C}} \setminus (\cup_{k=1}^n cl\mathbb{D}_k \cup G_1)$. Thus $\partial G_1 \cup \partial G_2 = \cup_{k=1}^n \mathbb{T}_k, \partial G_1 \cap \partial G_2 = W := \{w_1, ..., w_n\}$. The chosen orientation on $\mathbb{T}_k$ ("positive") leaves $\mathbb{D}_k$ to the left for every $k = 1, ..., n$. It generates the orientation on the boundaries G_1 and G_2. Suppose for simplicity $0 \in G_1$.

Let f be a given continuously differentiable function on ∂G_1. We consider the problem of finding a function $\phi \in \mathcal{C}_A(G_1) \cap \mathcal{C}^1(clG_1 \setminus W)$ satisfying the boundary condition

$$\Re \phi(t) = f(t), \ t \in \partial G_1. \tag{4.8.1}$$

This is a special case of the Riemann–Hilbert boundary value problem (see Section 2.9). Instead of using the general scheme for the solution of (4.8.1) we apply here

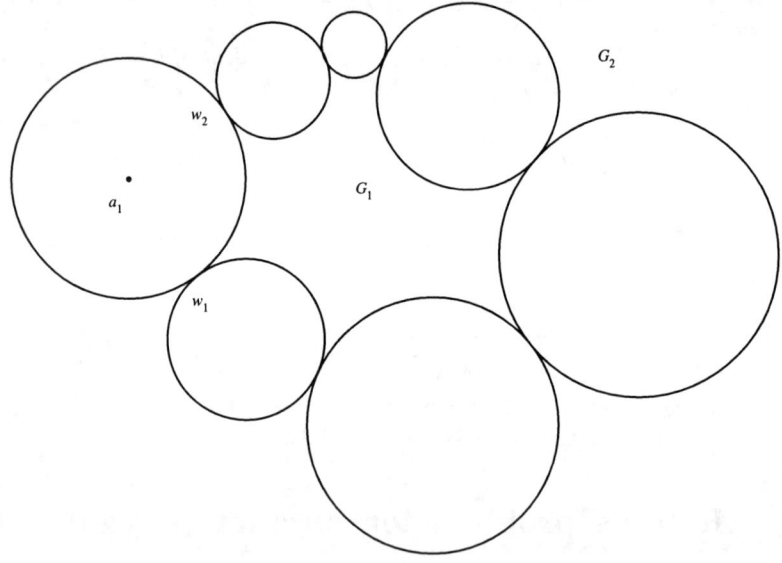

Figure 4.4: Circular polygon with zero angles

the method developed in Section 4.5, which allows finding an exact solution of the Riemann–Hilbert boundary value problem for a multiply connected circular domain.

Let us first continue the given function $f(t), t \in \partial G_1$ up to a function continuously differentiable on $\cup_{k=1}^{n} \mathbb{T}_k$, using the same notation for it, and consider the auxiliary problem

$$\Re\phi(t) = f(t), \ t \in \partial G_1, \tag{4.8.2}$$

in the class of functions $\phi \in \mathcal{C}_A(G_2) \cap \mathcal{C}^1(clG_2\backslash W)$. Since w_k, $k = 1, \ldots, n$ are wedge-points for both contours ∂G_1, and ∂G_2, the limit values of the functions $\phi|_{G_1}$ and $\phi|_{G_2}$ at these points can be in general different.

We associate with the function ϕ a collection of functions $\phi_k, k = 1, \ldots, n$, analytic in $\mathbb{D}_k$, continuously differentiable on $cl\mathbb{D}_k\backslash\{w_k, w_{k+1}\}$, and almost bounded at w_k, w_{k+1}. The latter means (cf., e.g., [91]) that ϕ_k can be represented in the form

$$\phi_k(z) = \phi_k^0(z) - \frac{1}{\pi i}\sum_{j=k}^{k+1}(\phi_k(w_j+0) - \phi_k(w_j-0))\log(z-w_j),$$

4.8. CIRCULAR POLYGONS WITH ZERO ANGLES.

where $\phi_k^0 \in \mathcal{C}_\mathcal{A}^1(\mathbb{D}_k)$. The functions ϕ_k are connected with ϕ via the following boundary condition

$$2\Re\phi_k(t) = \Re(\phi(t) - f(t)), \ t \in \mathbb{T}_k. \tag{4.8.3}$$

Let us consider now the $\mathbb{R}$-linear problem

$$\phi(t) = \phi_k(t) - \overline{\phi_k(t)} + f_k(t), \ t \in \mathbb{T}_k, \ k = 1, \ldots, n, \tag{4.8.4}$$

in the above prescribed classes of unknown functions ϕ, ϕ_k, $k = 1, \ldots, n$, and the functions $f_k(t)$ being the boundary functions of the solution of the Schwarz problem

$$\Re f_k(t) = f(t), \ t \in \mathbb{T}_k,$$

for the fixed disc $\mathbb{D}_k$.

Lemma 4.9 *The function ϕ is a solution of the problem (4.8.1), (4.8.2), if and only if the functions ϕ, ϕ_k related by (4.8.3) constitute a solution of problem (4.8.4).*

Proof. If $\phi(z)$ is a solution of (4.8.4), then it evidently satisfies relations (4.8.1) and (4.8.3) with relevant functions ϕ_k. Conversely, let ϕ be a solution of the problem (4.8.1), (4.8.2); then due to (4.8.3) the real parts of both sides of (4.8.4) coincide. We have to construct such function ϕ_k that the imaginary part of both sides of (4.8.4) is equal too, i.e., ϕ_k have to satisfy the following boundary condition:

$$2\Im\phi_k(t) = \Im(\phi(t) - f_k(t)), \ t \in \mathbb{T}_k. \tag{4.8.5}$$

It should be noted that the given functions (namely $\Im\phi$) can have jump discontinuities at the points w_k, w_{k+1}. Hence, following [91] we have the solution of the problem (4.8.5) in the following form

$$\phi_k(z) = \frac{1}{2\pi}\int_{\mathbb{T}_k} \frac{\Im(\phi(\tau) - f_k(\tau))}{\tau - z}d\tau - \frac{1}{4\pi}\int_{\mathbb{T}_k} \frac{\Im(\phi(\tau) - f_k(\tau))}{\tau - a_k}d\tau$$

$$= -\frac{1}{2\pi}\sum_{j=k}^{k+1}\Im[\phi(w_j + 0) - \phi(w_j - 0)]\log(z - w_j) + \phi_k^0(z),$$

$\phi_k^0(z) \in \mathcal{C}_\mathcal{A}(\mathbb{D}_k)$, the branch of the logarithmic function is fixed in $\widehat{\mathbb{C}}\setminus L_j$, where the curve $L_j \subset G_2 \cup \{w_j\}$ connects w_j and ∞. It is easily seen that ϕ, ϕ_k satisfy condition (4.8.4).

It completes the proof. ∎

For further constructions it is better to rewrite the right-hand side of the formula for ϕ_k in the form:

$$\phi_k(z) = \Phi_k(z) + p_k \log(z - w_k) + q_k \log(z - w_{k+1}), \quad z \in \mathbb{D}_k, \tag{4.8.6}$$

where p_k, q_k are real constants, $\Phi_k \in \mathcal{C}_A(\mathbb{D}_k) \cap \mathcal{C}^1(cl\mathbb{D}_k \setminus \{w_k, w_{k+1}\})$. In order to reduce (4.8.4) to a system of functional equations we introduce the function

$$\Omega(z) = \begin{cases} \phi_k(z) + \sum_{m \neq k} \overline{\phi_m\left(z^*_{(m)}\right)} + f_k(z), & z \in \mathbb{D}_k, \\ \phi(z) + \sum_{m=1}^n \overline{\phi_m\left(z^*_{(m)}\right)}, & z \in G_1 \cup G_2. \end{cases}$$

It follows from (4.8.4) that $\Omega(z)$ is continuous in $\mathbb{C}$ except the set W, where it is almost bounded [91]. By virtue of Liouville's theorem we have $\Omega(z) = constant$. Let us calculate this constant:

$$\Omega(z) = \Omega(0) = \phi(0) + \sum_{m=0} \overline{\phi_m\left(0^*_{(m)}\right)}, \quad z \in \mathbb{C}.$$

From the definition of $\Omega(z)$ in $\mathbb{D}_k$ and (4.8.6) we obtain

$$\Phi_k(z) = -\sum_{m \neq k}\left[\overline{\Phi_m\left(z^*_{(m)}\right)} - \overline{\Phi_m\left(0^*_{(m)}\right)}\right] - f_k(z) + \phi(0) + \overline{\Phi_k\left(0^*_{(k)}\right)} - \alpha_k(z), \tag{4.8.7}$$

where

$$\alpha_k(z) = p_k \log\left[(z - w_k) / \left(\overline{0^*_{(k)} - w_k}\right)\right] + q_k \log\left[(z - w_{k+1}) / \left(\overline{0^*_{(k)} - w_{k+1}}\right)\right]$$

$$- \sum_{m \neq k}\left[p_m \log \overline{\frac{z^*_{(m)} - w_m}{0^*_{(m)} - w_m}} + q_m \log \overline{\frac{z^*_{(m)} - w_{m+1}}{0^*_{(m)} - w_{m+1}}}\right]. \tag{4.8.8}$$

As the points $w_k, w_{k+1} \in \mathbb{T}_k$, so $(w_k)^*_{(k)} = w_k$ and $(w_{k+1})^*_{(k)} = w_{k+1}$. Hence

$$\log\left(\overline{z^*_{(k)} - w_k}\right) = \log\left(\overline{z^*_{(k)} - (w_k)^*_{(k)}}\right) = \log \frac{r_k^2(z - w_k)}{(z - a_k)(a_k - w_k)}.$$

Along similar lines

$$\log\left(\overline{z^*_{(k)} - w_{k+1}}\right) = \log\left(\overline{z^*_{(k)} - (w_{k+1})^*_{(k+1)}}\right) = \log \frac{r_k^2(z - w_{k+1})}{(z - a_k)(a_k - w_{k+1})}.$$

4.8. CIRCULAR POLYGONS WITH ZERO ANGLES.

It follows from (4.8.7) that the function $\alpha_k(z)$ has to be continuous in clD_k. From the other side the logarithms appearing in (4.8.8) have jumps along the curve connecting the points $z = w_k$, $z = a_k$ and $z = w_{k+1}$. This contradiction can be overcame only if $p_k = q_{k-1}$, $k = 1, 2, ..., n$. Let us write the functions $\alpha_k(z)$ in the form $\alpha_k(z) = \sum_{j=0}^{n} p_j H_k^j(z)$, where

$$H_k^k(z) = \log \frac{a_k(a_k - w_k)(z - a_{k-1})}{r_k^2 a_{k-1}}, \quad H_k^{k+1}(z) = \log \frac{a_k(a_k - w_{k+1})(z - a_{k+1})}{r_k^2 a_{k+1}},$$

$$H_k^j(z) = \log \left[\overline{\frac{\left(0_{(j)}^* - w_{(j)}^*\right)\left(0_{(j-1)}^* - w_{(j)}^*\right)}{\left(z_{(j)}^* - w_{(j)}^*\right)\left(z_{(j-1)}^* - w_{(j)}^*\right)}} \right], \quad j \neq k, k+1. \quad (4.8.9)$$

The function $\phi(z)$ is analytic in G_1 and G_2, continuous in clG_1 and clG_2. Hence by the Cauchy theorem:

$$\int_{\partial G_1} \phi(z)\,dz + \int_{\partial G_2} \phi(z)\,dz = 0.$$

Let us rewrite the last equality in the form

$$\sum_{k=1}^{n} \int_{\mathbb{T}_k} \phi(z)\,dz = 0. \quad (4.8.10)$$

From the definition of $\Omega(z)$ and representation (4.8.6) we have

$$\phi(z) = \phi(0) - \sum_{m=1}^{n} \left[\overline{\Phi_m\left(z_{(m)}^*\right)} - \overline{\Phi_m\left(0_{(m)}^*\right)} \right] \quad (4.8.11)$$

$$- \sum_{m=1}^{n} \left[p_m \log \left(\overline{z_{(m)}^* - w_m}\right) + p_{m+1} \log \left(\overline{z_{(m)}^* - w_{m+1}}\right) \right], \quad z \in clG_1 \text{ and } clG_2.$$

The definition of logarithmic functions in representation (4.8.6) implies that the function $\log\left(\overline{z_{(m)}^* - w_m}\right)$ is an analytic branch in $\overline{\mathbb{D}_k} \backslash \widehat{(w_m, a_m)}$, and the function $\log\left(\overline{z_{(m)}^* - w_{m+1}}\right)$ is an analytic branch in $\overline{\mathbb{D}_k} \backslash \widehat{(w_{m+1}, a_m)}$, where $\widehat{(w_m, a_m)} =: (L_m)_{(m)}^*$, $\widehat{(w_{m+1}, a_m)} =: (L_{m+1})_{(m)}^*$ are the lines connecting corresponding points. Using (4.8.11) we calculate the integral (4.8.10). Since the increments of the logarithms for $m = k$ in (4.8.11) are equal to πi, then $\sum_{k=1}^{n}(p_k + p_{k+1}) = 0$, or simply

$$\sum_{k=1}^{n} p_k = 0. \quad (4.8.12)$$

Let us represent system (4.8.7) in the form of operator equation

$$\Phi = \mathbf{A}\Phi + F \tag{4.8.13}$$

in the space $\mathcal{C}_\mathcal{A}(\cup_{k=0}^n \mathbb{D}_k)$, where the operator A is defined by the formula

$$\mathbf{A}\Phi(z) := -\sum_{m \neq k} \left[\overline{\Phi_m\left(z^*_{(m)}\right)} - \overline{\Phi_m\left(0^*_{(m)}\right)} \right], \ z \in cl\mathbb{D}_k \ (k=1,2,...,n).$$

Lemma 4.10 *The homogeneous equation $\Phi = \mathbf{A}\Phi$ corresponding to (4.8.13) has only a trivial solution.*

Lemma 4.11 *Non-homogeneous equation (4.8.13) has a unique solution in the class $\mathcal{C}_\mathcal{A}(\cup_{k=0}^n \mathbb{D}_k)$ for each $F \in \mathcal{C}_\mathcal{A}(\cup_{k=0}^n \mathbb{D}_k)$.*

Proof. Let $\Phi(z)$ be a solution of (4.8.13) in $\mathcal{C}_\mathcal{A}(\cup_{k=0}^n \mathbb{D}_k)$. Hence the functions $\Phi_k(z) := \Phi(z)$ are analytic in $\mathbb{D}_k$ and continuous in $cl\mathbb{D}_k$. Let us introduce the function

$$\psi(z) := -\sum_{m=1}^n \left[\overline{\Phi_m\left(z^*_{(m)}\right)} - \overline{\Phi_m\left(0^*_{(m)}\right)} \right] \tag{4.8.14}$$

analytic in $G_1 \cup G_2$, continuous in $cl(G_1 \cup G_2)$. From (4.8.13) we have

$$\psi(t) = \Phi_k(t) - \overline{\Phi_k(t)} - F(t) + \overline{\Phi_k\left(0^*_{(k)}\right)}, \ t \in \mathbb{T}_k.$$

Let us rewrite the later equality in the form

$$\Re \psi(t) = -\Re F(t) + \Re \Phi_k\left(0^*_{(k)}\right), \ t \in \mathbb{T}_k, \tag{4.8.15}$$

$$2\Im \Phi_k(t) = \Im \left[\psi(t) + F(t) + \Phi_k\left(0^*_{(k)}\right)\right], \ t \in \mathbb{T}_k. \tag{4.8.16}$$

Since the functions $F(t)$ and $\psi(t)$ are continuous on $\mathbb{T}_k$, hence from (4.8.15) we get $\Re \Phi_k\left(0^*_{(k)}\right) = \Re \Phi_{k+1}\left(0^*_{(k+1)}\right)$. Therefore, the constant $\Re \Phi_k\left(0^*_{(k)}\right)$ does not depend on k. If $F(t) \equiv 0$ then from (4.8.15) we have $\psi(z) = constant$. But $\psi(0) = 0$, hence $\psi(z) \equiv 0$. It follows from the definition (4.8.14) and the Decomposition theorem (cf., Section 2.4) that each function $\overline{\Phi_m\left(z^*_{(m)}\right)} - \overline{\Phi_m\left(0^*_{(m)}\right)}$ is a constant. Therefore, using the relation $\Phi = \mathbf{A}\Phi$ we have $\Phi_m(z) \equiv 0$ for each m.
Lemma 4.10 is proved.

4.8. CIRCULAR POLYGONS WITH ZERO ANGLES.

By virtue of Subsection 2.7.1 the inhomogeneous problem (4.8.15) with respect to $\psi(z)$ analytic in $G_1 \cup G_2$ has a unique solution up to the additive constant $(c + i\gamma)$. Hence problem (4.8.16) with respect to $\Phi_k(z)$ has a unique solution up to an arbitrary additive constant which vanishes in (4.8.13).

Lemma 4.11 is proved. ∎

Theorem 4.18 *Equation (4.8.13) has a unique solution in the space $\mathcal{C}_A(\cup_{k=0}^n \mathbb{D}_k)$. This solution can be found by the method of successive approximations converging in $\mathcal{C}_A(\cup_{k=0}^n \mathbb{D}_k)$.*

Proof. We use Lemma 4.11 and repeat argumentations of Lemma 4.2 (see Section 4.3). ∎

Applying the method of successive approximations to system (4.8.7) we find

$$\overline{\Phi_k\left(z^*_{(k)}\right)} - \overline{\Phi_k\left(0^*_{(k)}\right)} = \mathbf{B}_k h(z),$$

where $h(z) := h_k(z)$ for $|z - a_k| \leq r_k$, $h_k(z) = f_k(z) + \alpha_k(z)$, the operator $\mathbf{B}_k$ is defined by the formula:

$$\mathbf{B}_k h(z) := -\left[\overline{h_k\left(z^*_{(k)}\right)} - \overline{h_k\left(0^*_{(k)}\right)}\right] + \sum_{k_1 \neq k}\left[\overline{h_{k_1}\left(z^*_{(k_1 k)}\right)} - \overline{h_{k_1}\left(0^*_{(k_1 k)}\right)}\right]$$

$$- \sum_{k_1 \neq k}\sum_{k_2 \neq k_1}\left[\overline{h_{k_2}\left(z^*_{(k_2 k_1 k)}\right)} - \overline{h_{k_1}\left(0^*_{(k_2 k_1 k)}\right)}\right] + \ldots, \quad |z - a_k| \geq r_k.$$

(4.8.17)

It should be noted that this operator maps each function from $\mathcal{C}_A(\cup_{m=0}^n \mathbb{D}_m)$ to the function from $\mathcal{C}_A(\mathbb{D}_k)$. It follows from the definitions of $h_k(z)$ and $\alpha_k(z)$ that

$$\overline{\Phi_k\left(z^*_{(k)}\right)} - \overline{\Phi_k\left(0^*_{(k)}\right)} = \mathbf{B}_k f(z) + \sum_{j=1}^n p_j d_k(z), \qquad (4.8.18)$$

where $d_k(z) := \mathbf{B}_k H_j(z)$, $H_j(z) := H_k^j(z)$ for $|z - a_k| \leq r_k$. The functions $H_k^j(z)$ have the form (4.8.9). Substituting $z = 0^*_{(k)}$ into (4.8.7) we obtain

$$\begin{aligned}\Phi_k\left(0^*_{(k)}\right) &= \sum_{m \neq k}\left[\mathbf{B}_m f\left(0^*_{(k)}\right) + \sum_{j=1}^n p_j d_k\left(0^*_{(k)}\right)\right] \\ &- f_k\left(0^*_{(k)}\right) + \phi(0) + \overline{\Phi_k\left(0^*_{(k)}\right)} - \sum_{j=0}^n p_j H_k^j\left(0^*_{(k)}\right).\end{aligned}$$

(4.8.19)

The real parts of the relations (4.8.19) together with (4.8.12) constitute a real system of $(n+1)$ linear algebraic equations with respect to $(n+1)$ unknown values

$\Re\phi(0)$, $p_1, p_2, ..., p_n$. After solving to (4.8.19) and using (4.8.11) we arrive at the formula

$$\begin{aligned}\phi(z) &= \Re\phi(0) + i\Im\phi(0) - \sum_{m=1}^{n}\left[\mathbf{B}_m f(z) + \sum_{j=1}^{n} p_j \mathbf{B}_m H_j(z)\right] \\ &+ i\sum_{m=1}^{n}\left[p_m \log\left(\overline{z_{(m)}^* - w_m}\right) + p_{m+1}\log\left(\overline{z_{(m)}^* - w_{m+1}}\right)\right], \\ &\qquad z \in clG_1 \text{ and } clG_2,\end{aligned} \quad (4.8.20)$$

where $i\Im\phi(0)$ is an arbitrary imaginary constant. Let us prove that the system (4.8.12), $\Im$(4.8.19) (i.e., real part of (4.8.19)) always has a unique solution. The system (4.8.12), $\Re$(4.8.19) corresponding to the homogeneous system ($f = 0$ and $i\Im\phi(0) = 0$) has only a trivial solution, because the opposite case contradicts to representation (4.8.6) and the Decomposition theorem (see Section 2.4). So we have proved the following

Theorem 4.19 *The solution of problem (4.8.1), (4.8.2) has the form (4.8.20) where $i\Re\phi(0)$ is an arbitrary constant. The numbers $\Im\phi(0)$, $p_1, p_2, ..., p_n$ are defined from the uniquely solvable system of linear algebraic equations (4.8.12), $\Re$(4.8.19).*

Let us apply the last theorem to construct the conformal mapping $\omega(z)$. Assume that $f(t) = -\log|t|$. Then $f_k(z) = -\log z$, $z \in \mathbb{D}_k$, where the branch of the logarithm is fixed for every $k = 1, \ldots, n$ in the complex plane with the cut connecting the points $z = 0$ and $z = \infty$ belongs to $G_1 \cup \{w_k\} \cup G_2$. In (4.8.20) let us calculate

$$\sum_{m=1}^{n} \mathbf{B}_m f(z) = \sum_{m=1}^{n}\left[\overline{\log z_{(m)}^*} - \overline{\log 0_{(m)}^*}\right] - \sum_{m=1}^{n}\sum_{k_1 \neq k}\left[\overline{\log z_{(k_1 m)}^*} - \overline{\log 0_{(k_1 m)}^*}\right] + \ldots$$

$$= \log \prod_{m=1}^{n}\overline{\frac{z_{(m)}^*}{0_{(m)}^*}} \prod_{m=1}^{n}\prod_{k_1 \neq m}\overline{\frac{0_{(k_1 m)}^*}{z_{(k_1 m)}^*}} \prod_{m=1}^{n}\prod_{k_1 \neq m}\prod_{k_2 \neq k_1}\overline{\frac{z_{(k_2 k_1 m)}^*}{0_{(k_2 k_1 m)}^*}} \ldots \quad (4.8.21)$$

Ultimately, let us describe the finite algorithm to construct the conformal mapping in analytic form.

i) determine $H_j(z) := H_k^j(z)$ for $|z - a_k| \leq r_k$; the functions $H_k^j(z)$ has the form (4.8.9);

ii) find $d_k(z) := \mathbf{B}_k H_j(z)$ and $\mathbf{B}_k f(z)$ according to (4.8.17) and (4.8.21);

iii) solve the real system of $(n+1)$ linear algebraic equations

$$\sum_{j=1}^{n} p_j \sum_{m \neq k}\Re\left[\mathbf{D}_m^j\left(0_{(k)}^*\right) - H_m^j\left(0_{(k)}^*\right)\right] + \Re\phi(0) = \Re\left[f_k\left(0_{(k)}^*\right) - \sum_{m \neq k}\mathbf{B}_k f(0_{(k)}^*)\right],$$

$$k = 1, 2, ..., n; \quad \sum_{j=1}^{n} p_j = 0$$

with respect to $\Re\phi(0)$, $p_1, p_2, ..., p_n$. The conformal mapping has the form $w(z) = z \exp \phi(z)$, where the function $\phi(z)$ has the form (4.8.20); $i\Im\phi(0)$ is an arbitrary pure imaginary constant.

4.9 Generalized method of Schwarz and other methods.

4.9.1 Generalized method of Schwarz for a doubly connected domain.

In the present subsection we come back to the doubly connected domain considered in Section 4.1 in order to introduce the main idea of the generalized method of Schwarz (GMS). Let us consider the Dirichlet problem:

$$u(t) = f(t) \text{ on } |t - a_1| = r_1 \text{ and } |t - a_2| = r_2. \quad (4.9.1)$$

This problem has been solved by complex potentials in Section 4.1. In the present section we propose a slight modification of this method. Using the Decomposition Theorem we represent the unknown function in the form (see Fig.4.9.1)

$$u(z) = u_1(z) + u_2(z), \ z \in c\mathbb{D}, \quad (4.9.2)$$

where $u_j(z)$ is harmonic in $|z - a_j| > r_j$ and continuous in $|z - a_j| \geq r_j$ ($j = 1, 2$). Let us write (4.9.1) for $k = 1$ as

$$u_1(t) + u_2(t^*_{(1)}) = f(t) \text{ on } |t - a_1| = r_1. \quad (4.9.3)$$

The function $u_1(z) + u_2(z^*_{(1)})$ is harmonic in $|z - a_1| > r_1$ and continuous in $|z - a_1| \geq r_1$. Using the Green function for $|z - a_1| > r_1$ we harmonically continue $f(t)$ into $|z - a_1| > r_1$. Then (4.9.3) implies

$$u_1(z) = -u_2(z^*_{(1)}) + f_1(z) \text{ in } |z - a_1| \geq r_1, \quad (4.9.4)$$

where $f_1(t) = f(t)$ on $|t - a_1| = r_1$. Similar arguments for (4.9.1) with $k = 2$ yield the relation

$$u_2(z) = -u_1(z^*_{(2)}) + f_2(z) \text{ in } |z - a_2| \geq r_2. \quad (4.9.5)$$

The relations (4.9.4), (4.9.5) constitute a system of functional equations in a class of harmonic functions. The system (4.9.4), (4.9.5) is similar to the system (4.1.7), (4.1.8) with respect to complex potentials. The system (4.9.4), (4.9.5) can be solved by successive approximations. Let us consider now the method of successive approximations for (4.9.4), (4.9.5) from another point of view. The zero-order approximation for (4.9.4), (4.9.5)

$$u_1^{(0)}(z) = f_1(z), \quad |z - a_1| \geq r_1$$

is a solution to the Dirichlet problem with respect to $u_1^{(0)}(z)$ harmonic in the domain $|z - a_1| > r_1$ with the boundary data $f(t)$ (see Figure 4.9.1). So we do not take into account the second disc in the zero-order approximation. Instead of (4.9.2) we put $u^{(0)}(z) = u_1^{(0)}(z)$ in D as the zero-order approximation for the original problem (4.9.1). We call the Dirichlet problem for $|z - a_1| > r_1$ by the first problem, the Dirichlet problem for $|z - a_2| > r_2$ - by the second problem. We use the same names for the boundary conditions (4.9.1). The function $u^{(0)}(z)$ satisfies the first boundary condition (4.9.1) and does not satisfy the second one. The first-order approximation

$$u_2^{(1)}(z) = -u_1^{(0)}(z_{(2)}^*) + f_2(z), \quad |z - a_2| \geq r_2$$

is a solution to the Dirichlet problem for the domain $|z - a_2| > r_2$ with the boundary data $-u_1^{(0)}(t) + f_2(t)$ on $|t - a_2| = r_2$ (see Figure 4.9.1). Let us construct $u(z)$ according to (4.9.2)

$$u^{(1)}(z) = u_1^{(0)}(z) + u_2^{(1)}(z).$$

The function $u^{(1)}(z)$ satisfies the second boundary condition (4.9.1) and does not satisfy the first one. At first we solve the first problem with the boundary data $f(t)$ (see Figure 4.9.1). After we calculate the solution $u_1^{(0)}$ on the second circle, and solve the second problem with the corrected data $-u_1^{(0)}(t) + f_2(t)$ (see Figure 4.9.1). One can consider the result $u^{(1)}(z)$ as a perturbation of the second problem with the boundary data $f(t)$, $|t - a_2| = r_2$, by the first problem with the boundary data $f(t)$, $|t - a_1| = r_1$. sFurther, in order to correct $u^{(1)}(z)$ we calculate $u_2^{(1)}(t)$ on $|t - a_1| = r_1$ and solve the first problem with the boundary data $-u_2^{(1)}(t)$ (see Figure 4.9.1). This is the next approximation in (4.9.4). Then

$$u_1^{(2)}(z) = -u_2^{(1)}(z_{(1)}^*), \quad |z - a_1| \geq r_1, \quad u^{(2)}(z) = u_1^{(0)}(z) + u_2^{(1)}(z) + u_1^{(2)}(z), \quad z \in cl\mathbb{D}.$$

One can consider $u^{(2)}(z)$ as a perturbation of the first problem with the boundary data $f(t)$, $|t - a_1| = r_1$, by the second problem which involves the perturbation by the "pure" first problem. The function $u^{(2)}(z)$ satisfies the first boundary condition (4.9.1)

4.9. GENERALIZED METHOD OF SCHWARZ AND OTHER METHODS. 183

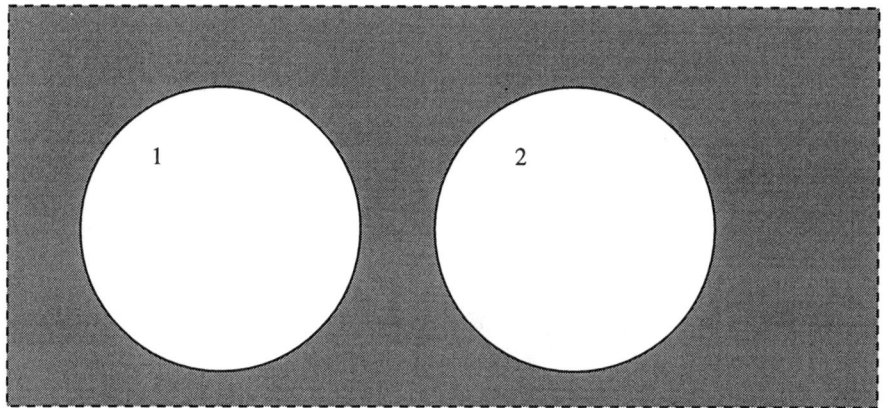

Figure 4.5: Doubly connected domain

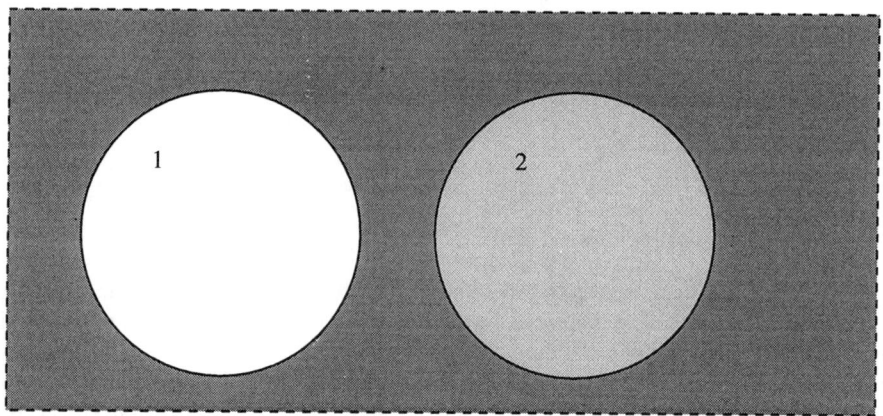

Figure 4.6: Even step

and does not satisfy the second one. To construct the next approximation $u^{(3)}(z)$ we take $u_2^{(3)}(z) = -u_1^{(2)}\left(z_{(2)}^*\right)$, $|z - a_2| \geq r_2$. Then $u^{(3)}(z) = u_1^{(0)}(z) + u_2^{(1)}(z) + u_1^{(2)}(z) + u_2^{(3)}(z)$, $z \in cl\mathbb{D}$. The function $u^3(z)$ satisfies the second boundary condition (4.9.1) and does not satisfy the first one. The next approximations are constructed by the

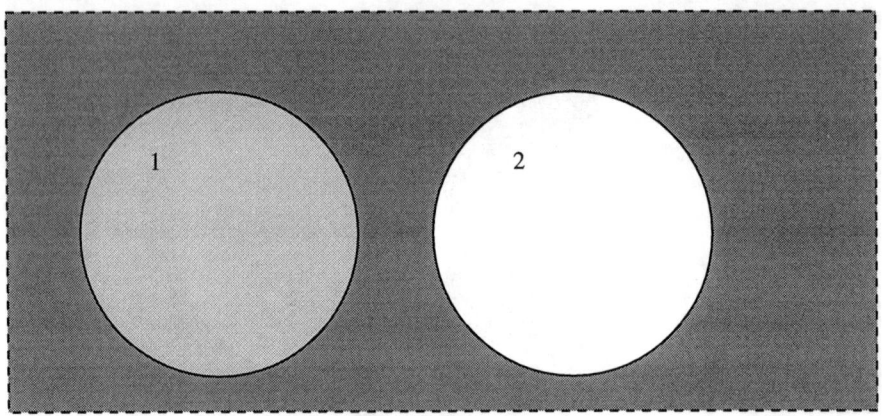

Figure 4.7: Odd step

same method. We have, for instance, for the even N

$$u^{(N)}(z) = u_1^{(0)}(z) + u_2^{(1)}(z) + u_1^{(2)}(z) + u_2^{(3)}(z) + \ldots + u_2^{(N)}(z), \ z \in cl\mathbb{D}. \quad (4.9.6)$$

Remark 4.8 *It is assumed in GMS [157] to write (4.9.6) in the form*

$$u^{(N)}(z) = u_1^{(0)}(z) - v_2^{(1)}(z) + u_1^{(2)}(z) - v_2^{(3)}(z) + \ldots - v_2^{(N)}(z), \ z \in cl\mathbb{D},$$

where $v_2^{(p)}(z) := -u_2^{(p)}(z)$, $p = 1, 3, 5, \ldots$.

We have described the classic generalized alternating method of Schwarz. We have shown on the example of a doubly connected domain that GMS for circular domains is reduced to the method of functional equations. However, there is a difference between the functional equations (4.9.4), (4.9.5) and (4.1.7), (4.1.8) which is concluded in the following. The functional equations (4.9.4), (4.9.5) are fulfilled in $|z - a_1| \geq r_1$ and $|z - a_2| \geq r_2$ and (4.1.7), (4.1.8) are fulfilled in $|z - a_1| \leq r_1$ and $|z - a_2| \leq r_2$. The difference in the classes (harmonic and analytic functions) is not so essential.

4.9.2 Generalized method of Schwarz for multiply connected domains and other methods.

We now discuss GMS for an arbitrary multiply connected domain D derived in Subsection 2.7.2. For the definiteness we take a multiply connected domain D containing

4.9. GENERALIZED METHOD OF SCHWARZ AND OTHER METHODS.

infinity and consider the Dirichlet problem

$$u(t) = f(t), \ t \in \partial D \tag{4.9.7}$$

in the space $h_-^1(\partial D)$ (harmonic in D and continuously differentiable in clD). Here D is bounded by the curves L_k ($k = 1, 2, ..., n$) orientated in a counterclockwise sense (see Figure 2.2 from Section 2.1). Let us fix a point $w \in clD$. Using the Decomposition Theorem from Section 2.4 we represent a function $u(z)$ harmonic in D and continuous in clD ($u \in c_-(\partial D)$ in designation of Section 2.2) by the sum

$$u(z) = \sum_{k=1}^{n} u_k(z) + \sum_{k=1}^{n} A_k \log|z - z_k| + A, \tag{4.9.8}$$

where $u_m(z)$ is the real part of a function analytic in $D_m^- := ext L_m$, $u_m(w) = 0$, A_m and A are real constants,

$$\sum_{k=1}^{n} A_k = 0. \tag{4.9.9}$$

The function $u_m(z)$ is harmonic in D_m^- and continuous in clD_m^-. We define the operator $\mathbf{H}_m$ on L_m as an operator solving the Dirichlet problem for D_k^- corrected by subtraction of the solution at w, i.e., $\mathbf{H}_m f \in h_-(L_m)$ and

$$\mathbf{H}_m f = \frac{1}{2\pi} \int_{L_m} f(\tau) \left[\frac{\partial G_m(z, \tau)}{\partial \nu} - \frac{\partial G_m(w, \tau)}{\partial \nu} \right] d\sigma, \tag{4.9.10}$$

where G_m is the Green function of D_m^- (see Subsection 2.7.1), $f \in \mathcal{H}(L_m)$. We understand by z a point in clD_m^-, by τ a point on L_m. We can define $\mathbf{W}_{mk}$ in such a way that

$$\mathbf{W}_{mk}\mathbf{H}_k f = \frac{1}{2\pi} \int_{L_k} f \frac{\partial G_k}{\partial \nu} ds \tag{4.9.11}$$

acts from $C(L_k)$ to $C(L_m)$. It is a compact operator from $C(L_k)$ to $C(L_m)$, since the kernel of the integral operator (4.9.11) is continuous and bounded on $L_k \times L_m$. If $f(\tau) = u_m(\tau)$ then (4.9.10) implies

$$u_m(z) = \frac{1}{2\pi} \int_{L_m} u_m(\tau) \left[\frac{\partial G_m(z, \tau)}{\partial \nu} - \frac{\partial G_m(w, \tau)}{\partial \nu} \right] d\sigma, \tag{4.9.12}$$

since $u_m(w) = 0$. Applying (4.9.8) and (4.9.12) to (4.9.7) on each L_k we obtain

$$u_k(t) + \frac{1}{2\pi} \sum_{m \neq k} \int_{L_m} u_m(\tau) \left[\frac{\partial G_m(t, \tau)}{\partial \nu} - \frac{\partial G_m(w, \tau)}{\partial \nu} \right] d\sigma \tag{4.9.13}$$

$$= f(t) - \sum_{m=1}^{n} A_m \log |t - z_m| - A, \ t \in L_k, \ k = 1, 2, ..., n.$$

Let us harmonically continue (4.9.13) from L_k to D. Let $f_k(z)$, $g_{km}(z)$ be solutions of the Dirichlet problems

$$f_k(t) = f(t), \ g_{km}(t) = \log |t - z_m|, \ t \in L_k,$$

for the domain D_k^- ($k, m = 1, 2, ..., n$), for instance, $f_k(z) = \frac{1}{2\pi} \int_{L_k} f(\tau) \frac{\partial G_k(t,\tau)}{\partial n} d\sigma.$
Then (4.9.13) becomes

$$u_k(z) + \frac{1}{2\pi} \sum_{m \neq k} \int_{L_m} u_m(\tau) \left[\frac{\partial G_m(z,\tau)}{\partial \nu} - \frac{\partial G_m(w,\tau)}{\partial \nu} \right] d\sigma \quad (4.9.14)$$

$$= f_k(z) - \sum_{m=1}^{n} A_m g_{km}(z) - A, \ z \in cl D, \ k = 1, 2, ..., n.$$

Substituting $z = w$ in (4.9.14) we arrive at the equalities

$$\sum_{m=1}^{n} A_m g_{km}(w) + A = f_k(w), \ k = 1, 2, ..., n, \quad (4.9.15)$$

which simultaneously with (4.9.9) constitute a linear algebraic system with respect to A_m, A. The homogeneous system corresponding to (4.9.15), (4.9.9) has only a zero solution, because in the opposite case we get a non-zero function $u(z)$ from (4.9.8) satisfying the Dirichlet problem for D with the zero-th data. Hence, we may assume that the right-hand parts of (4.9.13) and (4.9.14) are uniquely defined. Following [157] let us prove

Theorem 4.20 *The homogeneous system*

$$u_k(t) + \frac{1}{2\pi} \sum_{m \neq k} \int_{L_m} u_m(\tau) \left[\frac{\partial G_m(t,\tau)}{\partial \nu} - \frac{\partial G_m(w,\tau)}{\partial \nu} \right] d\sigma = 0, \ t \in L_k, \ k = 1, 2, \ldots, n, \quad (4.9.16)$$

has only a zero solution.

Proof. Let $v_k(z)$ ($k = 1, 2, ..., n$) be a certain solution of (4.9.16). We have

$$v_k(z) + \frac{1}{2\pi} \sum_{m \neq k} \int_{L_m} v_m(\tau) \frac{\partial G_m(z,\tau)}{\partial n} d\sigma = c_k, \ z \in cl D, \ k = 1, 2, ..., n, \quad (4.9.17)$$

4.9. GENERALIZED METHOD OF SCHWARZ AND OTHER METHODS.

where $c_k = \sum_{m \neq k} v_m(w)$. Let us consider the function

$$v(z) = \sum_{k=1}^{n} v_k(z), \qquad (4.9.18)$$

harmonic in D and continuous in $cl D$ ($v \in c_+(\partial D)$). It is evident that the function conjugate to $v_k(z)$ is single-valued in D_k^-, since the domain D_k^- is simply connected. Hence, the function conjugate to $v(z)$ is single-valued in D. Formula (4.9.17) shows that the function $v(z)$ takes the constant values c_k on L_k ($k = 1, 2, \ldots, n$). One can consider this condition on $v(z)$ as the modified Dirichlet problem (see Subsection 2.7.2)

$$\Re w(t) = c_k, \ t \in L_k, \ k = 1, 2, \ldots, n, \qquad (4.9.19)$$

where $v(z) = \Re w(z)$, $w(z) \in \mathcal{C}_A(D)$. The problem (4.9.19) has only the constant solution (see Subsection 2.7.2), $v(z) = c$. Then (4.9.19) yields the equalities

$$v_k(z) + \sum_{m \neq k} v_m(z) = c, \ z \in cl D, \ k = 1, 2, \ldots, n,$$

which imply that $v_k(z)$ is harmonic inside L_k. But $v_k(z)$ is harmonic in D_k^- and continuous in $cl D_k^-$ by definition. Hence it appears that $v_k(z)$ is harmonic in $\widehat{\mathbb{C}}$. According to Liouville's theorem $v_k(z) = C_k$ where C_k is a constant ($k = 1, 2, \ldots, n$). Setting $v_k(z) = C_k$ in (4.9.17) we obtain $v_k(z) \equiv 0$.

The theorem is proved. ∎

By virtue of Fredholm's alternative the non-homogeneous system (4.9.13) or (4.9.14) has a unique solution. It is possible to show that the system of integral equation (4.9.14) for a circular domain is transformed into a system of functional equations as we have made it for a doubly connected domain in Section 4.9.1.

The aim of the method of addition theorems [81] is to construct an appropriate basic system for $ext L_k$ (disc, ellipse and so on) and to find a relation between this basic system to another basic system for $ext L_m$ ($m \neq k$). Such an approach leads to an infinite linear algebraic system corresponding to GMS. Let us sketch the main idea of addition theorems for the Dirichlet problem (4.9.7).

Let us consider the Hilbert space l^2 of the sequences $\alpha := (\alpha_1, \alpha_2, \ldots, \alpha_k, \ldots)$ with the norm

$$\|\alpha\| := \left(\sum_{k=1}^{\infty} |\alpha_k|^2 \right)^{1/2}$$

Let the functions $w_l^k(z)$ ($l = 1, 2, \ldots$) for fixed $k = 1, 2, \ldots, n$ belong to $h_-(L_k)$ and generate an orthogonal system in the Hilbert space $l_-^2(L_k)$. Recall that $g \in l_-^2(L_k)$, if

g is harmonic in $D_k^- = ext L_k$ and the limit boundary values of g belong to $\mathcal{L}^2(L_k)$, i.e., $\exists \, \|g\| := \left(\int_{L_k} |g(t)|^2 \, dt \right)^{1/2}$ (see Section 2.2). Each function $u_m(z)$ from $l_-^2(L_k)$ is represented in the form of the series

$$u_m(z) = \sum_{l=1}^{\infty} X_l^m \omega_l^m(z). \qquad (4.9.20)$$

Since the Hilbert spaces $l_-^2(L_k)$ and l^2 are isometric, the sequence X_l^m belongs to l^2. Let the function $f(t) \in l^2(L_k)$, then

$$f(t) = \sum_{l=1}^{\infty} f_l^k \omega_l^k(t), \; t \in L_k, \qquad (4.9.21)$$

where $f_l^k \in l^2$. In view of (4.9.20) and the representation

$$u(z) = \sum_{k=1}^{n} u_k(z), \; z \in cl D,$$

with $u_k \in l_-^2(L_k)$, the unknown function $u(z)$ is represented in the form

$$u(z) = \sum_{m=1}^{n} \sum_{l=1}^{\infty} X_l^m \omega_l^m(z).$$

It follows from (4.9.7) and (4.9.21) that

$$\sum_{l=1}^{\infty} X_l^m \omega_l^m(t) + \sum_{m \neq k} \sum_{l=1}^{\infty} X_l^m \omega_l^m(t) = \sum_{l=1}^{\infty} f_l^k \omega_l^k(t) \text{ on each } L_k.$$

The formula

$$\omega_l^m(t) = \sum_{p=1}^{\infty} \gamma_{pl}^{km} \omega_p^k(t), \; t \in L_k, \; m = 1, 2, ..., n, \; m \neq k, \qquad (4.9.22)$$

with known γ_{pl}^{km} is called an *addition theorem*. Using the addition theorem we obtain the relations

$$\sum_{l=1}^{\infty} X_l^m \omega_l^m(t) + \sum_{m \neq k} \sum_{p=1}^{\infty} X_l^m \gamma_{lp}^{km} \omega_l^k(t) = \sum_{l=1}^{\infty} f_l^k \omega_l^k(t) \text{ on each } L_k.$$

4.9. GENERALIZED METHOD OF SCHWARZ AND OTHER METHODS.

If the second series converges absolutely, then it is possible to change the order of summation.

$$\sum_{l=1}^{\infty} \left(X_l^k + \sum_{m \neq k} \sum_{p=1}^{\infty} \gamma_{lp}^{km} X_p^m \right) w_l^k(t) = \sum_{l=1}^{\infty} f_l^k w_l^k(t), \ k = 1, 2, ..., n. \qquad (4.9.23)$$

Since the system $w_l^k(t)$ is orthogonal, we obtain the following infinite set of linear algebraic equations

$$X_l^k + \sum_{m \neq k} \sum_{p=1}^{\infty} \gamma_{lp}^{km} X_p^m = f_l^k, \ l = 1, 2, ..., \ k = 1, 2, ..., n \qquad (4.9.24)$$

from (4.9.21).

The method of addition theorems is a discrete variant of GMS under the condition of absolute convergence of the corresponding series. Let us apply the method of successive approximations to the set (4.9.24). As the zero-order approximation we assume $\left(X_l^k\right)^0 = f_l^k$. The corresponding solution from $l_-^2(L_k)$

$$u_k^0(z) = \sum_{l=1}^{\infty} \left(X_l^k\right)^0 w_l^k(z) = \sum_{l=1}^{\infty} f_l^k w_l^k(z)$$

coincides with the zero-order approximation of GMS. Once the $s-1$ approximation $\left(X_l^k\right)^{s-1}$ has been constructed, then the s-th approximation is constructed by the formula (4.9.24). This coincides with the construction of the function $u_k^s(z)$ of GMS. If GMS converges, then the method of successive approximation convergence applied to (4.9.24) converges in l^2, since l^2 and $l_-^2(L_k)$ are isometric. Therefore, the following theorem holds

Theorem 4.21 *The method of successive approximations applied to (4.9.24) converges in l^2 if and only if the corresponding GMS converges in $\mathcal{L}^2(\cup_{k=1}^n L_k)$.*

4.9.3 The modified method of Schwarz.

We have described the classic method of Schwarz in the previous section. We now modify this method and prove the uniform convergence of the modified method for an arbitrary multiply connected domain D. We use the classic logarithmic potentials [157]: the potential of the double layer $\mathbf{Q}_k \mu(z) := \frac{1}{2\pi} \int_{L_k} \mu \frac{\cos(r,n)}{r} d\sigma$, and of simple layer $\mathbf{P}_k \mu(z) := \frac{1}{2\pi} \int_{L_k} \mu \log \frac{1}{r} d\sigma$ for which

$$\mathbf{P}_k^+ \mu(t) - \mathbf{P}_k^- \mu(t) = \mu(t), \quad \frac{\partial \mathbf{P}_k^+ \mu}{\partial n} = \frac{\partial \mathbf{P}_k^- \mu}{\partial n},$$

$$\mathbf{Q}_k^+\mu(t) = \mathbf{Q}_k^-\mu(t), \quad \frac{\partial \mathbf{Q}_k^+\mu}{\partial n} - \frac{\partial \mathbf{Q}_k^-\mu}{\partial n} = \mu(t), \ t \in L_k.$$

Here $\mu \in C^1(L_k)$, $\mathbf{P}_k^+\mu(t) := \lim_{z \to t \ z \in D_k} \mathbf{P}_k^+\mu(z)$ and so on.

Lemma 4.12 *Given a point $w \in D$ and a function $g \in \mathcal{H}(L_k)$ for which*

$$\int_{L_k} \frac{\partial g}{\partial n} ds = 0. \qquad (4.9.25)$$

The problem

$$\frac{\partial u_k}{\partial n} - \frac{\partial \mathbf{H}_k u_k}{\partial n} = g, \ t \in L_k, \ \mathbf{H}_k u_k(w) = 0, \qquad (4.9.26)$$

has a unique solution u_k in the space $h_+^1(D_k)$ (harmonic in D_k and continuously differentiable in $cl D_k$). Here the operator $\mathbf{H}_k$ is defined by (4.9.9).

Proof. According to the definition of $\mathbf{H}_k$ the function $u^-(z) := \mathbf{H}_k u_k(z)$ belongs to $C_-(L_k)$ and $u^- = u_k$ on L_k. Therefore, problem (4.9.26) can be written as the problem of conjugation

$$u^- = u_k, \quad \frac{\partial u_k}{\partial n} = \frac{\partial u^-}{\partial n} + g \text{ on } L_k, \ \mathbf{H}_k u_k(w) = 0. \qquad (4.9.27)$$

In terms of complex potentials (4.9.27) becomes

$$\varphi_k(t) = \varphi^-(t) + g_1(t), \ t \in L_k, \qquad (4.9.28)$$

$$\varphi^-(w) = 0, \qquad (4.9.29)$$

where $\varphi_k \in \mathcal{H}_+(L_k)$, $\varphi^- \in \mathcal{H}_-(L_k)$, $g_1(t)$ is such a function that $\frac{\partial g_1(t)}{\partial s} = g(t)$ on L_k. We have the general solution of the jump problem (4.9.28)

$$\frac{1}{2\pi i} \int_{L_k} \frac{g_1(\tau)}{\tau - z} d\tau + c = \begin{cases} \varphi_k(z), \ z \in D_k, \\ \varphi^-(z), \ z \in D_k^-, \end{cases}$$

where $u_k(z) = \Re \varphi_k(z)$, $u^-(z) = \Re \varphi^-(z)$. The constant $c = \int_{L_k} \frac{g_1(\tau)}{\tau - w} d\tau$ is determined by (4.9.29).

The lemma is proved. ∎

If the function g from (4.9.30) does not satisfy condition (4.9.25), then we can modify the problem introducing an undetermined constant c_k. Then we obtain

4.9. GENERALIZED METHOD OF SCHWARZ AND OTHER METHODS.

Lemma 4.13 *The problem*

$$\frac{\partial u_k}{\partial n} - \frac{\partial \mathbf{H}_k u_k}{\partial n} = g + c_k, \ t \in L_k, \ \mathbf{H}_k u_k(w) = 0, \tag{4.9.30}$$

has a unique solution u_k in the space $h^1_+(D_k)$ with $c_k = -\int_{L_k} \frac{\partial g}{\partial n} ds$.

We now proceed to modify GMS. For the definiteness we consider the Dirichlet problem (4.9.7). Let us know its solution u. Using u we construct the functions $u_k \in h^1_-(L_k)$ by the boundary value problem

$$\frac{\partial u_k}{\partial n} - \frac{\partial \mathbf{H}_k u_k}{\partial n} = \frac{\partial u}{\partial n} + A_k \text{ on } L_k, \ A_k := \int_{L_k} \frac{\partial u}{\partial n} ds, \ \mathbf{H}_k u_k(w) = 0, \tag{4.9.31}$$

for each fixed $k = 1, 2, \ldots, n$. By virtue of Lemma 4.12 u_k can be constructed uniquely.

Remark 4.9 *The functions u_k were harmonic in D_k^- in the classic method of Schwarz from the previous section. In our version of the method, u_k are harmonic in D_k. This is the crucial point of our modification. Formally it does not matter what integral equation appears. But one can see below that this insignificant modification leads to convergence of the method for an arbitrary multiply connected domain.*

In order to deduce a system of integral equations we define the function $U(z)$ as follows

$$U(z) := \begin{cases} u_k(z) + \sum_{m \neq k} \mathbf{H}_m u_m(z) + f^+(z) - \sum_{m \neq k} A_m F_m^+(z), \ z \in D_k, \\ \qquad k = 1, 2, \ldots, n, \\ u(z) + \sum_{m=1}^n \mathbf{H}_m u_m(z) + f^-(z) - \sum_{m=1}^n A_m F_m^-(z), \ z \in D. \end{cases}$$

Here $f^\pm(z) \in h^1_\pm(\partial D)$ is the potential of a simple layer along ∂D satisfying the relations

$$f^+(t) - f^-(t) = f(t), \ \frac{\partial f^+}{\partial n} = \frac{\partial f^-}{\partial n}. \tag{4.9.32}$$

We have

$$F_k^+(t) - F_k^-(t) = 0, \ \frac{\partial F_k^+}{\partial n} - \frac{\partial F_k^-}{\partial n} = 1 \text{ on } L_k, \text{ for fixed } k, \tag{4.9.33}$$

where $F_k^\pm(z) \in h^1_\pm(L_k)$. Let us calculate the jump Δ_k of U on each L_k

$$\Delta_k = u(t) + f^-(t) - u_k(t) - f^+(t) - A_k \left(F_k^+(t) - F_k^-(t) \right).$$

Using (4.9.27), (4.9.32), (4.9.33) and the properties of $\mathbf{H}_k$ we conclude that $\Delta_k = 0$. The jump Δ'_k of the normal derivative of U on each L_k is calculated as follows

$$\Delta'_k = \frac{\partial u}{\partial n} + \frac{\partial}{\partial n}\left(\mathbf{H}_k u_k(t)\right) - \frac{\partial u_k}{\partial n} + \frac{\partial f^-}{\partial n} - \frac{\partial f^+}{\partial n} - A_k\left(\frac{\partial F_k^+}{\partial n} - \frac{\partial F_k^-}{\partial n}\right).$$

Using (4.9.31), (4.9.32), (4.9.33) we obtain $\Delta'_k = 0$. Equalities $\Delta_k = \Delta'_k = 0$ are equivalent to the Analytic Continuation Principle for the corresponding complex potentials (see (2.12.1) with $\lambda_k = 1$ and (2.12.5) with $\rho_k = 1$). Hence, the function $U(z)$ is harmonic in $\hat{\mathbb{C}}$. It follows from Liouville's theorem that $U(z)$ is a constant which we denote by A. By the definition of $U(z)$ in D_k we have

$$u_k(z) = -\sum_{m \neq k} \mathbf{H}_m u_m(z) - f^+(z) + \sum_{m \neq k} A_m F_m^+(z) + A, \ z \in cl D_k, \ k = 1, 2, \ldots, n. \tag{4.9.34}$$

The relations (4.9.34) can be considered as a counterpart of the system of functional equations (4.9.4), (4.9.5). Moreover, (4.9.34) corresponds to (4.9.14) from the method of Schwarz in the classic form.

We have to find the functions $u_k \in h_+(L_k)$ and the real constants $A, A_1, A_2, \ldots, A_n$ satisfying the condition

$$\sum_{m=1}^{n} A_m = 0. \tag{4.9.35}$$

Moreover, $u_k(z)$ satisfy the additional conditions

$$\mathbf{H}_k u_k(w) = 0, \ k = 1, 2, \ldots, n. \tag{4.9.36}$$

At the beginning we solve (4.9.34) by the method of successive approximations. After we find $A, A_1, A_2, \ldots, A_n$ from (4.9.35), (4.9.36).

Theorem 4.22 *Given $f(z) \in h_+(L_k)$. The system*

$$u_k(z) = -\sum_{m \neq k} \mathbf{H}_m u_m(z) + f(z), \ z \in cl D_k, \ k = 1, 2, \ldots, n \tag{4.9.37}$$

with respect to $u_k \in h_+(L_k)$ has a unique solution. This solution can be found by the method of successive approximations.

Proof. According to Theorem 2.3 and Corollary 2.2 from Section 2.3 it is sufficiently to prove that the complex system

$$u_k(z) = -\lambda \sum_{m \neq k} \mathbf{H}_m u_m(z), \ z \in cl D_k, \ k = 1, 2, \ldots, n, \tag{4.9.38}$$

4.9. GENERALIZED METHOD OF SCHWARZ AND OTHER METHODS.

has only a zero solution. Let us introduce the function

$$u(z) = -\lambda \sum_{m=1} \mathbf{H}_m u_m(z), \quad z \in cl\, D. \tag{4.9.39}$$

The following boundary conditions hold true:

$$u(t) = (1-\lambda)\, u_k(t), \quad \frac{\partial u}{\partial n}(t) = \frac{\partial u_k}{\partial n} - \lambda \frac{\partial}{\partial n}\left(\mathbf{H}_k u_k\right)(t), \quad t \in L_k, \quad k=1,2,...,n. \tag{4.9.40}$$

Let $u = \psi_1 + i\psi_2$, $u_k = v_k + iw_k$ be the complex solution of (4.9.40), and let $\lambda = \nu + i\mu$ be a complex number satisfying the condition $|\lambda| \leq 1$. We multiply the first equality (4.9.40) by $\overline{u(t)} = (1-\overline{\lambda})\,\overline{u_k(t)}$, integrate it, and apply Green's formulae

$$\iint_G \left[\left(\frac{\partial V}{\partial x}\right)^2 + \left(\frac{\partial V}{\partial y}\right)^2\right] dx\,dy = \int_{\partial G} V \frac{\partial V}{\partial n} ds, \tag{4.9.41}$$

$$\int_{\partial G} \left(V_1 \frac{\partial V_2}{\partial n} - V_2 \frac{\partial V_2}{\partial n}\right) ds = 0, \quad V, V_1, V_2 \in h_+(\partial G),$$

for each D_k and sum from 1 to n. We obtain

$$-\int_{\partial D}\left(\psi_1 \frac{\partial \psi_1}{\partial n} + \psi_2 \frac{\partial \psi_2}{\partial n}\right) ds - i \int_{\partial D}\left(\psi_1 \frac{\partial \psi_2}{\partial n} - \psi_2 \frac{\partial \psi_1}{\partial n}\right) ds$$

$$= (1-\overline{\lambda}) \sum_{k=1}^{n} \left[\int_{L_k}\left(v_k \frac{\partial v_k}{\partial n} + w_k \frac{\partial w_k}{\partial n}\right) ds + i \int_{L_k}\left(v_k \frac{\partial w_k}{\partial n} - w_k \frac{\partial v_k}{\partial n}\right) ds\right]$$

$$-\lambda(1-\overline{\lambda}) \sum_{k=1}^{n} \left[\int_{L_k^-}\left(\mathbf{H}_k v_k \frac{\partial \mathbf{H}_k v_k}{\partial n} + \mathbf{H}_k w_k \frac{\partial \mathbf{H}_k w_k}{\partial n}\right) ds\right.$$

$$\left.+i \int_{L_k}\left(\mathbf{H}_k v_k \frac{\partial \mathbf{H}_k w_k}{\partial n} - \mathbf{H}_k w_k \frac{\partial \mathbf{H}_k v_k}{\partial n}\right) ds\right] = (1-\overline{\lambda}) \sum_{k=1}^{n} \int_{L_k}\left(v_k \frac{\partial v_k}{\partial n} + w_k \frac{\partial w_k}{\partial n}\right) ds$$

$$-\lambda(1-\overline{\lambda}) \sum_{k=1}^{n} \int_{L_k^-}\left(\mathbf{H}_k v_k \frac{\partial \mathbf{H}_k v_k}{\partial n} + \mathbf{H}_k w_k \frac{\partial \mathbf{H}_k w_k}{\partial n}\right) ds.$$

Let us note that the normal derivatives $\partial/\partial n$ have opposite signs on L_k and L_k^-. Using (4.9.41) we introduce the non-negative quantities

$$a := \mathbf{J}[D]\psi_1 + \mathbf{J}[D]\psi_2, \quad b := \sum_{k=1}^{n} \left(\mathbf{J}[D_k] v_k + \mathbf{J}[D_k] w_k\right),$$

$$c := \sum_{k=1}^{n} \left(\mathbf{J}[D_k] \mathbf{H}_k v_k + \mathbf{J}[D_k] \mathbf{H}_k w_k \right),$$

where

$$\mathbf{J}[G]V := \int_{\partial G} V \frac{\partial V}{\partial n} ds = \int\int_G \left[\left(\frac{\partial V}{\partial x} \right)^2 + \left(\frac{\partial V}{\partial y} \right)^2 \right] dx dy.$$

Then we have

$$a + (1-\nu)b + (|\lambda|^2 - \nu)c = 0, \; \mu(b-c) = 0.$$

For $\lambda \neq 1$ these equalities are possible only if $a = b = c = 0$. This implies that u and u_k are constants. Applying (4.9.36) we conclude that $u(z) \equiv u_k(z) \equiv 0$ by (4.9.39). If $\lambda = 1$, then the second equality (4.9.40) and Lemma 4.12 imply that $u_k(z) \equiv 0$.

This proves the theorem. ∎

The desired function $u(z)$ has the form

$$u(z) = -\sum_{m=1}^{n} \mathbf{H}_m u_m(z) - f^-(z) + \sum_{m=1}^{n} A_m F_m^-(z) + A, \; z \in D, \qquad (4.9.42)$$

where $u_m(z)$ are found from (4.9.34) by the method of successive approximations; A_m and A are found from the finite set of linear algebraic equations (4.9.35), (4.9.36).

4.10 Notes and Comments.

4.1.° The topic *Iterative functional equations* in classes of analytic functions is a traditional branch of complex analysis. One can find the general theory of the simple functional equations (4.1.9) and others in [46, 58, 132, 161, 162, 167, 172, 179, 204, 248] and references cited therein. There are many different methods to solve (4.1.9). Following [132] we apply the method of successive approximations in Theorem 4.5.

4.2.° In Section 4.2.1 we follow the results of L. G. Mikhajlov [156]. The difference between Section 4.2.1 and [156] is in the choice of spaces: Theorems 4.6 and 4.7 were proved by L. G. Mikhajlov [156] for the space $\mathcal{L}_p(L)$. He also used another form of boundary condition (4.2.1), namely

$$(\mathbf{W}_1 \overline{\psi^-})(t) = b(t)\overline{\psi^-(t)} + (\mathbf{R}\overline{\psi^-})(t),$$

with $\mathbf{R}$ being a nonlinear operator.

Estimates and exact formulae for norms of the singular integral along the unit circle in the Hölder spaces were calculated in [19] (see also Section 2.13).

4.10. NOTES AND COMMENTS.

M. Kuczma considered the functional equation (4.1.11), when the coefficient $G(z)$ has a pole at z_0. He showed [132, p.164] that equation (4.1.11) may have no solution in $\mathcal{C}_A(\mathbb{U})$ and may even have no local solution analytic near z_0. The complete solution of (4.1.11) with meromorphic coefficients has been given in [126], [161]. The local functional equation

$$z^\chi \phi(z) = \sum_{k=1}^{n} G_k(z)\phi(s_k z) + g(z),$$

with $\chi > 0$ and $|s_k| < 1$, has been solved in [204]. We apply the method derived in [174, 204] to equation (4.2.17).

4.3 – 4.4.° Here we follow [188] and [201]. There is a remarkable relationship between the Green function and the Bergman kernel function $K(z, \zeta)$:

$$K(z, \overline{\zeta}) = -\frac{2}{\pi} \frac{\partial^2 G(z, \zeta)}{\partial z \partial \overline{\zeta}}.$$

See [37, p.60]. Therefore, we can construct $K(z, \overline{\zeta})$ for a multiply connected circular domain in closed form. There are other important functions inherent to a domain, for instance, Robin's function [257]. We think that they can be constructed by the method of functional equations for circular domains in closed form.

4.5.° Following [188] and [201] we solve the Riemann–Hilbert problem for the unbounded domain $\mathbb{D} = \widehat{\mathbb{C}} \setminus \cup_{k=0}^{n} cl\mathbb{D}_k$ by the method of functional equations. One can see here and it was known (see Chapter 1 and [91]) that the connectivity of the domain n is an important parameter of the problem as well as the index of the problem χ.

4.6.° H. Poincaré [228] created the series (4.6.1) and investigated its absolute convergence using algebraic-geometric ideas. When $q = 1$ the series (4.6.1) can be either absolutely convergent or absolutely divergent. It depends on the properties of $\mathcal{G}$. The main method of the proof of absolute convergence is concluded in the estimation of the θ_2-series (4.6.1) by the number series $\sum_{j=1}^{\infty} |c_j|^{-2}$, where c_j is a coefficient of $\gamma_j(z)$. The c_j are the radii of the so-called isometric circles $|c_j z + d_j| = 1$. See [84] and [138]. If the series $\sum_{j=1}^{\infty} |c_j|^{-2}$ converges, then (4.6.1) for $q = 1$ converges absolutely and uniformly in each compact subset of B. Absolute convergence immediately yields the automorphic property of the Poincaré θ_2-series, since it is possible to change the order of summation. Necessary and sufficient conditions for absolute and uniform convergence of the series have been found in [7], [8] in terms of the Hausdorff dimension of $\Lambda(\mathcal{G})$. This result is based on the study of the series $\sum_{j=1}^{\infty} |c_j|^{-2}$.

Let us note that absolute and uniform convergence was not studied separately in the previous works [7, 8, 84, 228]. Following [200] we discuss only uniform convergence.

This approach allows us to extend knowledge about the Poincaré series.

Historically such series were the crucial point in construction of the harmonic measures and the Schwarz operator.

4.7.° This result is new. We want to stress a difference between the methods for such mixed problems used in Section 4.7 and Section 3.7.

4.8.° Here we follow [202]. For collection see Chapter 1.

4.9.° If we apply the method of successive approximations to (4.9.14) then we obtain GMS for the multiply connected domain D. A natural question one asks is the following. Does this method converge? S. G. Mikhlin [157] took the point $w = \infty$ and proved convergence of GMS for an arbitrary doubly connected domain and for multiply connected domains, when the curves L_k are sufficiently far from each other. In the latest case it is easily seen that the kernel of the system (4.9.14) is small and the method of successive approximations can be applied. As a result S. G. Mikhlin has had absolutely convergent GMS. G. M. Goluzin [102] proved absolute convergence for an arbitrary triply connected circular domain. There are other works devoted to applied problems, where GMS is formally applied (see, for instance, [246]).

The formula (4.9.42) for $u(z)$ is similar to (4.9.2) from the method of functional equations. However, $u_m(z)$ now involves integrals instead of the compositions of the functions. Moreover, these integrals involve the Green functions for D_m^- in kernel. To construct such a Green function is a separate problem. Therefore, we do not insist here that the Dirichlet problem (4.9.7) is solved in closed form. Roughly speaking the final result can be presented as follows. If the solution of the Dirichlet problem for each domain D_m^- is constructed, then we propose a simple algorithm based on successive approximations to construct the solution of the Dirichlet problem for the domain D. The results of Subsection 4.9.3 are based on [197, 198].

Following [189] we formulate the method of addition theorems in Subsection 4.9.2 as a discrete form of GMS. In general the aim of the method of addition theorems is to construct an appropriate system $\omega_l^k(z)$ and to determine the coefficients γ_{pl}^{km} in (4.9.22). We do not study this constructive question here and refer to [81]. We can only verify that it is possible to apply addition theorems in such a way that the corresponding infinite set of linear algebraic equations can be solved by convergent successive approximations. We also show a way how to do it. This way is based on the modified GMS, because of Theorem 4.21.

The method of functional equation coincides with the modified GMS for circular domains. These methods are also interpreted as the classic method of images [40]. The method of images, when used in combination with addition theorems, allows us to calculate the induced electrostatic moments on a set of dielectric cylinders (discs in the two-dimensional interpretation). There are decompositions algorithms and

4.10. NOTES AND COMMENTS.

multigrid methods applied for numerical solution to the partial differential equations [252]. Decomposition algorithms can be considered as preconditioning methods when a large linear system is divided into smaller problems whose solutions are used to produce a preconditioner.

We do not include in the book results concerning boundary value problems on Riemann surfaces beginning from the seminal paper [294], where E. I. Zverovich has solved the $\mathbb{C}$-linear problem on an arbitrary closed Riemann surface in terms of the principal functionals. Let us consider a particular case of the Riemann surface, the double $\mathcal{D}$ of a multiply connected domain D. The principal functionals of $\mathcal{D}$ are expressed by harmonic measures constructed in Section 4.4. Hence, we can solve a $\mathbb{C}$-linear problem on $\mathcal{D}$. It was done in [196] by reducing to the problem on the complex plane

$$\varphi(t) = G(t)\overline{\psi(t)} + g(t), \ t \in \partial D,$$

first considered by B. Bojarski [42]. If $|G(t)| = 1$ and $\overline{G(t)}g(t) + \overline{g(t)} = 0$ on ∂D, then $\omega(z) := \frac{1}{2}(\varphi(z) + \psi(z))$, $z \in cl D$, solves the Riemann–Hilbert problem

$$\omega(t) = G(t)\overline{\omega(t)} + g(t), \ t \in \partial D.$$

The principal functionals of certain Riemann surfaces generalized the double have been constructed in [199].

Chapter 5

Nonlinear Problems of Mechanics

Methods of complex analysis constitute one of the powerful tools of continuum mechanics. Classic literature on the subject deals mainly with the linear models. Nowadays there are appearent results concerning nonlinear models of mechanics (theoretical as well as numerical). For details see references in Chapter 1.

Here we describe some of the nonlinear type problems. Main attention is paid to the constructive analysis of the composite materials with a finite number of inclusions.

5.1 Steady heat conduction of nonlinear composite materials

In this section we consider the two-dimensional steady heat conduction described by the quasi-linear equations

$$\nabla \mathbf{q} = 0, \quad \mathbf{q} = \lambda(T)\nabla T, \tag{5.1.1}$$

where $\nabla := (\partial/\partial x, \partial/\partial y)$, $T = T(x,y)$ is the temperature distribution on the plane (x,y), $\mathbf{q} = \mathbf{q}(x,y)$ is the heat flux, $\lambda(T)$ is the thermal conductivity in general depending on T. We assume that $\lambda(T)$ is a continuous positive function on $[0, +\infty)$, $\lambda(0) := \lim_{T \to +0} \lambda(T) > 0$. The discontinuity set of $\lambda(T)$ is finite. These assumptions on $\lambda(T)$ are of a physical nature of the heat conduction. The relations (5.1.1) can be written as equation

$$\nabla [\lambda(T)\nabla T] = 0. \tag{5.1.2}$$

Remark 5.1 *If $\lambda(T) = $ constant, then (5.1.2) becomes the Laplace equation with respect to the potential T. This case refered to as the well studied branch of the applications of the complex analysis (see Section 2.12).*

By virtue of the properties of $\lambda(\xi)$ we conclude that the function

$$f(\xi) := \int_0^\xi \lambda(\xi)d\xi \qquad (5.1.3)$$

satisfies the following conditions:
i) $f : [0, +\infty) \to [0, +\infty)$,
ii) f is an increasing function,
iii) $f(0) = 0$,
iv) $f(\xi)$ is a continuous piece-wise smooth function on $[0, +\infty]$ with finite possible points of discontinuity.

One can see that there exists an inverse function f^{-1} satisfying the same properties i) - iv).

Different boundary conditions can be posed for equation (5.1.2). Let D be a domain in $\mathbb{C}$ (simply or multiply connected), $L := \partial D$. One can consider two classical boundary value problems for (5.1.2).

Dirichlet problem. Given a continuous function $g(x,y)$ on L, find a function $T(x,y)$ satisfying (5.1.2) $((x,y) \in D)$, continuous in $cl D$ with the boundary condition

$$T = g. \qquad (5.1.4)$$

Neumann problem. Given a continuous function $g(x,y)$ on L, find a function $T(x,y)$ satisfying (5.1.2) $((x,y) \in D)$, continuously differentiable in $cl D$ with the boundary condition

$$\frac{\partial T}{\partial n} = g \qquad (5.1.5)$$

Let us generalize now these problems. Let L consist of mutually disjoint simple, smooth, closed curves L_k $(k = 1, 2, ..., n)$ dividing the complex plane $\widehat{\mathbb{C}}$ into two domains D^+ and D^-. The domain D^- is a union of the simply connected disjoint domains D_k bounded by L_k, and the domain D^+ contains the infinite point (see Figure 2.2). The chosen orientation on L_k (called "positive") leaves D_k to the left.

We consider the following *Conjugation problem.* Given a Hölder continuous function $g_2(x,y)$ on L $(g_2 \in \mathcal{H}(L))$, find two functions $T^+(x,y)$, $T^-(x,y)$ such that
1) $T^{\pm}(x,y)$ satisfy (5.1.2) in domains $D^{\pm}$ respectively,
2) $T^{\pm}(x,y)$ is continuously differentiable in the closures of the domains considered, satisfying the boundary conditions

$$T^+ = T^-, \quad \lambda\left(T^+\right)\frac{\partial T^+}{\partial n} = \lambda_k\left(T^-\right)\frac{\partial T^-}{\partial n} + g_2 \text{ on } L, \qquad (5.1.6)$$

5.1. STEADY HEAT CONDUCTION. NONLINEAR COMPOSITES.

where the function λ_k has the same properties as λ.

Let us now transform (5.1.2) to the Laplace equation, and see what kind of problems we have to consider for it instead of (5.1.4), (5.1.5), and (5.1.6). For this purpose we introduce the new pair of unknown functions $u^\pm(x,y) := f(T^\pm(x,y))$ with $f(\xi)$ defined by (5.1.3). Then the equation (5.1.2) becomes the Laplace equation

$$\Delta u = 0, (x,y) \in D^+ \cup D^-. \tag{5.1.7}$$

At the same time the boundary condition (5.1.4) is transformed to

$$u = f(g(x,y)) \text{ on } L. \tag{5.1.8}$$

Therefore the quasi-linear problem (5.1.2), (5.1.4) has been reduced to the linear problem (5.1.7), (5.1.8). Along similar lines the quasi-linear Neumann problem (5.1.2), (5.1.5) is reduced to a linear one.

We now proceed to study the conjugation problem (5.1.2), (5.1.6). Without loss of generality one can assume that $\lambda(0) = 1$. Introduce the functions $f_k(\xi) := \int_0^\xi \lambda_k(\xi)d\xi$ having the same properties i) - iv) as f. Put $\mu_k := \lambda_k(0) > 0$. The function $F_k(\xi) := f \circ f_k^{-1}(\mu_k \xi)$ satisfies the properties i) - iv). It follows immediately from the relation $F_k(f_k(\mu_k \xi)) := f(\xi)$.

Introduce the functions

$$\mu(x,y) := \begin{cases} \mu^+(x,y) := 1, \ (x,y) \in D^+, \\ \mu_k^-(x,y) := \mu_k, \ (x,y) \in cl D_k, \ k = 1, 2, \ldots, n, \end{cases}$$

$$F(x,y,\xi) := \begin{cases} F^+(x,y,\xi) := F(\xi), \ (x,y) \in D^+, \\ F_k^-(x,y,\xi) := F_k(\xi), \ (x,y) \in cl D_k, \ k = 1, 2, \ldots, n, \end{cases}$$

and the new unknown functions

$$u^\pm(x,y) := \begin{cases} f(T(x,y)), \ (x,y) \in D^+, \\ \mu_k^{-1} f_k(T(x,y)), \ (x,y) \in cl D_k, \ k = 1, 2, \ldots, n. \end{cases}$$

For the functions $u^\pm$ satisfying the Laplace equation (5.1.7) in $D^\pm$ respectively we consider the following conjugation problem

$$u^+ = F(t, u^-) + g_1, \quad \frac{\partial u^+}{\partial n} = \mu \frac{\partial u^-}{\partial n} + g_2, \text{ on } L, \tag{5.1.9}$$

where $F(t, u^-(t)) := \lim_{z \to t} F_k(u(z))$, $t \in L_k$ ($k = 1, 2, \ldots, n$); $g_1, g_2 \in \mathcal{H}(L)$. As usual we put $z = x + iy$ and identify, for instance, the functions $F(x, y, \xi)$ and $F(z, \xi)$. Problem (5.1.6) is reduced to problem (5.1.9) with $g_1 = 0$. The reason for the appearance of g_1 will be explained at the end of the section.

Since $u^\pm$ are harmonic in D^+ and D_k, we have

$$\int_L \frac{\partial u^+}{\partial n} ds = 0 \text{ and } \int_{L_k} \frac{\partial u^-}{\partial n} ds = 0, \ k = 1, 2, \ldots, n.$$

Using the second relation (5.1.9) we obtain the necessary solvability condition on g_2:

$$\int_L g_2(t) ds = 0. \tag{5.1.10}$$

This relation has to be valid even for the initial statement of problem (5.1.6) too.

Introduce two complex potentials $\phi^+(z) := u^+ + iv^+$ in $cl D^+$ and $\phi^-(z) := u^- + iv^-$ in $cl D_k$, where the unknown function $v^\pm$ are conjugate harmonic function to $u^\pm$. The functions v^+ is in general multi-valued in D^+. Denoting by A_k the period of $\frac{1}{2\pi i}\phi^+(t)$ with respect to L_k, we have

$$A_k := \frac{1}{2\pi i} \int_{L_k} d\phi^+(t) = \frac{1}{2\pi} \int_{L_k} dv^+(t), \ k = 1, 2, \ldots, n. \tag{5.1.11}$$

Using the Cauchy–Riemann equations we can write the second condition (5.1.9) in the form

$$\frac{\partial v^+}{\partial s} = \mu \frac{\partial v^-}{\partial s} + g_2,$$

where s is the natural parameter of L_k. It follows from the relation $\int_{L_k} dv^-(t) = 0$ that $\int_{L_k} dv^+(t) = \int_{L_k} g_2(t) ds$. Then (5.1.11) yields

$$A_k = \frac{1}{2\pi} \int_{L_k} g_2(t) ds. \tag{5.1.12}$$

Therefore, according to the Decomposition Theorem $\phi^+(z)$ in D^+ is represented in the form

$$\phi^+(z) = u^+ + i\tilde{v}^+ + \sum_{k=1}^n A_k \log(z - z_k) + i\gamma,$$

where z_k is an arbitrary point from D_k, A_k has the form (5.1.12), and $\tilde{v}^+$ is a single-valued function harmonic in D^+. One can assume that $\gamma = 0$, since a complex

5.1. STEADY HEAT CONDUCTION. NONLINEAR COMPOSITES.

potential is defined up to an arbitrary additive constant. Moreover, (5.1.10) and (5.1.12) imply that $\sum_{k=1}^{n} A_k = 0$. Hence, $\phi^+(z)$ is bounded at infinity.

Thus, conditions (5.1.9) can be written as (see Section 2.12 for details)

$$\Re\phi^+(t) = F(t, \Re\phi^-(t)) + g_1, \quad \Im\phi^+(t) = \mu \operatorname{Im} \phi^-(t) + g_3(t), \quad t \in L, \qquad (5.1.13)$$

where $g_3(t)$ is a primitive of g_2. One can combine two real relations (5.1.13) into a complex one

$$\phi^+(t) = F\left[t, \frac{1}{2}\left(\phi^-(t) + \overline{\phi^-(t)}\right)\right] + \frac{\mu(t)}{2}\left(\phi^-(t) - \overline{\phi^-(t)}\right) + g_4(t), \quad t \in L,$$

where $g_4(t) := g_1(t) + ig_3(t)$. Subtracting $\frac{1}{2}(\mu(t) + 1)\phi^-(t)$ from the parts of the last equality we obtain

$$\begin{aligned}\phi^+(t) - \tfrac{1}{2}(\mu(t)+1)\phi^-(t) &= F\left[t, \tfrac{1}{2}\left(\phi^-(t) + \overline{\phi^-(t)}\right)\right] \\ &\quad - \tfrac{1}{2}\phi^-(t) - \tfrac{\mu(t)}{2}\overline{\phi^-(t)} + g_4(t), \quad t \in L.\end{aligned} \qquad (5.1.14)$$

It is convenient to introduce new unknown function

$$\omega(z) := \frac{1}{2}(\mu(z)+1)\phi^-(z) \text{ in } D^- = \bigcup_{k=1}^{n} D_k.$$

Then (5.1.14) becomes

$$\begin{aligned}\phi^+(t) - \omega(t) &= F\left[t, \tfrac{1-\rho(t)}{2}\left(\omega(t) + \overline{\omega(t)}\right)\right] \\ &\quad - \tfrac{1-\rho(t)}{2}\omega(t) - \tfrac{1+\rho(t)}{2}\overline{\omega(t)} + g_4(t), \quad t \in L,\end{aligned} \qquad (5.1.15)$$

where $\rho(z) := \rho_k$ in clD_k, $\rho_k := (\mu_k - 1)(\mu_k + 1)^{-1}$. Let us represent the function $F(\xi)$ in the form

$$F(t, \xi) = \xi + G(t, \xi), \qquad (5.1.16)$$

where the function $G(t, \xi)$ satisfies the following property

$$|G(t, \xi_1) - G(t, \xi_2)| \leq \gamma_0 |\xi_1 - \xi_2|, \text{ for all } \xi_1, \xi_2 \geq 0, \ t \in L, \qquad (5.1.17)$$

and γ_0 is a positive constant.

Let us consider the projectors $\mathbf{P} := \frac{1}{2}\mathbf{I} + \frac{1}{2}\mathbf{S}$ and $\mathbf{Q} := \frac{1}{2}\mathbf{I} - \frac{1}{2}\mathbf{S}$, where $\mathbf{I}$ is the identity operator, $\mathbf{S}$ is the singular integral operator on L (see Section 2.6). Applying $\mathbf{Q}$ to (5.1.15) and using (5.1.16) we arrive at the singular integral equation

$$\omega = \mathbf{Q}\rho\overline{\omega} - \mathbf{Q}G\left[t, \frac{1-\rho}{2}(\omega + \overline{\omega})\right] + h \text{ on } L, \qquad (5.1.18)$$

with respect to $\omega \in \mathcal{H}(L)$. Here $h := \mathbf{Q}g_4$.

Theorem 5.1 *If*

$$\|\mathbf{Q}\|(\rho + (1-\rho)\gamma_0) \leq q < 1, \qquad (5.1.19)$$

then equation (5.1.18) has a unique solution in a space of the Hölder continuous functions. This solution can be found by the method of successive approximations.

Proof is based on the Banach theorem (see Section 2.3). We have

$$\left\|\rho\mathbf{Q}\overline{z_1} + \mathbf{Q}G(t,(1-\rho)(z_1+\overline{z_1})) - \rho\mathbf{Q}\overline{z_2} + \mathbf{Q}G\left(t, \frac{1-\rho}{2}(z_2+\overline{z_2})\right)\right\|$$

$$\leq \|\mathbf{Q}\|\left(\rho\|\overline{z_1} - \overline{z_2}\| + (1-\rho)\gamma_0\|\Re(z_1 - z_2)\|\right)$$

$$\leq \|\mathbf{Q}\|(\rho + (1-\rho)\gamma_0)\|z_1 - z_2\|.$$

If (5.1.19) holds then the Banach theorem can be applied to equation (5.1.18), and it has a unique solution.

The theorem is proved. ■

The integral equation (5.1.18) has been constructed with respect to the function ω analytic in D^-. We now construct a similar equation with respect to the function ϕ^+ analytic in D^+. As in the previous case we begin with the conjugation condition (5.1.6), which can be written in the following form

$$u^- = F^{-1}(t, u^+), \quad \frac{\partial u^-}{\partial n} = \mu^{-1}\frac{\partial u^+}{\partial n} + \mu^{-1}g_2, \ t \in L, \qquad (5.1.20)$$

where F^{-1} is inverse to $F(t,\xi)$ with respect to ξ. The condition (5.1.9) implies equation (5.1.14). Similar arguments yield

$$\phi^-(t) - \tfrac{1}{2}(\mu^{-1}+1)\phi^+(t) = F^{-1}\left[t, \tfrac{1}{2}\left(\phi^+(t) + \overline{\phi^+(t)}\right)\right]$$

$$- \tfrac{1}{2}\phi^+(t) - \tfrac{\mu^{-1}}{2}\overline{\phi^+(t)} + i\mu^{-1}g_3(t), \ t \in L, \qquad (5.1.21)$$

5.1. STEADY HEAT CONDUCTION. NONLINEAR COMPOSITES.

from (5.1.20). Let us represent the function $F(\xi)$ in the form

$$F^{-1}(t,\xi) = \xi - H(t,\xi), \tag{5.1.22}$$

where the function $H(t,\xi)$ satisfies the property (5.1.17). Then (5.1.21) becomes

$$\frac{2}{\mu^{-1}+1}\phi^-(t) - \phi^+(t) = -\frac{\mu^{-1}-1}{\mu^{-1}+1}\overline{\phi^+(t)}$$

$$-\frac{2}{\mu^{-1}+1}H\left[t, \frac{1}{2}\left(\phi^+(t) + \overline{\phi^+(t)}\right)\right] + \frac{2ig_3(t)}{1+\mu}, \ t \in L. \tag{5.1.23}$$

Let us note that the function

$$\frac{2}{\mu^{-1}+1}\phi^-(t) = \frac{2\mu_k}{1+\mu_k}\phi_k(t), \ t \in D_k$$

can be analytically continued into D_k for each $k = 1, 2, \ldots, n$. Thus $\mathbf{P}\left(\frac{2}{\mu^{-1}+1}\phi^-\right) = 0$, where $\mathbf{P} := \frac{1}{2}\mathbf{I} + \frac{1}{2}\mathbf{S}$ is the standard projector, and we have from (5.1.23) the following singular integral equation

$$\phi^+ = \mathbf{P}\rho\overline{\phi^+} - \mathbf{P}(1+\rho)G\left[t, \frac{1}{2}(\phi^+ + \overline{\phi^+})\right] + h \text{ on } L, \tag{5.1.24}$$

where $h := \mathbf{P}\frac{2ig_3(t)}{1+\mu}$.

The same arguments as in Theorem 5.1 give us the following result

Theorem 5.2 *If*

$$\|\mathbf{P}\|\left(\rho + (1+\rho)\gamma\right) \le q < 1,$$

then equation (5.1.24) has a unique solution in a space of the Hölder continuous functions. This solution can be found by the method of successive approximations.

We now proceed to explain which way the model problem (5.1.6) can be applied to the determination of the temperature distribution of plane composite materials. Consider a finite plane domain G, $\partial G =: L_0$ divided into domains D^+ and D_k, $(k = 1, 2, \ldots, n)$ as it is represented in Figure 2.1. Suppose that the domains D^+ and D_k are occupied by isotropic materials with different conductivities $\lambda(T)$ and $\lambda_k(T)$, respectively, where $T = T(x, y)$ is the temperature distribution in G. Let the temperature distribution be known on L_0 :

$$T = g \text{ on } L_0, \tag{5.1.25}$$

where g is a given Hölder continuous function. If we assume that the thermal contact between D^+ and $D_k (k = 1, 2, \ldots, n)$ is perfect, then the limit values of the temperature and the normal component to L_k of the heat flux have to be equal:

$$T^+ = T^-, \quad \mathbf{q}_n^+ = \mathbf{q}_n^- \text{ on } L_k, \; k = 1, 2, \ldots, n.$$

Using (5.1.1) we obtain the conjugation conditions

$$T^+ = T^-, \; \lambda \left(T^+ \right) \frac{\partial T^+}{\partial n} = \lambda_k \left(T^- \right) \frac{\partial T^-}{\partial n} \text{ on } L_k, \; k = 1, 2, \ldots, n. \tag{5.1.26}$$

So we arrive at the mixed problem (5.1.25), (5.1.26) for the nonlinear equation (5.1.6). Let us reduce this problem to problem (5.1.9).

Similarly one can reduce problem (5.1.25), (5.1.26) to the following one

$$u = \widetilde{g} \text{ on } L_0, \; u^+ = F(t, u^-), \; \mu \frac{\partial u^-}{\partial n} = \frac{\partial u^+}{\partial n} \text{ on } L_k, \; k = 1, 2, \ldots, n, \tag{5.1.27}$$

with respect to the function u sectionally harmonic in G. Let $z = z(w)$ be a conformal mapping of the unit disk $\mathbb{U}$ onto G. Then problem (5.1.27) transforms to the same problem with $L_0 = \mathbb{T}$. Let us consider the auxiliary Dirichlet problem $u_0 = \widetilde{g}$ on $\mathbb{T}$ with respect to the function u_0 harmonic in $\mathbb{U}$ and continuous on $cl\mathbb{U}$. The function u_0 is written in closed form by (2.7.2). Introduce the function $U(z) := u(z) - u_0(z)$ in $cl\mathbb{D}$ and $U(z) := -u(z^*) + u_0(z^*)$ in $cl\mathbb{D}^*$, where $z^* := 1/\overline{z}$, $(D^+)^*$ is symmetric to D^+ with respect to $\mathbb{T}$. The function $U(z)$ is harmonic in $D^+ \cup \mathbb{T} \cup (D^+)^*$, since $U = 0$ on $\mathbb{T}$. We can also continue the conjugation condition (5.1.27) from L_k to $(L_k)^*$. By result we arrive at the problem

$$U^+ = F(u^-) - u_0, \; \mu \frac{\partial u^-}{\partial n} = \frac{\partial u^+}{\partial n} - \frac{\partial u_0}{\partial n} \text{ on } L_k \text{ and } (L_k)^*, \; k = 1, 2, \ldots, n,$$

which is precisely a problem of type (5.1.9).

5.2 Linearized problem.

The linear part of the operator from (5.1.18) is $\mathbf{Q}\rho\overline{\omega}$, since $G(x) = o(x)$, as $x \to +0$. Solution to the linearized equation $\omega = \mathbf{Q}\rho\overline{\omega} + h$ is important, because the linear equation is the benchmark of the Newton method to solve the nonlinear equation (5.1.18). We are not discussing here the Newton method described in [114, p. 679]. In the present section we solve the $\mathbb{R}$-linear problem for a multiply connected circular domain, a mixed problem for a multiply connected circular domain and a mixed problem for a multiply connected symmetric circular domain. All these problems can be considered as special problems of the mechanics of composite materials.

5.2.1 Linearized problem for circular domains.

In the present section we solve a linearized problem corresponding to (5.1.18) in the case of the circular domain $D^+ = \mathbb{D}$.

Given a function $h(t)$, Hölder continuous on $\mathbb{T}_k$, and constants ρ_k ($k = 1, 2, \ldots, n$), find the functions $\phi(z)$, $\phi_k(z)$ analytic in $\mathbb{D}$, $\mathbb{D}_k$, respectively, continuous in the closures of the domains considered with the $\mathbb{R}$- linear conjugation condition (see Section 2.12)

$$\phi(t) = \phi_k(t) - \rho_k \overline{\phi_k(t)} - h(t), \quad |t - a_k| = r_k, \quad k = 1, 2, \ldots, n. \quad (5.2.1)$$

Problem (5.2.1) corresponds to the linearized equation $\omega = \mathbf{Q}\rho\overline{\omega} + h$, when $\rho = \rho_k$ on $|t - a_k| = r_k$, $\omega(z) = \phi_k(z)$ in $|z - a_k| \leq r_k$. We will assume that $-1 < \rho_k < 1$. This corresponds to the most interesting case from the physical point of view.

We follow Section 4.4 in order to reduce problem (5.2.1) to a system of functional equations. First represent $h(t)$ on $L = \cup_{k=1}^n \mathbb{T}_k$ as the difference $h(t) = h^+(t) - h^-(t)$, $t \in L$, by the Sokhotsky–Plemelj formulae (see Section 2.6). Here $h^\pm \in \mathcal{C}_\pm(L)$ and it is convenient to put $h^-(w) = 0$ instead of $h^-(\infty) = 0$. As usual w is a fixed point of $\mathbb{D}\setminus\{\infty\}$.

Introduce the function

$$\Phi(z) := \begin{cases} \phi_k(z) + \sum_{m \neq k} \rho_m \overline{\left[\phi_m\left(z^*_{(m)}\right) - \phi_m\left(w^*_{(m)}\right)\right]} - \rho_k \overline{\phi_k\left(w^*_{(k)}\right)} - h^+(z), \\ \qquad |z - a_k| \leq r_k, \\ \phi(z) + \sum_{m=0}^n \rho_m \overline{\left[\phi_m\left(z^*_{(m)}\right) - \phi_m\left(w^*_{(m)}\right)\right]} - h^-(z), \quad z \in \mathbb{D}. \end{cases}$$

Here the sum $\sum_{m \neq k}$ contains the terms $m = 1, 2, \ldots, n$; $m \neq k$. Taking into account the Analytic Continuation Principle and Liouville's theorem we conclude that $\Phi(z)$ is a constant: $\Phi(z) = Q - \sum_{m=0}^n \rho_m \overline{\phi_m\left(w^*_{(m)}\right)}$, where $Q = \phi(w)$. The definition of $\Phi(z)$ yields the system of functional equations

$$\phi_k(z) = -\sum_{m \neq k} \rho_m \overline{\left[\phi_m\left(z^*_{(m)}\right) - \phi_m\left(w^*_{(m)}\right)\right]} + \rho_k \overline{\phi_k\left(w^*_{(k)}\right)} + f_k(z) + Q, \quad (5.2.2)$$

$$|z - a_k| \leq r_k, \quad k = 0, 1, \ldots, n,$$

with respect to the functions $\phi_k(z) \in \mathcal{C}_A(\mathbb{D}_k)$. The function $\phi(z)$ is related to the auxiliary functions by the formula

$$\phi(z) = Q + \sum_{m=0}^n \rho_m \overline{\left[\phi_m\left(z^*_{(m)}\right) - \phi_m\left(w^*_{(m)}\right)\right]}. \quad (5.2.3)$$

Theorem 5.3 *A general solution of the system is*

$$\psi_k(z) = -\sum_{m \neq k} \rho_m \overline{\psi_m\left(z^*_{(m)}\right)} + g(z), \qquad (5.2.4)$$

where $g \in C_A\left(\cup_{k=1}^n \mathbb{D}_k\right)$ *has the form*

$$\psi_k(z) = \left[g(z) - g\left(w^*_{(k)}\right)\right] - \sum_{k_1 \neq k} \rho_{k_1} \overline{\left[g\left(z^*_{(k_1)}\right) - g\left(w^*_{(k_1 k)}\right)\right]}$$

$$+ \sum_{k_1 \neq k} \sum_{k_2 \neq k_1} \rho_{k_1} \rho_{k_2} \left[g\left(z^*_{(k_2 k_1)}\right) - g\left(w^*_{(k_2 k_1 k)}\right)\right] - \ldots + c_k$$

$$= \sum_{m=0}^{\infty} \sum_{k_1 \neq k} \sum_{k_2 \neq k_1} \cdots \sum_{k_m \neq k_{m-1}} \rho_{k_1} \rho_{k_2} \cdots \rho_{k_m} \qquad (5.2.5)$$

$$\times (-1)^m \mathbf{C}^m \left[g\left(z^*_{(k_m \ldots k_2 k_1)}\right) - g\left(w^*_{(k_m \ldots k_1 k)}\right)\right] + c_k,$$

where c_k *are constants,* $\mathbf{C}$ *is the operator of complex conjugation. The zero term ($m = 0$) of the series is defined as* $\left[g(z) - g\left(w^*_{(k)}\right)\right]$. *Series (5.2.5) converges absolutely and uniformly in* $|z - a_k| \leq r_k$.

Proof follows from the convergence of the series (5.2.5) with $\rho_{k_l} = 1$ (see Section 4.3). ∎

Let us apply Theorem 5.3 to (5.2.2). By virtue of (5.2.2), (5.2.3) and (5.2.5) we obtain

$$\phi_k(z) = \sum_{m=0}^{\infty} \sum_{k_1 \neq k} \sum_{k_2 \neq k_1} \cdots \sum_{k_m \neq k_{m-1}} \rho_{k_1} \rho_{k_2} \cdots \rho_{k_m} \qquad (5.2.6)$$

$$\times (-1)^m \mathbf{C}^m \left[h^+\left(z^*_{(k_m \ldots k_2 k_1)}\right) - h^+\left(w^*_{(k_m \ldots k_1 k)}\right)\right] + c_k,$$

$$\phi(z) = \sum_{m=0}^{\infty} \sum_{k=0}^{n} \sum_{k_1 \neq k} \sum_{k_2 \neq k_1} \cdots \sum_{k_m \neq k_{m-1}} \rho_{k_1} \rho_{k_2} \cdots \rho_{k_m} \rho_k$$

$$\times (-1)^{m+1} \mathbf{C}^{m+1} \left[h^+\left(z^*_{(k_m \ldots k_1 k)}\right) - h^+\left(w^*_{(k_m \ldots k_1 k)}\right)\right] + Q,$$

where $c_k = \phi_k\left(w^*_{(k)}\right)$. We now proceed to determine the undetermined constant appearing in (5.2.6). The initial problem (5.2.1) has the solution depending on an

5.2. LINEARIZED PROBLEM.

arbitrary additive constant. We assume Q to be that arbitrary constant. Substituting $z = w^*_{(k)}$ in (5.2.6) we obtain the values

$$R_{mk} := \overline{\phi_m\left(w^*_{(mk)}\right)} - \overline{\phi_m\left(w^*_{(m)}\right)}, \quad k, m = 1, 2, \ldots, n; \ k \neq m.$$

Then by substituting $z = w^*_{(k)}$ in (5.2.2) we have the $\mathbb{R}$-linear set of algebraic equations

$$c_k - \rho_k \overline{c_k} = Q + R_k, \quad k = 1, 2, \ldots, n,$$

where $R_k := f_k\left(w^*_{(k)}\right) - \sum_{m \neq k}^n \rho_m R_{mk}$. After solving these equations we obtain

$$c_k = \frac{\Re(R_k + Q)}{1 - \rho_k} + i\frac{\Im(R_k + Q)}{1 + \rho_k}, \quad k = 1, 2, \ldots, n. \tag{5.2.7}$$

So we obtain the solution of (5.2.2) in analytic form (5.2.6), (5.2.7). The original function $\phi(z)$ has the form (5.2.3).

5.2.2 A mixed problem for a circular domain.

We consider a multiply connected circular domain $\mathbb{D}$ bounded by external circle $\mathbb{T}_0 := \{t \in \mathbb{C} : |t - a_0| = r_0\}$ and internal circles $\mathbb{T}_k := \{t \in \mathbb{C} : |t - a_k| = r_k\}$ restricted by the discs $\mathbb{D}_k := \{z \in \mathbb{C} : |z - a_k| < r_k\}$ (see Figure 4.3). Let us consider the steady heat conduction governed by the Laplace equation of the materials occupying the domains $\mathbb{D}$, $\mathbb{D}_k$ with the thermal conductivities $\lambda = 1$, λ_k, respectively. Let the contact between materials be perfect and the temperature distribution $u(z)$ be given on the external circle $\mathbb{T}_0$.

A mathematical model of the problem considered follows. Given a function $f(t) \in \mathcal{H}(\mathbb{T}_0)$, find functions $u(z)$ and $u_k(z)$ harmonic in $\mathbb{D}$ and $\mathbb{D}_k$, continuously differentiable in the closures of the domains considered ($u \in h^1_+(\partial \mathbb{D})$, $u_k \in h^1_-(\mathbb{T}_k)$) with the boundary conditions

$$u(t) = u_k(t), \quad \frac{\partial u}{\partial n}(t) = \lambda_k \frac{\partial u_k}{\partial n}(t), \quad |t - a_k| = r_k, \ k = 1, 2, \ldots, n, \tag{5.2.8}$$

$$u(t) = f(t), \quad |t - a_0| = r_0. \tag{5.2.9}$$

(5.2.8) models the perfect contact between materials; (5.2.9) models the external temperature distribution. Problem (5.2.8), (5.2.9) can be written in terms of the complex potentials (see Section 2.12)

$$\phi(t) = \phi_k(t) - \rho_k \overline{\phi_k(t)}, \quad |t - a_k| = r_k, \ k = 1, 2, \ldots, n, \tag{5.2.10}$$

$$\Re\phi(t) = f(t), \quad |t - a_0| = r_0, \tag{5.2.11}$$

where $\rho_k := (\lambda_k - 1)/(\lambda_k + 1)$, $\phi \in \mathcal{C}_A^1(\mathbb{D})$, $\phi_k \in \mathcal{C}_A^1(\mathbb{D}_k)$. Harmonic and analytic functions are related by the equalities

$$u(z) = \Re\phi(z), \quad \frac{\lambda_k + 1}{2} u_k(z) = \Re\phi_k(z).$$

Following Section 2.12 we write (5.2.11) as the $\mathbb{R}$-linear problem

$$\phi(t) = \phi_0(t) - \overline{\phi_0(t)} + f^+(t), \quad |t - a_0| = r_0, \tag{5.2.12}$$

where the function

$$f^+(z) = \frac{1}{\pi i} \int_{\mathbb{T}_0^-} \frac{f(\tau)d\tau}{\tau - z} - \frac{1}{2\pi i} \int_{\mathbb{T}_0^-} \frac{f(\tau)d\tau}{\tau - a_0} \tag{5.2.13}$$

belongs to $\mathcal{C}_A(\mathbb{D}_0^-)$ and satisfies the boundary condition

$$\Re f^+(t) = f(t), \quad |t - a_0| = r_0,$$

the unknown function ϕ_0 belongs to $\mathcal{C}_A(\mathbb{D}_0^-)$, and $\Re\phi_0(\infty) = 0$. Here $\mathbb{D}_0^-$ is defined as follows $\mathbb{D}_0^- := \left\{ z \in \widehat{\mathbb{C}} : |z - a_0| > r_0 \right\}$. Let us rewrite conditions (5.2.10), (5.2.12) as

$$\phi(t) = \phi_k(t) - \rho_k \overline{\phi_k(t)} + f_k(t), \quad |t - a_k| = r_k, \quad k = 0, 1, \ldots, n, \tag{5.2.14}$$

where $f_k(t) \equiv 0$ for $k = 1, 2, \ldots, n$, $f_0(z) := f^+(z)$, $\rho_0 := 1$.

The problem (5.2.14) in Section 5.2.1 is reduced to the following system of functional equations

$$\phi_k(z) = -\sum_{m \neq k} \rho_m \left[\overline{\phi_m\left(z_{(m)}^*\right)} - \phi_m\left(w_{(m)}^*\right) \right] + \rho_k \overline{\phi_k\left(w_{(k)}^*\right)} + f_k(z) + Q, \tag{5.2.15}$$

$$|z - a_k| \leq r_k, \quad k = 0, 1, \ldots, n.$$

The function $\phi(z)$ has the form

$$\phi(z) = Q + \sum_{m=0}^{n} \rho_m \left[\overline{\phi_m\left(z_{(m)}^*\right)} - \phi_m\left(w_{(m)}^*\right) \right].$$

Applying Theorem 5.3 we obtain

$$\phi_k(z) = \sum_{m=0}^{\infty} \sum_{k_1 \neq k} \sum_{k_2 \neq k_1} \cdots \sum_{k_m \neq k_{m-1}} \rho_{k_1} \rho_{k_2} \cdots \rho_{k_m} \tag{5.2.16}$$

$$\times (-1)^m \mathbf{C}^m \left[f_{k_m}\left(z_{(k_m \ldots k_2 k_1)}^*\right) - f_{k_m}\left(w_{(k_m \ldots k_1 k)}^*\right) \right] + c_k,$$

5.2. LINEARIZED PROBLEM.

$$\phi(z) = \sum_{m=0}^{\infty}\sum_{k=0}^{n}\sum_{k_1\neq k}\sum_{k_2\neq k_1}\cdots\sum_{k_m\neq k_{m-1}} \rho_{k_1}\rho_{k_2}\cdots\rho_{k_m}\rho_k \qquad (5.2.17)$$
$$\times (-1)^{m+1}\,\mathbf{C}^{m+1}\left[f_{k_m}\left(z^*_{(k_m\ldots k_1 k)}\right) - f_{k_m}\left(w^*_{(k_m\ldots k_1 k)}\right)\right] + Q,$$

where $c_k = \phi_k\left(w^*_{(k)}\right)$. We now proceed to determine the undetermined constant appearing in (5.2.16). Initial problem (5.2.14) has the solution depending on an arbitrary additive constant (say Q is that constant). Substituting $z = w^*_{(k)}$ in (5.2.16) we obtain the values

$$R_{mk} := \overline{\phi_m\left(w^*_{(mk)}\right)} - \overline{\phi_m\left(w^*_{(m)}\right)}, \ k, m = 0, 1, \ldots, n; \ k \neq m.$$

Then by substituting $z = w^*_{(k)}$ in (5.2.15) we have the $\mathbb{R}$-linear set of algebraic equations

$$c_k - \rho_k \overline{c_k} = Q + R_k, \ k = 0, 1, \ldots, n, \qquad (5.2.18)$$

where $R_k := f_k\left(w^*_{(k)}\right) - \sum_{m\neq k}^{n}\rho_m R_{mk}$. After solving these equations we obtain

$$c_k = \frac{\Re(R_k + Q)}{1 - \rho_k} + i\frac{\Im(R_k + Q)}{1 + \rho_k}, \ k = 1, 2, \ldots, n, \ c_0 = \frac{i}{2}\Im(R_0 + Q) + \alpha,$$

where α is an arbitrary real constant. Moreover, (5.2.18) yields the following additional condition, $\Re(R_0 + Q) = 0$. The final formulae for the constants are the following

$$Q = -\Re R_0 + iC, \qquad (5.2.19)$$

$$c_k = (1 - \rho_k)^{-1}\Re(R_k - R_0) + i(1 + \rho_k)^{-1}(\Im R_k + C), \ k = 1, 2, \ldots, n,$$

where C is an arbitrary real constant vector.

We now proceed to study the vector-matrix problem which is formally written as (5.2.10), (5.2.11):

$$\phi(t) = \phi_k(t) - \rho_k\overline{\phi_k(t)}, \ |t - a_k| = r_k, \ k = 1, 2, \ldots, n, \qquad (5.2.20)$$

$$\Re\phi(t) = f(t), \ |t - a_0| = r_0. \qquad (5.2.21)$$

However, now we assume that ρ_k is a constant matrix $N \times N$, $f(t)$ is a given vector of the dimension N with the Hölder continuous components, and the components of the unknown vector-functions ϕ, ϕ_k of the dimension N belong to $\mathcal{C}^1_A(\mathbb{D})$, $\in C^1_A(\mathbb{D}_k)$, respectively.

Theorem 5.4 *Let the matrix ρ_k be represented in the form*

$$\rho_k = p_k^{-1} s_k \overline{p}_k, \qquad (5.2.22)$$

or in the form $\rho_k = p_k^{-1} s_k p_k$, where p_k is a non-degenerated matrix, s_k is a diagonal matrix with elements s_{kj} satisfying the inequality $|s_{kj}| < 1$. Then the general solution of problem (5.2.20), (5.2.21) has the form (5.2.16), (5.2.17), where ρ_{km} are given matrix, f_{km} are the vector-functions: $f_k(t) \equiv 0$ for $k = 1, 2, \ldots, n$, $f_0(z) := f^+(z)$, and $f^+(z)$ has the form (5.2.13). The constant vectors c_k, Q have the form (5.2.19); C is an arbitrary real vector.

Proof repeats the proof in the scalar case and it can be easily obtained. ∎

Lemma 5.1 *Let a constant μ satisfy the inequality $|\mu| < 1$ and matrices ρ_k are represented in the form (5.2.22). Then the vector-matrix problem*

$$\phi(t) = \phi_k(t) - \mu \rho_k \overline{\phi_k(t)} + \gamma_k, \quad |t - a_k| = r_k, \ k = 0, 1, \ldots, n, \qquad (5.2.23)$$

has only constant solutions in the class $\mathcal{C}_A(\partial(\mathbb{D}))$.

Proof. Let us introduce the function

$$U(z, \overline{z}) := \begin{cases} \phi(z), & z \in \mathbb{D} \\ \phi_k(z) - \mu \rho_k \overline{\phi_k(z)} + \gamma_k, & z \in \mathbb{D}_k, \ k = 0, 1, \ldots, n, \end{cases}$$

satisfying the differential equation

$$U_{\overline{z}} - Q(z) \overline{U_z} = 0, \ z \in \cup_{k=1}^n \mathbb{D}_k \cup \mathbb{D}, \qquad (5.2.24)$$

where

$$Q(z) := \begin{cases} 0, & z \in \mathbb{D} \\ -\mu \rho_k, & z \in \mathbb{D}_k, \ k = 0, 1, \ldots, n. \end{cases}$$

We now demonstrate that equation (5.2.24) is of elliptic type. It is evident in $\mathbb{D}$. In each domain $\mathbb{D}_k$ we make the change $V(z, \overline{z}) := p_k U(z, \overline{z})$; then (5.2.24) becomes

$$V_{\overline{z}} - s_k \overline{V_z} = 0. \qquad (5.2.25)$$

The matrices s_k are diagonal; hence (5.2.25) can be separated onto scalar equations of an elliptic type, because the moduli of the elements of s_k are less than the unit.

5.2. LINEARIZED PROBLEM.

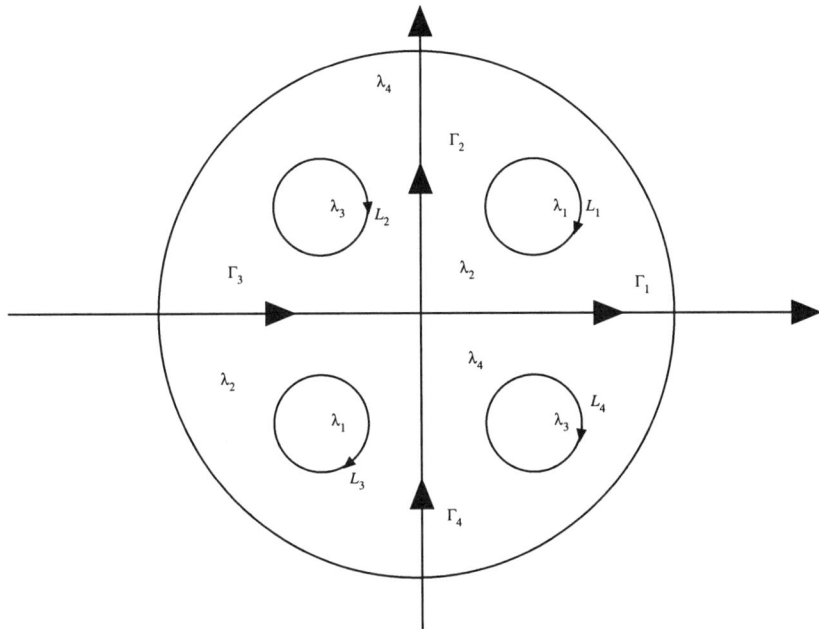

Figure 5.1: Section of composite material. Components have the conductivities λ_1, λ_2, λ_3, λ_4

Therefore, (5.2.24) is of elliptic type. We have the equal limit values of $U(z,\bar{z})$ on each circle L_k. Then following [43] we can consider (5.2.24) in a class of generalized functions for which Liouville's theorem holds and, hence, $U(z,\bar{z})$ is a constant vector. We have no other possibility for $U(z,\bar{z})$, since any classic solution of an elliptic equation is a generalized solution which is in our case unique.

The lemma is proved. ∎

5.2.3 A mixed boundary value problem.

Consider steady heat conduction of the two-dimensional composite material occupying the disc $\mathbb{U}(0, r_0)$ (see Figure 5.1). We assume that the disc is compound for different materials of the conductivities λ_1, λ_2, λ_3, λ_4, respectively. The external temperature distribution is given on the circle $\mathbb{T}_0 := \mathbb{T}(0, r_0)$.

The mathematical model of the problem is the following. Given a Hölder continuous vector $g(t)$, $t \in \mathbb{T}(0, r_0)$, find a vector-function $h(z)$ analytic in

$$B := \mathbb{U}(0, r_0) \setminus (\cup_{j=1}^{4} \Gamma_j \cup \cup_{j=1}^{4} \mathbb{T}_j)$$

(see Figure 5.1), continuous in the closures of the domains considered except the points $(0,0)$, $(0, \pm r_0)$, $(\pm r_0, 0)$, where integrable singularities are admitted [91]. Let us denote $h^+(t)$ and $h^-(t)$ the limit values of $h(z)$ on ∂B introduced in accordance with the orientation of the curves (see Figure 5.1). The boundary values $h^+(t)$ and $h^-(t)$ satisfy the following $\mathbb{R}$-linear conjugation conditions

$$h^+(t) = \frac{\lambda_2 + \lambda_4}{2\lambda_2} h^-(t) + \frac{\lambda_2 - \lambda_4}{2\lambda_2} \overline{h^-(t)}, \; t \in \Gamma_1 \cup \Gamma_4, \tag{5.2.26}$$

$$h^+(t) = \frac{\lambda_4 + \lambda_2}{2\lambda_4} h^-(t) + \frac{\lambda_4 - \lambda_2}{2\lambda_4} \overline{h^-(t)}, \; t \in \Gamma_2 \cup \Gamma_2, \tag{5.2.27}$$

$$h^+(t) = \frac{\lambda_2 + \lambda_1}{2\lambda_2} h^-(t) + \frac{\lambda_2 - \lambda_1}{2\lambda_2} \overline{h^-(t)}, \; t \in \mathbb{T}_1 \cup \mathbb{T}_3, \tag{5.2.28}$$

$$h^+(t) = \frac{\lambda_4 + \lambda_3}{2\lambda_4} h^-(t) + \frac{\lambda_4 - \lambda_3}{2\lambda_4} \overline{h^-(t)}, \; t \in \mathbb{T}_2 \cup \mathbb{T}_4, \tag{5.2.29}$$

and the boundary condition

$$\Re h(t) = g(t), \; t \in \mathbb{T}_0. \tag{5.2.30}$$

Remark 5.2 *As one can see in the present subsection we use a slightly other designation for the limit boundary values of the functions on curves. The indices "+" and "−" are used here to derive boundary values in accordance with orientation of the curves. They are not assigned to functions as in all other subsections. Such designations are more convenient for the compound contour used here.*

For the sake of simplicity of calculations we additionally suggest that $g(t)$ is an even vector-function, i.e., $g(t) = g(-t)$. Then using symmetry we conclude that the heat field is symmetric too, and we are looking for only a symmetric solution, i.e., $h(z) = h(-z)$.

We shall reduce problem (5.2.26) - (5.2.30) to a vector-matrix $\mathbb{R}$-linear problem for the domain $\mathbb{D} := \mathbb{U}(0, r_0) \setminus (\cup_{j=1}^{4} \mathbb{T}_j)$. First we introduce the following vector-function

$$\Phi(z) = (\Phi_1(z), \Phi_2(z), \Phi_3(z), \Phi_4(z))^T := \left(h(z), \overline{h(\overline{z})}, h(-z), \overline{h(-\overline{z})}\right)^T, \tag{5.2.31}$$

5.2. LINEARIZED PROBLEM.

satisfying the symmetry condition

$$\Phi_1(z) \equiv \overline{\Phi_2(\bar{z})} \equiv \Phi_3(-z) \equiv \overline{\Phi_4(-\bar{z})}.$$

Let us rewrite the conditions (5.2.26) - (5.2.30) in terms of $\Phi(z)$. Consider, for instance, the curve Γ_1. If $z \to t^- \in \Gamma_1$, then $-z \to -t \in \Gamma_3$, $-\bar{z} \to -t^+ \in \Gamma_1$. Hence, the following relations are valid

$$\Phi_1^+(t) = h^+(t), \ \Phi_2^+(t) = \overline{h^-(t)}, \ \Phi_3^+(t) = h^-(-t), \ \Phi_4^+(t) = \overline{h^+(-t)},$$

$$\Phi_1^-(t) = h^-(t), \ \Phi_2^-(t) = \overline{h^+(t)}, \ \Phi_3^-(t) = h^+(-t), \ \Phi_4^-(t) = \overline{h^-(-t)},$$

and the conjugation condition on Γ_1 becomes

$$\Phi^+(t) = G_1 \Phi^-(t), \ t \in \Gamma_1,$$

where

$$G_1 = \frac{1}{\lambda_2 + \lambda_4} \begin{pmatrix} 2\lambda_4 & \lambda_2 - \lambda_4 & 0 & 0 \\ -(\lambda_2 - \lambda_4) & 2\lambda_2 & 0 & 0 \\ 0 & 0 & 2\lambda_4 & \lambda_2 - \lambda_4 \\ 0 & 0 & -(\lambda_2 - \lambda_4) & 2\lambda_2 \end{pmatrix}.$$

Along similar lines we obtain the conjugation conditions on the other curves

$$\Phi^+(t) = G_k \Phi^-(t), \ t \in \Gamma_k, \tag{5.2.32}$$

$$\Phi^+(t) = T_k \Phi^-(t) + H_k \overline{\Phi^-(t)}, \ t \in \mathbb{T}_k, \ k = 1, 2, 3, 4, \tag{5.2.33}$$

$$\Re \Phi(t) = q(t), \ t \in \mathbb{T}_0, \tag{5.2.34}$$

where

$$G_2 = \frac{1}{\lambda_2 + \lambda_4} \begin{pmatrix} 2\lambda_2 & 0 & 0 & -(\lambda_2 - \lambda_4) \\ 0 & 2\lambda_4 & \lambda_2 - \lambda_4 & 0 \\ 0 & -(\lambda_2 - \lambda_4) & 2\lambda_2 & 0 \\ \lambda_2 - \lambda_4 & 0 & 0 & 2\lambda_4 \end{pmatrix},$$

$$G_3 = \frac{1}{\lambda_2 + \lambda_4} \begin{pmatrix} 2\lambda_2 & -(\lambda_2 - \lambda_4) & 0 & 0 \\ \lambda_2 - \lambda_4 & 2\lambda_4 & 0 & 0 \\ 0 & 0 & 2\lambda_2 & -(\lambda_2 - \lambda_4) \\ 0 & 0 & \lambda_2 - \lambda_4 & 2\lambda_2 \end{pmatrix},$$

$$G_4 = \frac{1}{\lambda_2 + \lambda_4} \begin{pmatrix} 2\lambda_4 & 0 & 0 & \lambda_2 - \lambda_4 \\ 0 & 2\lambda_2 & -(\lambda_2 - \lambda_4) & 0 \\ 0 & \lambda_2 - \lambda_4 & 2\lambda_4 & 0 \\ -(\lambda_2 - \lambda_4) & 0 & 0 & 2\lambda_2 \end{pmatrix},$$

$$T_1 = \begin{pmatrix} \frac{\lambda_2+\lambda_1}{2\lambda_2} & 0 & 0 & 0 \\ 0 & \frac{\lambda_4+\lambda_3}{2\lambda_4} & 0 & 0 \\ 0 & 0 & \frac{\lambda_2+\lambda_1}{2\lambda_2} & 0 \\ 0 & 0 & 0 & \frac{\lambda_4+\lambda_3}{2\lambda_4} \end{pmatrix}, \quad H_1 = \begin{pmatrix} \frac{\lambda_2-\lambda_1}{2\lambda_2} & 0 & 0 & 0 \\ 0 & \frac{\lambda_4-\lambda_3}{2\lambda_4} & 0 & 0 \\ 0 & 0 & \frac{\lambda_2-\lambda_1}{2\lambda_2} & 0 \\ 0 & 0 & 0 & \frac{\lambda_4-\lambda_3}{2\lambda_4} \end{pmatrix},$$

$$T_3 = T_1; H_3 = H_1,$$

$$T_2 = \begin{pmatrix} \frac{\lambda_4+\lambda_3}{2\lambda_4} & 0 & 0 & 0 \\ 0 & \frac{\lambda_2+\lambda_1}{2\lambda_2} & 0 & 0 \\ 0 & 0 & \frac{\lambda_4+\lambda_3}{2\lambda_4} & 0 \\ 0 & 0 & 0 & \frac{\lambda_2+\lambda_1}{2\lambda_2} \end{pmatrix}, \quad H_2 = \begin{pmatrix} \frac{\lambda_4-\lambda_3}{2\lambda_4} & 0 & 0 & 0 \\ 0 & \frac{\lambda_2-\lambda_1}{2\lambda_2} & 0 & 0 \\ 0 & 0 & \frac{\lambda_4-\lambda_3}{2\lambda_4} & 0 \\ 0 & 0 & 0 & \frac{\lambda-\lambda_1}{2\lambda_2} \end{pmatrix},$$

$$T_4 = T_2; H_4 = H_2,$$

$$q(t) = \left(g(t), g(\overline{t}), g(t), g(\overline{t})\right)^T.$$

We now try to remove the $\mathbb{C}$-linear conditions (5.2.32). Let us introduce the vector-function

$$F(z) := \begin{cases} \Phi(z), & \Re z > 0, \ \Im z > 0, \\ G_2^{-1}\Phi(z), & \Re z < 0, \ \Im z > 0, \\ G_2^{-1}G_3\Phi(z), & \Re z < 0, \ \Im z < 0, \\ G_2^{-1}G_3G_4\Phi(z), & \Re z > 0, \ \Im z < 0. \end{cases} \tag{5.2.35}$$

We shall write the $\mathbb{C}$-linear conditions (5.2.32) in terms of $F^\pm(t)$. We have $F^+(t) = G_2^{-1}\Phi^+(t)$, $F^-(t) = \Phi^-(t)$, $t \in \Gamma_2$. It follows from $\Phi^+(t) = G_2\Phi^-(t)$, $t \in \Gamma_2$, that $F^+(t) = F^-(t)$, $t \in \Gamma_2$. The same condition holds on Γ_3 and Γ_4: $F^+(t) = F^-(t)$, $t \in \Gamma_3 \cup \Gamma_4$. The condition on Γ_1 takes the form

$$F^+(t) = G_1 G_4^{-1} G_3^{-1} G_2 F^-(t), \ t \in \Gamma_1. \tag{5.2.36}$$

It is easy to check that $G_3^{-1} = G_1$, $G_4^{-1} = G_2$. Then (5.2.35) becomes

$$F^+(t) = GF^-(t), \ t \in \Gamma_1, \tag{5.2.37}$$

where

$$G = (G_1 G_2)^2 := \begin{pmatrix} B_1 & B_2 \\ B_2 & B_1 \end{pmatrix}^2,$$

5.2. LINEARIZED PROBLEM.

$$B_1 := \frac{1}{(\lambda_2 + \lambda_4)^2} \begin{pmatrix} (\lambda_2 - \lambda_4)^2 & -2\lambda_4(\lambda_2 - \lambda_4) \\ 2\lambda_2(\lambda_2 - \lambda_4) & (\lambda_2 - \lambda_4)^2 \end{pmatrix},$$

$$B_2 := \frac{1}{(\lambda_2 + \lambda_4)^2} \begin{pmatrix} (\lambda_2 - \lambda_4)^2 & 2\lambda_4(\lambda_2 - \lambda_4) \\ -2\lambda_2(\lambda_2 - \lambda_4) & (\lambda_2 - \lambda_4)^2 \end{pmatrix}.$$

We have

$$G = \begin{pmatrix} B_1 & B_2 \\ B_2 & B_1 \end{pmatrix}^2 = \begin{pmatrix} B_1^2 + B_2^2 & 2B_1B_2 \\ 2B_1B_2 & B_1^2 + B_2^2 \end{pmatrix}. \tag{5.2.38}$$

The matrices B_1 and B_2 satisfy the following relations $B_1B_2 = B_2B_1$, $B_1 + B_2 = I$, where I is the unit 2×2 matrix.

It follows from (5.2.35) that $F^-(t) = G_2^{-1}G_3G_4\Phi^-(t)$, $t \in \Gamma_1$. Moreover, we are looking for a symmetric solution for which $h(z) = h(-z)$. Hence, $\Phi_1(z) = \Phi_3(z)$, $\Phi_2(z) = \Phi_4(z)$ and we can introduce the functions $u(z)$ and $v(z)$ in such a way that

$$U(t) = \begin{pmatrix} u(t) \\ v(t) \end{pmatrix}, \; G_2^{-1}G_3G_4\Phi^-(t) =: \begin{pmatrix} U(t) \\ U(t) \end{pmatrix} \Leftrightarrow F^-(t) = \begin{pmatrix} U(t) \\ U(t) \end{pmatrix}, \; t \in \Gamma_1.$$

Using the properties of the matrix G we arrive at the following equation on Γ_1:

$$GF^-(t) = \begin{pmatrix} B_1^2 + B_2^2 & 2B_1B_2 \\ 2B_1B_2 & B_1^2 + B_2^2 \end{pmatrix} \begin{pmatrix} U(t) \\ U(t) \end{pmatrix} = \begin{pmatrix} (B_1 + B_2)^2 U(t) \\ (B_1 + B_2)^2 U(t) \end{pmatrix} = \begin{pmatrix} U(t) \\ U(t) \end{pmatrix}.$$

Hence, we can write $F^+(t) = F^-(t)$, $t \in \Gamma_1$, instead of (5.2.35).

The $\mathbb{R}$-linear conditions (5.2.33) in terms of $F^\pm(t)$ become

$$F^+(t) = V_k F^-(t) + W_k \overline{F^-(t)}, \; t \in \mathbb{T}_k, \; k = 1,2,3,4, \tag{5.2.39}$$

where $V_1 = T_1$, $V_2 = G_2^{-1}T_2G_2$, $V_3 = G_2^{-1}G_3T_3G_3^{-1}G_2$, $V_4 = G_2^{-1}G_3G_4T_3G_4^{-1}G_3^{-1}G_2$, $W_1 = H_1$, $W_2 = G_2^{-1}H_2G_2$, $W_3 = G_2^{-1}G_3H_3G_3^{-1}G_2$, $W_4 = G_2^{-1}G_3G_4H_3G_4^{-1}G_3^{-1}G_2$.
The boundary condition (5.2.34) in terms of $F^\pm(t)$ becomes

$$\Re F(t) = f(t), \; t \in \mathbb{T}_0, \tag{5.2.40}$$

where

$$f(t) := \begin{cases} q(t), & \Re t > 0, \; \Im t > 0, \\ G_2^{-1}q(t), & \Re t < 0, \; \Im t > 0, \\ G_2^{-1}G_3q(t), & \Re t < 0, \; \Im t < 0, \\ G_2^{-1}G_3G_4q(t), & \Re t > 0, \; \Im t < 0, \end{cases} \; t \in \mathbb{T}_0.$$

Thus we have reduced problem (5.2.32), (5.2.33), (5.2.34) to problem (5.2.39), (5.2.40) has been solved in the previous subsection. In order to use Theorem 5.4 we should use the designations of Subsection 5.2.2. Let us denote

$$\phi(z) := F(z),\ z \in \mathbb{U}(0,r_0)\setminus\cup_{j=1}^{4}\mathbb{D}_j,\ \phi_k(z) := V_k F(z),\ z \in \mathbb{D}_k,\ k=1,2,3,4. \quad (5.2.41)$$

Then (5.2.39), (5.2.40) become

$$\phi(t) = \phi_k(t) - \rho_k \overline{\phi_k(t)},\ |t - a_k| = r_k,\ k=1,2,3,4, \quad (5.2.42)$$

$$\Re\phi(t) = f(t),\ |t| = r_0, \quad (5.2.43)$$

where $\rho_k := -W_k V_k^{-1}$.

Let us check the conditions of Theorem 5.4. We calculate

$$\rho_1 = -W_1 V_1^{-1} = -\begin{bmatrix} \frac{\lambda_2-\lambda_1}{\lambda_2+\lambda_1} & 0 & 0 & 0 \\ 0 & \frac{\lambda_4-\lambda_3}{\lambda_4+\lambda_3} & 0 & 0 \\ 0 & 0 & \frac{\lambda_2-\lambda_1}{\lambda_2+\lambda_1} & 0 \\ 0 & 0 & 0 & \frac{\lambda_4-\lambda_3}{\lambda_4+\lambda_3} \end{bmatrix} = s_1,$$

$$\rho_2 = -W_2 V_2^{-1} = -G_2^{-1} H_2 G_2 G_2^{-1} T_2^{-1} G_2 = G_2^{-1}\left(-H_2 T_2^{-1}\right) G_2,$$

$$s_2 = -H_2 T_2^{-1} = -\begin{bmatrix} \frac{\lambda_4-\lambda_3}{\lambda_4+\lambda_3} & 0 & 0 & 0 \\ 0 & \frac{\lambda_2-\lambda_1}{\lambda_2+\lambda_1} & 0 & 0 \\ 0 & 0 & \frac{\lambda_4-\lambda_3}{\lambda_4+\lambda_3} & 0 \\ 0 & 0 & 0 & \frac{\lambda_2-\lambda_1}{\lambda_2+\lambda_1} \end{bmatrix},$$

$$\begin{aligned}\rho_3 &= -W_3 V_3^{-1} = -G_2^{-1} G_3 H_3 G_3^{-1} G_2 G_2^{-1} G_3 T_3^{-1} G_3^{-1} G_2 \\ &= \left(G_3^{-1} G_2\right)^{-1}\left(-H_3 T_3^{-1}\right)\left(G_3^{-1} G_2\right),\end{aligned}$$

$$s_3 = -H_3 T_3^{-1} = -H_1 T_1^{-1} = s_1,$$

$$\begin{aligned}\rho_4 &= -W_4 V_4^{-1} = -G_2^{-1} G_3 G_4 H_4 G_4^{-1} G_3^{-1} G_2 G_2^{-1} G_3 G_4 T_4^{-1} G_4^{-1} G_3^{-1} G_2 \\ &= \left(G_4^{-1} G_3^{-1} G_2\right)^{-1}\left(-H_4 T_4^{-1}\right)\left(G_4^{-1} G_3^{-1} G_2\right),\end{aligned}$$

$$s_4 = -H_4 T_4^{-1} = -H_2 T_2^{-1} = s_2.$$

The moduli of the diagonal elements of s_k, $k=1,2,3,4$, are less than the unit: $\left|\frac{\lambda_2-\lambda_1}{\lambda_2+\lambda_1}\right| < 1$, $\left|\frac{\lambda_4-\lambda_3}{\lambda_4+\lambda_3}\right| < 1$, since the thermal conductivities of the materials λ_k, $k=1,2,3,4$, are positive. Therefore, the matrices ρ_k satisfy the conditions of Theorem 5.4 and the solution of the problem (5.2.42), (5.2.43) is given in Theorem 5.4 in closed form. The original unknown function $h(z)$ is determined by (5.2.41), (5.2.35) and (5.2.31).

5.3 Constructive solution to integral equations.

The integral equations (5.1.18) can be written in the form

$$w(z) = \mathbf{A}w(z) + h(z), \quad z \in clD^-, \tag{5.3.1}$$

where $\mathbf{A}w = \mathbf{A}_1 w + \mathbf{A}_2 w$, $\mathbf{A}_1 w := \mathbf{Q}\rho\overline{w}$, $\mathbf{A}_2 w := -\mathbf{Q}G\left[t, \frac{1}{2}(1-\rho)(w+\overline{w})\right]$, since the functions w, $\mathbf{A}w$ and h are analytically continued into D^-.

Applying the residue calculus in the present section we propose a simple algorithm to calculate explicitly the integral operators involved in $\mathbf{A}$.

If condition (5.1.19) holds, then the operator $\mathbf{A}$ is contractive, i.e., $\|\mathbf{A}z_1 - \mathbf{A}z_2\| \leq \|z_1 - z_2\|$; hence the function w can be found by the successive approximations

$$w^0 = h, \quad w^{k+1} = \mathbf{A}w^k + h, \quad k = 0, 1, 2, \ldots. \tag{5.3.2}$$

Instead of (5.3.2) we consider the following iterative method

$$\Omega^0 = -R_0, \quad \Omega^{k+1} = \mathbf{A}R_k + h, \quad k = 0, 1, 2, \ldots, \tag{5.3.3}$$

where the function $R_k(z)$ is a rational approximation of the function $w^k(z)$ satisfying the condition $\|w^k - R_k\| < 2^{-k}\varepsilon$. Let us estimate the value

$$\|w^k - \Omega^k\| = \|\mathbf{A}w^{k-1} - \mathbf{A}R_{k-1}\| \leq \|w^{k-1} - R_{k-1}\| \leq$$

$$\|w^{k-1} - \Omega^{k-1}\| + \|\Omega^{k-1} - R_{k-1}\| \leq \|w^{k-1} - \Omega^{k-1}\| + 2^{-k}\varepsilon < \ldots < 2\varepsilon.$$

Therefore, method (5.3.3) may be applied to the approximate solution to (5.3.1).

At the beginning let us consider examples of the operator $\mathbf{A}_1$.

1. Let $D^+ = \mathbb{U}$, then the function $\overline{w}(\tau) = w(1/\overline{\tau})$ is analytically continued into $|z| < 1$ and, hence,

$$\mathbf{A}_1 w(\frac{1}{z}) = \frac{\rho}{2\pi i} \int_{|\tau|=1} \overline{w(1/\overline{\tau})} \frac{d\tau}{\tau - z} = \overline{w(\infty)}, \text{ when } |z| < 1.$$

2. Let D^- consist of mutually disjoint disks $|z - a_k| < r_k$, $k = 1, 2, \ldots, n$. Let $w_k(z) := w(z)$ in $|z - a_k| \leq r_k$. Then

$$\mathbf{A}_1 w(z) = \sum_{m \neq k} \frac{\rho_m}{2\pi i} \int_{\mathbb{T}_m} \overline{w_m(\tau^*_{(m)})} \frac{d\tau}{\tau - z} + \frac{\rho_k}{2\pi i} \int_{\mathbb{T}_k} \overline{w_k(\tau^*_{(k)})} \frac{d\tau}{\tau - z},$$

$$|z - a_k| < r_k, \quad k = 1, 2, \ldots, n.$$

If $z \in D_k$ then $z^*_{(k)} \in \widehat{\mathbb{C}}\backslash D_m$ for $m \neq k$ and $\overline{\omega_m(z^*_{(m)})}$ is analytic in $|z - a_m| > r_m$. Therefore,
$$\mathbf{A}_1\omega(z) = -\sum_{m\neq k}\rho_m\overline{\omega_m(z^*_{(m)})} + \rho_k\overline{\omega_k(a_k)}, \quad |z - a_k| < r_k.$$
This form corresponds to equation (5.2.2).

3. Let D^- be a simply connected domain bounded by a curve L determined by the function $\alpha(w) = w + \sum_{k=0}^{n}\alpha_k w^{-k}$ which conformally maps $|w| > 1$ onto D^-. The operator $\mathbf{Q}\overline{\omega}(z)$, $z \in D^-$, has the form
$$\mathbf{Q}\overline{\omega}(z) = \frac{1}{2\pi i}\int_L \overline{\omega(\tau)}\frac{d\tau}{\tau - z} = \frac{1}{2\pi i}\int_{|t|=1}\overline{\omega(\alpha(1/\bar{t}))}\frac{\alpha'(t)dt}{\alpha(t) - \alpha(w)},$$
where $\tau = \alpha(t)$, $|t| = 1$. The function $\phi(t) := \overline{\omega(\alpha(1/\bar{t}))}$ is analytically continued into $|w| < 1$. Hence, the last integral is calculated by residue method
$$\mathbf{Q}\overline{\omega}(z) = -\frac{n}{2\pi i}\int_{|t|=1}\varphi(t)\frac{dt}{t} + \frac{1}{2\pi i}\int_{|t|=1}\varphi(t)K(t,w)dt, \quad (5.3.4)$$
since
$$\frac{1 - \sum_{k=1}^{n}k\alpha_k t^{-k-1}}{t - w + \sum_{k=1}^{n}\alpha_k(t^{-k} - w^{-k})} = -\frac{n}{t} + \frac{1}{t - w} + K(t,w).$$
The rational function $K(t,w)$ has poles at the roots of the equation $\alpha(t) - \alpha(w) = 0$ with respect to t with fixed w in $\mathbb{U}$. We are interested in the roots in $\mathbb{U}$ except $t = w$. In order to derive these roots we study the function
$$z = \alpha(w) = w + \sum_{k=0}^{n}\alpha_k w^{-k} = w^{-n}\left(w^{n+1} + \sum_{k=0}^{n}\alpha_k w^{n-k}\right).$$
The polynomial equation $w^{n+1} + \sum_{k=0}^{n}\alpha_k w^{n-k} = zw^n$ has $n + 1$ roots $w = \beta_j(z)$ ($j = 1, 2, \ldots, n + 1$). Then $\alpha(w)$ maps the extended complex w-plane onto a Riemann surface R of the function $\alpha(w)$. The surface R may be considered as $n + 1$ planes R_j slit along curves γ_j ($j = 1, 2, ..., n + 1$) and identified by a prescribed rule. The roots $\beta_j(z)$ can be considered as branches of the inverse function to $z = \alpha(w)$ on R_j. Let us denote by Γ_j the curves in the w-plane corresponding to the curve γ_j on R_j, i.e., $\alpha(\Gamma_j) = \gamma_j$.[1] Let us denote by L_j the curve l on R_j. Then $\beta_j(\Gamma_j) = \mathbb{T}$. The curves γ_j have the following properties:

i) each γ_j lies in the interior of L_j, since $w = \beta_j(z)$ conformally maps $D^-_j := \{ext(L_j) \text{ on } R_j\}$ onto the schlicht domain $|w| > 1$. A conformal mapping is a continuous function; hence, it cannot have jumps on D^-_j;

---
[1] It follows from [136] that Γ_j are circles.

5.3. CONSTRUCTIVE SOLUTION TO INTEGRAL EQUATIONS.

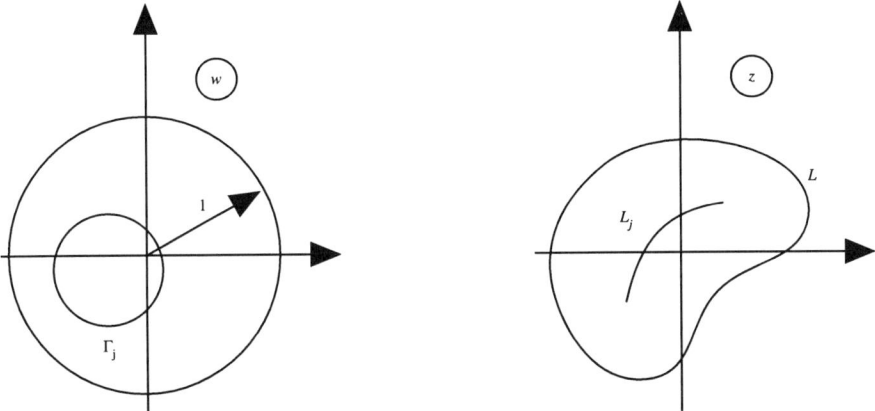

Figure 5.2: Nonconcentric circular domain

ii) on the Riemann surface R each curve L_j embraces all γ_m and L_m for $m = 1, 2, \ldots, n = 1;\ m \neq j$.

Taking into account these properties we conclude that equation $\alpha(t) - \alpha(w) = 0$ with respect to t with fixed w in $\mathbb{U}$ has $n + 1$ roots $t_j = \beta_j(\alpha(w))$ analytic in $\mathbb{U}$. We eliminate the root $t_{n+1} = w$. Then for each fixed $j = 1, 2, \ldots, n$ the set $\beta_j(D_m^-)$ lies in the interior of Γ_j for $m \neq j$. Therefore, by the Residue Calculus we have

$$\frac{1}{2\pi i} \int_{|t|=1} \varphi(t) K(t, w) dt = \sum_{j=1}^{n} G_j(w) \varphi[\beta_j(w)],$$

where the functions $G_j(w)$ analytic in $|w| < 1$ are expressed through $K(t, w)$. The functions $\beta_j(w)$ transform $cl\,\mathbb{U}$ into $\mathbb{U}$. Hence, (5.3.4) takes the form

$$\mathbf{P}^{-}\overline{\omega}(z) = -n\varphi(0) + \sum_{j=1}^{n} G_j(w)\varphi[\beta_j(\alpha(z))] = -n\overline{\omega(\infty)} + \sum_{j=1}^{n} G_j[\alpha(z)]\overline{\omega\left[\alpha\left(1/\overline{\beta_j(z)}\right)\right]},$$

where $\alpha\left(1/\overline{\beta_j(z)}\right)$ transforms $cl D^{-}$ into D^{-}.

Example 5.1 Let $\alpha(w) = w + \alpha_1/w$, where $0 < \alpha_1 < 1$. Then L is an ellipse. The operator $\mathbf{Q}\overline{\omega}$ becomes

$$\mathbf{Q}\overline{\omega}(z) = \frac{1}{2\pi i} \int_L \overline{\omega(\tau)} \frac{d\tau}{\tau - z} = \frac{1}{2\pi i} \int_{|t|=1} \varphi(t) \frac{1 - \alpha_1/t^2}{t - w + \alpha_1(t^{-1} - w^{-1})} dt$$

$$= \frac{1}{2\pi i}\int_{|t|=1}\varphi(t)\frac{t^2-\alpha_1}{t\left(t-w\right)\left(t-\alpha_1 w^{-1}\right)}dt = -\varphi(0)+\varphi(\alpha_1 w^{-1}),\ |w|>1.$$

Using the definition of φ we obtain

$$\mathbf{Q}\overline{w}(z) = -\overline{w(\infty)} + w\left[\frac{1}{2\alpha_1}\left(\overline{z}+\sqrt{\overline{z}^2-4\alpha_1}\right) + \frac{2\alpha_1^2}{\overline{z}+\sqrt{\overline{z}^2-4\alpha_1}}\right]$$

$$= -\overline{w(\infty)} + w\left[\frac{\overline{z}}{2}\left(\frac{1}{\alpha_1}+\alpha_1\right) + \frac{\sqrt{\overline{z}^2-4\alpha_1}}{2}\left(\frac{1}{\alpha_1}-\alpha_1\right)\right].$$

In this example $\beta_1(z) = \frac{1}{2}\left(z+\sqrt{z^2-4\alpha_1}\right)$, $\beta_2(z) = \frac{1}{2}\left(z-\sqrt{z^2-4\alpha_1}\right)$ are the roots of the equation $w^2+zw-\alpha_1 = 0$. The Riemann surface R consists of two planes R_1 and R_2, slit along the segment $\gamma := (-2\sqrt{\alpha_1}, 2\sqrt{\alpha_1})$ of the real axes. The upper bound of the segment $\gamma_1 := \gamma$ on R_1 is glued with the lower bound of the segment $\gamma_2 := \gamma$ on R_2, and conversely, the lower bound of γ_1 is glued with the upper bound of γ_2. The set $\alpha(\gamma_j)$ is the circle $|w| = 2\sqrt{\alpha_1}$; the half-axes of the ellipse L are $1\pm\alpha_1$. The roots of the equation $\alpha(t)-\alpha(w)=0$ with respect to t are $t_1 = \beta_1\left[\alpha(w)\right] = \alpha_1/w$ and $t_2 = w$. The shift $\alpha\left(1/\overline{\beta_1(z)}\right)$ has the form $\alpha\left(1/\overline{\beta_1(z)}\right) = \frac{\overline{z}}{2}\left(\frac{1}{\alpha_1}+\alpha_1\right) + \frac{\sqrt{\overline{z}^2-4\alpha_1}}{2}\left(\frac{1}{\alpha_1}-\alpha_1\right).$

We now sketch a method of calculation of $\mathbf{A}_2\omega$, where ω is a rational function. Let $L = \mathbb{T}$, $G(t,\xi)$ be a rational function on ξ. Then the integral

$$\mathbf{A}_2\omega(z) = -\frac{1}{2\pi i}\int_{|\tau|=1} G\left[t,\frac{1-\rho}{2}\left(\omega(\tau)+\overline{\omega(1/\overline{\tau})}\right)\right]\frac{d\tau}{\tau-z}$$

has a rational function as an integrand. So it can be calculated by residue method.

These examples show us that in many important cases it is not necessary to apply methods of calculation involving computation of integrals, but it is sufficient to compute a composition of some analytic functions and approximate the result by an appropriate rational function.

5.4 A quasi-linear problem and its application to composite materials with reactive inclusions

From the beginning of Chapter 5 we discuss the steady heat conduction of the materials, when the thermal conductivity λ depends on the temperature. This assumption

5.4. COMPOSITE MATERIALS WITH REACTIVE INCLUSIONS.

implies the quasi-linear equation (5.1.2) and corresponding nonlinear boundary value problems. In the present section we consider another type of the physical problem, when a quasi-linear conjugation problem of the analytic function theory appears.

We consider the domains D^+ and D^- divided by a Lyapunov's curve L. Let D^+ and D^- be occupied by isotropic homogeneous materials with the parameters $\lambda_1 > 0$ and $\lambda = 1$, respectively; the parameter λ_1 depends on the potential. Let sources and sinks modelled by the known complex potential [236]

$$Q(z) = \sum_{m=1}^{n} q_m (z - w_m)^{-\alpha_m} \qquad (5.4.1)$$

lie in the domain D^-, i.e., $w_m \in D^-$.

To demonstrate the physical sense of the problem let us discuss it in the context of electrical conductivity coupled with the heat effect. Assume that the electrical field is described by a potential satisfying the Laplace equation in D^+ and D^-. The electrical field yields heating the materials. The electrical conductivity of many materials depends on the temperature. We assume that we have such a material in D^+. If the temperature is distributed sufficiently rather in D^+, then λ_1 is an undetermined constant depending on the electrical potential via the temperature. In order to get a well-posed problem we assume that we know this dependence. If the contact between materials is perfect, then

$$u^+ = u^-, \quad \frac{\partial u^+}{\partial n} = \lambda_1 \frac{\partial u^-}{\partial n} \quad \text{on } L, \qquad (5.4.2)$$

where $u^+(z)$, $u^-(z)$ are the electrical potentials in D^+ and D^-, respectively. Such physical problems are usually studied by coupled nonlinear partial differential equations involving the electric potential and the temperature distribution. It is hard to solve such problems under general assumptions. Actually we simplify the physical couple problem and propose an easy algorithm.

The mathematical model of the problem consists of the following: find functions $\phi^+(z)$ and $\phi^-(z)$ analytic in D^+ and D^-, respectively, continuous in the closures of the domains considered (in designation of Section 2.2 $\phi^\pm \in \mathcal{C}_\pm(L)$) with the conjugation condition

$$\phi^-(t) = \phi^+(t) + \rho\overline{\phi^+(t)} + \rho\overline{Q(t)}, \; t \in L, \qquad (5.4.3)$$

where $\phi^+(\infty) = 0$, $\rho := (\lambda_1 - 1)(\lambda_1 + 1)^{-1}$. The unknown positive constant λ_1 is related to the unknown function $\phi^-(z)$ by a certain relation. It is convenient to write this relation in the form

$$\rho = \mathbf{B}\phi^- + b_0, \qquad (5.4.4)$$

where $\mathbf{B} : \mathcal{C}_+(L) \to \mathbb{R}_+$ is a given functional, $\mathbb{R}_+$ is the set of positive numbers. For the sake of simplicity we assume that $\mathbf{B}$ is a linear bounded operator. Problem (5.4.3), (5.4.4) can be referred to as a quasi-linear one, since the coefficients depend on the unknown function.

Let us explain why conditions (5.4.2) is written in the form (5.4.3). Introduce the complex potentials $\psi^+(z)$ in D^+ and $\psi^-(z)$ in D^- in such a way that $u^+(z) = \Re\psi^+(z)$, $u^-(z) = 2(\lambda_1+1)^{-1}\Re\psi^-(z)$. The function $\psi^-(z)$ is represented in the form

$$\psi^-(z) = \phi^+(z) + Q(z), \tag{5.4.5}$$

where $\phi^+ \in \mathcal{C}_-(L)$, and the given function $Q(z)$ has the form (5.4.1). Following Section 2.12 and Section 5.1 we rewrite (5.4.2) as the $\mathbb{R}$-linear condition

$$\psi^+(t) = \psi^-(t) + \rho\overline{\psi^-(t)}, \ t \in L. \tag{5.4.6}$$

Substituting (5.4.5) in (5.4.6) and introducing the function $\phi^-(z) := \psi^+(z) - Q(z)$ analytic in D^+ we arrive at condition (5.4.3).

We have the inequality $|\rho| < 1$, because the parameter λ_1 is always positive. Following Section 4.2 we reduce (5.4.3) to the integral equation

$$\mu = \frac{\rho}{2}\overline{(-\mu + \mathbf{S}\mu + Q)} \text{ on } L, \tag{5.4.7}$$

with respect to $\mu \in \mathcal{H}(L)$. Here $\mathbf{S}$ is the singular integral from Section 2.6, $\mu(t) = \phi^-(t) - \phi^+(t)$ on L. If $\rho(1 + \|\mathbf{S}\|) < 2$, then equation (5.4.7) has the unique solution

$$\mu(t) = \sum_{k=0}^{\infty} \rho^{k+1} \mu_k(t) \text{ on } L,$$

where

$$\mu_0 = \overline{Q}, \ \mu_1 = \frac{1}{2}\overline{(-\mu_0 + \mathbf{S}\mu_0 + Q)}, \ \mu_2 = \frac{1}{2}\overline{(-\mu_1 + \mathbf{S}\mu_1 + Q)}, \ldots$$

The functions $\phi^\pm(z)$ have the form

$$\phi^\pm(z) = \sum_{k=0}^{\infty} \rho^{k+1} (\mathbf{S}\mu_k)(z), \ z \in D^\pm. \tag{5.4.8}$$

Substituting $\phi^-(z)$ from (5.4.8) to (5.4.4) we arrive at the number equation

$$\rho = \sum_{k=0}^{\infty} \sigma_k \rho^{k+1} \tag{5.4.9}$$

with respect to ρ. Here $\sigma_k := \mathbf{B}\mathbf{S}\mu_k + b_0$, $k = 0, 1, \ldots$. We have the following

5.5. STEADY HEAT CONDUCTION ON CONFIGURATIONS.

Theorem 5.5 *Problem (5.4.3), (5.4.4) is solvable if and only if equation (5.4.9) is solvable and the solution belongs to the segment $(-1,+1)$. If $\rho \in (-1,+1)$ is a root of (5.4.9), then $\phi^{\pm}(z)$ has the form (5.4.8).*

Example 5.2 *Let L be the unit circle $\mathbb{U}$, and the functional $\mathbf{B}$ has the form $\mathbf{B} = \max_{|t|=1}|\phi^{-}(t)|$, $b_0 = 0$. It is easily seen that the $\mathbb{R}$-linear problem (5.4.3) has the unique solution*

$$\phi^{-}(z) = -\rho^2 Q(z), \quad \phi^{+}(z) = -\rho \overline{Q\left(\frac{1}{\overline{z}}\right)}. \tag{5.4.10}$$

Then (5.4.4) becomes

$$\rho = q\rho^2, \tag{5.4.11}$$

where $q := \max_{|t|=1}|Q(t)|$. The roots of (5.4.11) are $\rho = 0$ and $\rho = q^{-1}$. If $q > 1$, then $\phi^{\pm}(z)$ has the form (5.4.10).

We discuss now a generalization of problem (5.4.3), (5.4.4), when the domain D^- consists of the mutually disjoint simply connected domains D_k ($k = 1, 2, \ldots, n$) which model reactive inclusions, i.e., the conductivity λ_k of D_k depends on the potential in D_k and it is an undetermined constant. This statement of the problem is important in applications, when the sizes of the inclusions are sufficiently small and we may approximate nonlinear behavior of the conductivity by the functionals

$$\rho_k = \mathbf{B}\left[(1 - \rho_k)\phi_k(z)\right], \quad k = 1, 2, \ldots, n. \tag{5.4.12}$$

Here $\rho_k := (\lambda_k - 1)(\lambda_k + 1)^{-1}$; the multiplier $(1 - \rho_k) = 2(\lambda_k + 1)^{-1}$ appears in (5.4.12), because the potential $u_k(z) = 2(\lambda_k + 1)^{-1}\Re\phi_k(z)$. The complex potentials are related by the $\mathbb{R}$-linear condition

$$\phi(t) = \phi_k(t) - \rho_k\overline{\phi_k(t)} - Q(t), \quad t \in L_k, \quad k = 1, 2, \ldots, n, \tag{5.4.13}$$

where $Q(t)$ is a given function. If we know a solution of (5.4.13), then substituting it into (5.4.12) we obtain a set of equations with respect to ρ_k, $k = 1, 2, \ldots, n$. The $\mathbb{R}$-linear problem (5.4.13) has been solved in Section 5.2 for circular domains. At first the formulae from Section 5.2 are applied. Then we obtain a set of equations with respect to the constants ρ_k, $k = 1, 2, \ldots, n$ following from (5.4.12).

5.5 Steady heat conduction on special surfaces and configurations.

We consider the steady heat conduction described by the quasi-linear equation (5.1.2) on a surface and surfaces glued along curves which we shall call the configuration. The

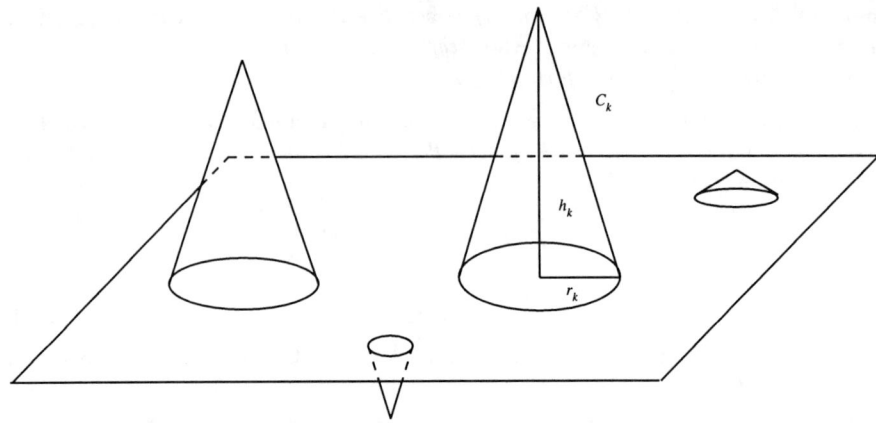

Figure 5.3: Special surface: plane with cones

operator ∇ from (5.1.2) can be considered on a surface. See [107, 236]. For general surfaces it is possible to discuss equation (5.1.2) as an elliptic equation and to apply the theory of the generalized analytic functions [30, 31, 34, 35, 227, 236, 268, 282].

5.5.1 A special surface.

In the present section we consider the surface S consisting of a plane multiply connected circular domain $\mathbb{D}$ on the complex plane $\widehat{\mathbb{C}}$ and cones C_k (Figure 5.3). The edges of $\mathbb{D}$ and C_k are glued along the circle $\mathbb{T}_k$.

Let us consider a fixed cone C_k (Figure 5.4). Each point P on C_k can be determined by two ways. The first one: P is determined by projection z on the disc $|z - a_k| < r_k$, where a_k is the projection of the vertex of C_k on the plane. The second way is the following. Let us cut the cone C_k along a generator, develop the cut cone on a plane and introduce a natural complex coordinate ζ. We introduce on the ζ-plane such a topology that the edges of the cut are identified. Transform C_k on the ζ-plane onto the z-plane by the conformal mapping

$$z = \lambda_k^{-\lambda_k}\zeta^{\lambda_k} + a_k \Leftrightarrow \zeta = \lambda_k(z - a_k)^{1/\lambda_k}, \qquad (5.5.1)$$

where $\lambda_k := \frac{2\pi}{\theta_k} = \frac{l_k}{r_k}$. The constant λ_k satisfies the inequality $1 < \lambda_k < +\infty$. The coordinate ζ on the cone C_k is local. We can construct the global conformal structure on the surface S by the following way. The domain $\mathbb{D}$ lies on the complex plane $\widehat{\mathbb{C}}$

5.5. STEADY HEAT CONDUCTION ON CONFIGURATIONS.

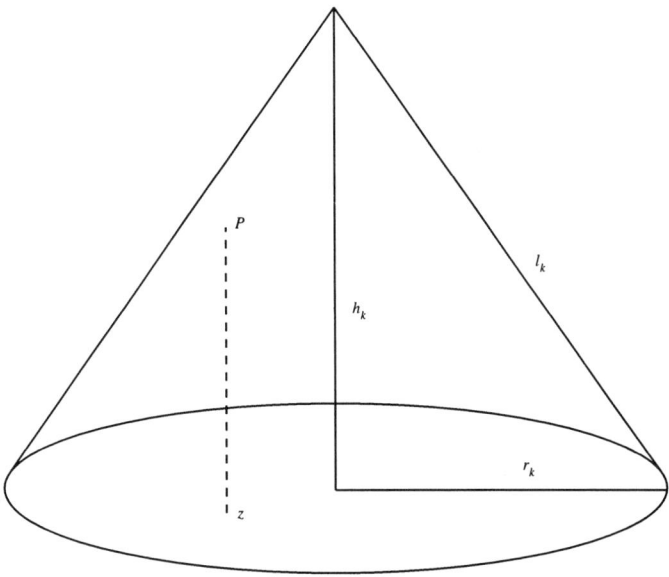

Figure 5.4: Cone

with natural coordinates. The coordinates on C_k are introduced by (5.5.1). Each point of S has one and only one complex coordinate z.

We write equation (5.1.2) in the coordinates introduced. This equation can be written only on smooth parts of the surface. Hence, (5.1.2) holds in $\mathbb{D}$ and $\mathbb{D}_k$, where $\nabla = (\partial/\partial x, \partial/\partial y)$ is written in the Cartesian coordinates, $z = x + iy$. Following Section 5.1 we introduce the function $u(x, y) := f(T(x, y))$ harmonic in $\mathbb{D}$ and $\mathbb{D}_k$. If the heat contact between the smooth parts of S is perfect, then the temperature distribution $T(x, y)$ satisfies the condition

$$T^+ = T^-, \quad \mathbf{q}_n^+ = \mathbf{q}_n^-, \qquad (5.5.2)$$

where, for instance, $T^+ := \lim_{z \to t} T(x, y)$, where $z = x + iy \in \mathbb{D}$, $t = x + iy \in \mathbb{T}_k$, q_n^+ is the normal heat flux on $\mathbb{T}_k$ from $\mathbb{D}$. We have

$$\mathbf{q}_n^+ = \lambda(T)\frac{\partial T^+}{\partial n} = \frac{\partial u^+}{\partial n}, \qquad (5.5.3)$$

$$\mathbf{q}_n^- = \lambda(T)\frac{\partial T^-}{\partial n}\lambda_k = \lambda_k \frac{\partial u^-}{\partial n} \text{ on } \mathbb{T}_k. \tag{5.5.4}$$

The normal vector on $\mathbb{T}_k$ coincides with the radius vector. The equality (5.5.3) expresses the heat flux on the plane and the equality (5.5.4) expresses the heat flux on the cone C_k in the z-coordinates. The multiplier λ_k appears in (5.5.4) because the real heat flux on the cone directed along the radius is greater λ_k times than its projection on $\mathbb{D}_k$. See Figure 5.4 and compare the coordinates z and ζ on C_k. The relations (5.5.2), (5.5.3), (5.5.4) yield the linear conjugation problem

$$u = u_k, \quad \frac{\partial u}{\partial n} = \lambda_k \frac{\partial u_k}{\partial n} \text{ on } \mathbb{T}_k, \ k = 1, 2, \ldots, n, \tag{5.5.5}$$

where $u_k := u$ in $cl\mathbb{D}_k$. Problem (5.5.5) is reduced to a $\mathbb{R}$-linear problem and is solved in closed form in Section 5.2.

Remark 5.3 *The described method can be also applied to surfaces having smooth components conformally isomorphic to plane domains.*

5.5.2 Configurations.

We consider steady heat conduction of the isotropic two-dimensional plates connected by a prescribed rule and generate a configuration $\mathcal{K}$. The simplest configuration is displayed in the Figure 5.5. First we need coordinates on $\mathcal{K}$. At the beginning let us explain how to introduce the coordinates in the simplest case, when we have three rectangles glued along the segment Γ (see Figure 5.5). Generate the sets $\mathsf{A}_1 := \{1; 2\}$, $\mathsf{A}_2 := \{1; 3\}$ and $\mathcal{S} := \{K(\mathsf{A}_1), K(\mathsf{A}_2)\}$, where $K(\mathsf{A}_1)$ and $K(\mathsf{A}_2)$ are the surfaces $K(\mathsf{A}_1) := D_1 \cup \Gamma \cup D_2$ and $K(\mathsf{A}_2) := D_1 \cup \Gamma \cup D_3$. The complex structure on $K(\mathsf{A}_1)$ is built in such a way that $K(\mathsf{A}_1)$ is a rectangle on the complex plane, where Γ is a segment of the real axes and D_1 and D_2 lie in the upper and lower half-planes, respectively (see Figure 5.5). The complex structure on $K(\mathsf{A}_2)$ is built in the same way. Moreover, these structures are compatible, i.e., D_1 in both structures has the same coordinates.

Let us consider the general case. Let the plates modeled by two-dimensional domains D_k ($k = 1, 2, \ldots, n$) endowed by a conformal structure intersect along a contour $\Gamma := \cap_{k=1}^n D_k$. The configuration $\mathcal{K}$ consists of D_k ($k = 1, 2, \ldots, n$) and Γ. Let us generate the set $\mathcal{S} := \{K(\mathsf{A}_1), K(\mathsf{A}_2), \ldots, K(\mathsf{A}_N)\}$, where A is a set of the numbers $k_1, k_2, \ldots, k_s$, where k_j can assume the values $k = 1, 2, \ldots, n$, and $K(\mathsf{A}_j)$ are surfaces consisting of D_k. These surfaces $K(\mathsf{A}) := \cup_{n \in \mathsf{A}} D_n \cup \Gamma$ are constructed in such a way that the set $\Gamma \cap K(\mathsf{A})$ for fixed A consists only of a finite set of the points. The

5.5. STEADY HEAT CONDUCTION ON CONFIGURATIONS. 229

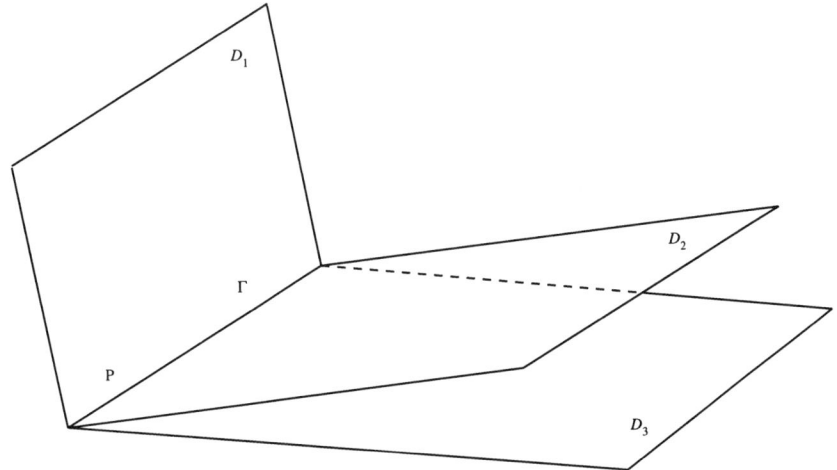

Figure 5.5: The simplest configuration

set of the surfaces S for the configuration $\mathcal{K}$ is called the *complete set* if each point of $\mathcal{K}$ belongs to a surface of the set S. We consider such configurations for which a complete set of orientated surfaces exists. Moreover, surfaces $K(\mathbf{A}_j)$ and $K(\mathbf{A}_l)$ of the complete set considered have the same conformal structures on $K(\mathbf{A}_j) \cap K(\mathbf{A}_l)$ for all j and l, if $K(\mathbf{A}_j) \cap K(\mathbf{A}_l) \neq \emptyset$.

Consider the steady heat conduction on $\mathcal{K}$. The temperature distribution satisfies the Laplace equation on each component D_k of $\mathcal{K}$. The boundary condition (Dirichlet, Neumann or mixed) is stated as in classical problems. We study the conjugation conditions on Γ. Let P be a point of Γ with the coordinates (x_0, y_0) in which m components of D_k meet (Figure 5.5). Let $u_k(x, y)$ be the temperature distribution on D_k ($k = 1, 2, ..., m$). If the contact between the plates is perfect, then all temperatures coincide at (x_0, y_0):

$$u_1(x_0, y_0) = u_2(x_0, y_0) = ... = u_m(x_0, y_0), \ (x_0, y_0) \in \Gamma. \tag{5.5.6}$$

The heat flux satisfies the condition

$$\frac{\partial u_1(x_0, y_0)}{\partial n} = \pm \frac{\partial u_2(x_0, y_0)}{\partial n} \pm \frac{\partial u_3(x_0, y_0)}{\partial n} \pm ... \pm \frac{\partial u_m(x_0, y_0)}{\partial n}, \ (x_0, y_0) \in \Gamma. \tag{5.5.7}$$

The sign in (5.5.7) is chosen in accordance with the conformal structure of D_k. The harmonic functions $u_k(x, y)$ define the complex potentials $\Phi_k(z)$, where $z = x + iy$.

For the sake of simplicity we consider only simply connected domains D_k. Then $\operatorname{Re} \Phi_k(z) = u_k(x,y)$. Equations (5.5.6) and (5.5.7) can be written in the form

$$\Re\Phi_1(t) = \Re\Phi_2(t) = \ldots = \Re\Phi_m(t), \tag{5.5.8}$$

$$\Im\Phi_1(t) = \Im\Phi_2(t) + \ldots + \Im\Phi_m(t), \ t \in \Gamma.$$

For the definiteness we take the sign plus. Using the relations

$$\Re\Phi_k(t) = \frac{1}{2}\left[\Phi_k(t) + \overline{\Phi_k(t)}\right], \ \Im\Phi_k(t) = \frac{1}{2i}\left[\Phi_k(t) - \overline{\Phi_k(t)}\right]$$

we rewrite (5.5.8) in the form

$$\begin{cases} \Phi_1(t) = \Phi_2(t) + \frac{1}{2}\left[\Phi_3(t) + \overline{\Phi_3(t)}\right] + \ldots + \frac{1}{2}\left[\Phi_m(t) + \overline{\Phi_m(t)}\right], & t \in \Gamma \\ \ldots \\ \Phi_1(t) = \Phi_m(t) + \frac{1}{2}\left[\Phi_2(t) + \overline{\Phi_2(t)}\right] + \ldots + \frac{1}{2}\left[\Phi_{m-1}(t) + \overline{\Phi_{m-1}(t)}\right], & t \in \Gamma. \end{cases} \tag{5.5.9}$$

In some cases the conditions (5.5.9) are reduced to a $\mathbb{R}$-linear conjugation problem. We study (5.5.9) in special cases characterizing the problem.

1. Let us consider two plates. The first one is modeled by an infinite plane; the second one is modeled by a bounded domain D_3. Let the second plate lie on the first one. We assume that these plates are thermally isolated except the curve Γ, the boundary of D_3, where the contact between the plates is perfect. Hence the first plate is divided onto two parts modeled by the domains $D_2 := D_3$ and $D_1 := \widehat{\mathbb{C}} \backslash cl D_2$. In this case (5.5.9) becomes

$$\Phi_1(t) = \Phi_2(t) + \frac{1}{2}\left[\Phi_3(t) - \overline{\Phi_3(t)}\right], \tag{5.5.10}$$

$$\Phi_1(t) = \Phi_3(t) + \frac{1}{2}\left[\Phi_2(t) - \overline{\Phi_2(t)}\right], \ t \in \Gamma.$$

For definiteness put $\Phi_1(\infty) = 0$. Let sources and sinks of the heat be on the configuration. We separate them in the complex potentials

$$\Phi_k(z) = \Psi_k(z) + Q_k(z), \ k = 1, 2, 3, \tag{5.5.11}$$

where $\Psi_k(z), Q_k(z) \in \mathcal{H}_A(D_k)$, $\Psi_k(z)$ is an unknown function, and $Q_k(z)$ is a given function. Substitute (5.5.11) in (5.5.10)

$$\Psi_1(t) = \Psi_2(t) + \frac{1}{2}\left[\Psi_3(t) - \overline{\Psi_3(t)}\right] + c_1(t), \tag{5.5.12}$$

5.5. STEADY HEAT CONDUCTION ON CONFIGURATIONS.

$$\Psi_1(t) = \Psi_3(t) + \frac{1}{2}\left[\Psi_2(t) - \overline{\Psi_2(t)}\right] + c_2(t), \ t \in \Gamma,$$

where, for instance, $c_1(t) = Q_2(t) + \frac{1}{2}\left[Q_3(t) - \overline{Q_3(t)}\right]$. It is easily seen that the functions $c_1(t)$ and $c_2(t)$ are related by the equality $\Im(c_1(t) - c_2(t)) = 0$.

The mathematical essence of problem (5.5.12) is not changed, if we identify the domains D_2 and D_3, i.e., to consider problem (5.5.12) on the plane with respect to $\Psi_1(z)$, $\Psi_2(z)$; $\Psi_3(z)$ analytic respectively in D_1; D_2. Subtracting the second equality (5.5.12) from the first one we obtain the Schwarz problem

$$\Re[\Psi_2(t) - \Psi_3(t)] = c_1(t) - c_2(t), \ t \in \Gamma, \qquad (5.5.13)$$

with respect to $\Psi_2(z) - \Psi_3(z)$ analytic in D_2. Let $g(z)$ be a vanishing at infinity solution of (5.5.13). Then $\Psi_3(z) = \Psi_3(z) - g(z)$, $z \in D_2$, and (5.5.12) becomes the $\mathbb{R}$-linear problem

$$\Psi_1(t) = \frac{3}{2}\Psi_2(t) - \frac{1}{2}\overline{\Psi_2(t)} + c_2(t) - g(t), \ t \in \Gamma. \qquad (5.5.14)$$

Let us consider the example when $\Gamma = \mathbb{T}(0;1)$ is the unit circle, the source and sink of the equal capacity q are located at the points $z = 0$ ($0 \in D_2$) and $z = \infty$, respectively. Then the solution to (5.5.10) has the form

$$\Phi_1(z) = q\log z, \ |z| \leq 1, \ \Phi_2(z) = q\log z, \ |z| \geq 1, \ \Phi_2(z) = 0, \ |z| \geq 1.$$

Therefore, the temperature distribution is

$$u(x,y) = \begin{cases} q\log\sqrt{x^2+y^2}, \ (x,y) \in D_1, D_2 \\ 0, \ (x,y) \in D_3. \end{cases}$$

2. Let us consider the heat conduction of the infinite plate and half-infinite cylinders perpendicular to the plane (see Figure 5.6). Let the plate and the cylinders be in perfect contact along a Lyapunov curve Γ. We introduce the usual complex coordinate on the plate. The complex coordinate on the cylinder is introduced as on a strip with the glued opposite sides.

In this case we arrive at problem (5.5.12) with $\Psi_1 \in \mathcal{C}_{\mathcal{A}}(D_1)$, $\Psi_2 \in \mathcal{C}_{\mathcal{A}}(D_2)$, $\Psi_3 \in \mathcal{C}_{\mathcal{A}}(D_3)$, where $D_1 := ext\Gamma$ ($\infty \in D_1$), $D_2 := int\Gamma$, D_3 is the strip modeled by the cylinder. Let us introduce the new unknown function $F_3(z) := \Psi_3(\exp z)$, analytic in the unit disc $\mathbb{U}(0;1)$. Then (5.5.12) becomes

$$\Psi_1(t) = \Psi_2(t) + \frac{1}{2}\left[F_3(\alpha(t)) - \overline{F_3(\alpha(t))}\right] + c_1(t), \qquad (5.5.15)$$

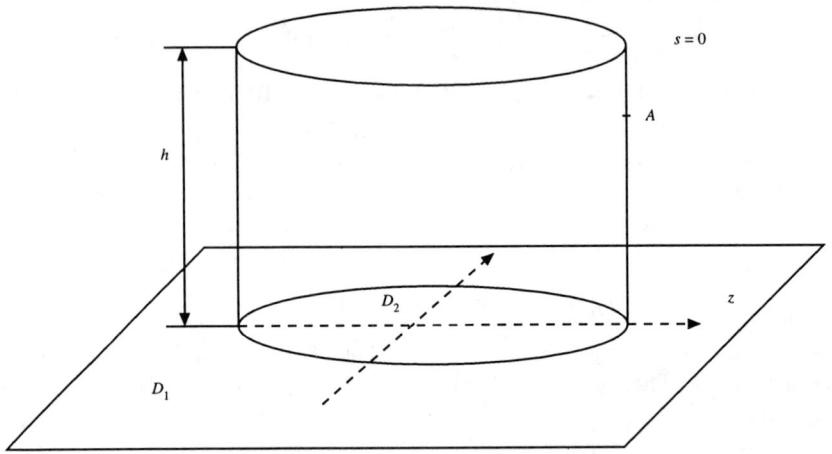

Figure 5.6: Configuration: cylinder on plane

$$\Psi_1(t) = F_3(\alpha(t)) + \frac{1}{2}\left[\Psi_2(t) - \overline{\Psi_2(t)}\right] + c_2(t), \ t \in \Gamma,$$

where $\alpha(t)$ is a homeomorphism preserving the natural parametrization of Γ. If $\Gamma = \mathbb{T}(0;1)$ is the unit circle, then $\alpha(t) \equiv t$, and (5.5.15) is reduced to (5.5.14).

3. Let us consider the same configuration as that in the previous example. Suppose only that $\Gamma = \mathbb{T}(0;1)$ is the unit circle, and the cylinder D_3 has a finite height with the circular boundary L, where the temperature distribution is given as a Hölder continuous function $f(t)$. The steady heat conduction of the configuration considered is reduced to the boundary value problem

$$\Psi_1(t) = \Psi_2(t) + \frac{1}{2}\left[F_3(t) - \overline{F_3(t)}\right], \ |t| = 1,$$

$$\Psi_1(t) = F_3(t) + \frac{1}{2}\left[\Psi_2(t) - \overline{\Psi_2(t)}\right] + c_2(t), \ |t| = 1, \qquad (5.5.16)$$

$$F_3(t) + \overline{F_3(t)} = f(t), \ |t| = r, \ \Psi_1(\infty) = 0.$$

We now reduce problem (5.5.16) to a functional equation.

We have $t = 1/\bar{t}$ on $|t| = 1$. Hence, the second condition (5.5.16) can be written in the form

$$\Psi_1(t) + \frac{1}{2}\overline{\Psi_2\left(\frac{1}{\bar{t}}\right)} = F_3(t) - \frac{1}{2}\Psi_2(t), \ |t| = 1. \qquad (5.5.17)$$

5.5. STEADY HEAT CONDUCTION ON CONFIGURATIONS.

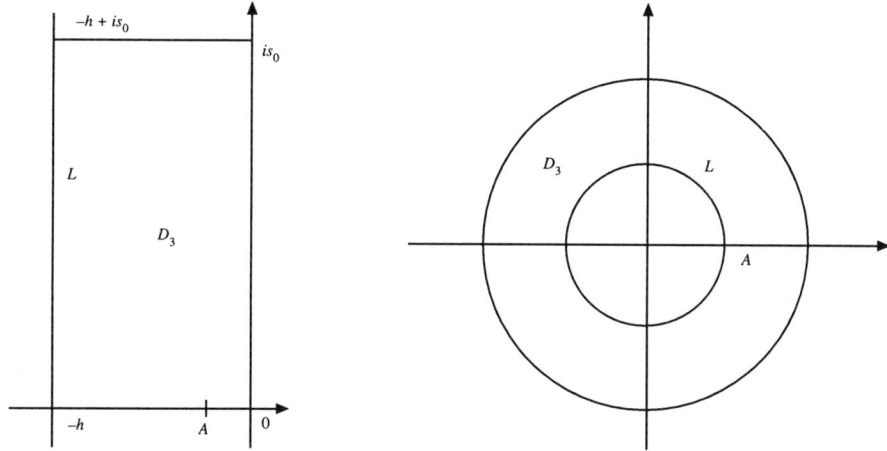

Figure 5.7: Complex structure on the cylinder

A function analytic in $r < |z| < 1$ is in the left-hand part of (5.5.16). By the Analytic Continuation Principle we introduce the function

$$\Omega(z) := \begin{cases} \Psi_1(z) + \frac{1}{2}\overline{\Psi_2\left(\frac{1}{\bar{z}}\right)}, & |z| > 1, \\ F_3(z) - \frac{1}{2}\Psi_2(z), & r < |z| < 1, \end{cases}$$

analytic in $|z| > r$. Let us find the limit boundary values of $\Omega(z)$ on $|t| = r$:

$$\Omega(t) = F_3(t) - \frac{1}{2}\Psi_2(t).$$

Using the third condition (5.5.16) we eliminate $F_3(t)$, then

$$\Omega(t) + \overline{\Omega(t)} = f(t) + \frac{1}{2}\left[\Psi_2(t) - \overline{\Psi_2(t)}\right], \quad |t| = r. \tag{5.5.18}$$

Using the relation $t = r^2/\bar{t}$ on $|t| = r$ we rewrite (5.5.18) in the form

$$\Omega(t) - \frac{1}{2}\overline{\Psi_2\left(\frac{r^2}{\bar{t}}\right)} + f^-(t) = -\overline{\Omega\left(\frac{r^2}{\bar{t}}\right)} + \frac{1}{2}\Psi_2(t) + f^+(t), \quad |t| = r, \tag{5.5.19}$$

where $f(t) = f^+(t) - f^-(t)$, $|t| = r$, is the representation of the Hölder function $f(t)$ in the form of difference of the functions $f^\pm \in \mathcal{H}_\pm(\mathbb{T}(0;r))$ by the Sokhotsky–Plemelj formulae (see Section 2.6). A function analytic in $|z| > r$ is in the left hand part of (5.5.19); a function analytic in $|z| < r$ is in the right hand part. Hence, by Liouville's theorem this relation determines a constant. This constant is equal to zero, since $\Omega(\infty) - \frac{1}{2}\overline{\Psi_2(0)} = 0$. Therefore,

$$\Omega(t) - \frac{1}{2}\overline{\Psi_2\left(\frac{r^2}{t}\right)} + f^-(t) = 0, \quad |z| > r.$$

Using the definition of $\Omega(z)$ we obtain

$$\Psi_1(z) + \frac{1}{2}\left[\overline{\Psi_2\left(\frac{1}{z}\right)} + \overline{\Psi_2\left(\frac{r^2}{z}\right)}\right] + f^-(t) = 0, \quad |z| > 1. \tag{5.5.20}$$

The first condition (5.5.16) implies that

$$\Psi_1(t) + \overline{\Psi_1(t)} = \Psi_2(t) + \overline{\Psi_2(t)}, \quad |t| = 1.$$

Hence,

$$\Psi_1(t) - \overline{\Psi_2\left(\frac{1}{t}\right)} = \Psi_2(t) - \overline{\Psi_1\left(\frac{1}{t}\right)}, \quad |t| = 1.$$

By virtue of Liouville's theorem we have

$$\Psi_1(z) - \overline{\Psi_2\left(\frac{1}{z}\right)} = c, \quad |z| \geq 1, \tag{5.5.21}$$

where c is a pure imaginary constant. Without loss of generality we can put that $c = 0$, since a complex potential is determined up to a pure imaginary constant. Eliminate $\Psi_1(z)$ from (5.5.20) and (5.5.21), change z by $1/\overline{z}$ and take the complex conjugation. As a result we obtain the functional equation

$$\Psi_2(z) = \frac{1}{3}\Psi_2(r^2 z) - \frac{2}{3}\overline{f^-\left(\frac{1}{\overline{z}}\right)}, \quad |z| \leq 1, \tag{5.5.22}$$

with respect to $\Psi_2 \in \mathcal{H}_A(\mathbb{U}(0;1))$ solved in Section 4.1:

$$\Psi_2(z) = -\frac{2}{3}\sum_{k=0}^{\infty}\frac{1}{3^k}\overline{f^-\left(\frac{1}{r^{2k}\overline{z}}\right)}, \quad |z| \leq 1.$$

The functions $\Psi_1(z)$ and $F_3(z)$ are determined by the relations (5.5.20) and (5.5.16).

5.6. AN ELASTIC PROBLEM FOR COMPOSITE MATERIALS.

Example 5.3 *Let the function $f(t)$ on L be given as a function in the natural parameter $f(s) = 2\cos\frac{2\pi}{s_0}s$, where $s_0 := |L|$ is the length of L. We consider the cylinder as a rectangle of the complex plane w with the glued opposite sides (see Figure 5.7). After conformal mapping of this rectangle onto an annulus we obtain the boundary value problem (5.5.16), where $r = \exp\left(\frac{2\pi}{s_0}h\right)$, h is the height of the cylinder, $f(t) = \frac{t}{r} + \frac{r}{t}$, and the point $s = 0$ of the circle L is mapped to the point $z = 1$. The functional equation (5.5.22) becomes*

$$\Psi_2(z) = \frac{1}{3}\Psi_2(r^2 z) + \frac{2}{3}rz, \quad |z| \le 1.$$

Its solution is $\Psi_2(z) = \frac{2r}{3-r^2}z$. It is easy to calculate the rest of the functions $\Psi_1(z) = \frac{2r}{3-r^2}z^{-1}$, $F_3(z) = \frac{2}{3-r^2}\left(\frac{3}{z} - z\right)$. Hence, the temperature distribution has the form

$$u(x,y) = \begin{cases} \frac{2r}{3-r^2}x, & (x,y) \in D_2, \\ \frac{2r}{3-r^2}\frac{x}{x^2+y^2}, & (x,y) \in D_1, \\ \frac{2r}{3-r^2}\cos\frac{2\pi}{s_0}y_1 \left[3\exp\left(-\frac{2\pi}{s_0}x_1\right) - \exp\left(\frac{2\pi}{s_0}x_1\right)\right] & (x_1,y_1) \in D_3. \end{cases}$$

5.6 An elastic problem of the mechanics of composite materials.

In the present section we consider a linear boundary value problem for a circular multiply connected domain $\mathbb{D}$ (see Figure 4.3) appearing in the stationary plane elasticity in the framework of Muskhelishvili's approach [206]. Let the domains $\mathbb{D}$, $\mathbb{D}_k$ be occupied by materials with elastic constants κ and μ, κ_k and μ_k, respectively. The stresses in $\mathbb{D}_k$ are expressed by the tensor

$$\begin{pmatrix} \sigma_{xx} & \sigma_{xy} \\ \sigma_{xy} & \sigma_{yy} \end{pmatrix}.$$

The component of the tensor can be determined by the Kolosov–Muskhelishvili formulae [206]

$$\sigma_{xx} + \sigma_{yy} = 4\Re\phi'_k(z), \quad \sigma_{xx} - \sigma_{yy} + 2i\sigma_{xy} = -2\left[\overline{z\phi''_k(z)} + \overline{\psi'_k(z)}\right],$$

where the functions $\phi_k(z)$ and $\psi_k(z)$ belong to $\mathcal{C}^2_{\mathcal{A}}(\mathbb{D}_k)$. The normal forces on $\mathbb{T}_k$ are given by the expression

$$\mathbf{T}_n^- = \phi_k(t) + t\overline{\phi'_k(t)} + \overline{\psi_k(t)}. \tag{5.6.1}$$

The displacements in $\mathbb{D}_k$ are expressed by the complex potentials:

$$U = \frac{1}{2\mu_k}\left[\kappa_k \phi_k(z) - z\overline{\phi'_k(z)} - \overline{\psi_k(z)}\right]. \tag{5.6.2}$$

Similar representations for the stresses and displacements hold in $cl\mathbb{D}$:

$$\mathbf{T}_n^+ = \phi(t) + t\overline{\phi'(t)} + \overline{\omega(t)}, \ t \in \partial\mathbb{D}, \tag{5.6.3}$$

$$U = \frac{1}{2\mu}\left[\kappa\phi(z) - z\overline{\phi'(z)} - \overline{\omega(z)}\right], \ z \in cl\mathbb{D},$$

where

$$\phi(z) = -P\log z + \varphi(z), \ \omega(z) = Q\log z + \psi(z), \tag{5.6.4}$$

P and Q are given complex constants, and $\varphi(z)$ and $\psi(z)$ belong to $\mathcal{C}_\mathcal{A}^2(\mathbb{D})$. We assume that a branch of $\log z$ is chosen in such a way that the cut connecting $z = 0$ and $z = \infty$ lies in $\mathbb{D}$.

We assume that the contact between different materials is perfect, i.e.,

$$\mathbf{T}_n^+ = \mathbf{T}_n^-, \ U^+ = U^- \text{ on } \partial\mathbb{D}, \tag{5.6.5}$$

where, for instance, $U^+(t) := \lim_{z \to t, \ z \in \mathbb{D}} U(z)$. Using the relation (5.6.1) - (5.6.4) we write the boundary condition (5.6.5) in the form

$$\phi_k(t) + t\overline{\phi'_k(t)} + \overline{\psi_k(t)} = \varphi(t) + t\overline{\varphi'(t)} + \overline{\omega(t)} + f(t),$$

$$\frac{1}{\mu_k}\left[\kappa_k \phi_k(t) - t\overline{\phi'_k(t)} - \overline{\psi_k(t)}\right] = \frac{1}{\mu}\left[\kappa\varphi(t) - t\overline{\varphi'(t)} - \overline{\omega(t)}\right] + g(t), \tag{5.6.6}$$

$$|t - a_k| = r_k, \ k = 1, 2, \ldots, n,$$

where $f(t)$ and $g(t)$ are given functions. It is worthwhile to study (5.6.6) with arbitrary given doubly differentiable functions $f(t)$ and $g(t)$, because problems with point forces, thermoelastic problems and others are reduced to (5.6.6) with various $f(t)$ and $g(t)$.

Let us introduce the new unknown functions

$$\Phi_k(z) = \left(\frac{r_k^2}{z - a_k} + \overline{a_k}\right)\phi'_k(z) + \psi_k(z), \ |z - a_k| \leq r_k,$$

$$\Psi_k(z) = \left(\frac{r_k^2}{z - a_k} + \overline{a_k}\right)\varphi'(z) + \psi(z), \ z \in cl\mathbb{D}, \ k = 1, 2, \ldots, n.$$

5.6. AN ELASTIC PROBLEM FOR COMPOSITE MATERIALS.

The functions $\Psi_k(z)$ are related by the identities

$$\Psi_k(z) = \Psi_l(z) + \lambda_{kl}(z)\varphi'(z), \quad z \in cl\mathbb{D}, \tag{5.6.7}$$

where

$$\lambda_{kl}(z) := \frac{r_k^2}{z - a_k} + \overline{a_k} - \frac{r_l^2}{z - a_l} - \overline{a_l}, \quad l, k = 1, 2, \ldots, n.$$

The condition (5.6.6) becomes

$$\phi_k(t) + \overline{\Phi_k(t)} = \varphi(t) + \overline{\Psi_k(t)} + f(t),$$

$$\frac{\mu}{\mu_k}\left[\kappa_k \phi_k(t) - \overline{\Phi_k(t)}\right] = \kappa\varphi(t) - \overline{\Psi_k(t)} + g(t), \quad |t - a_k| = r_k.$$

Eliminating $\overline{\Psi_k(t)}$ and $\varphi(t)$ we obtain

$$\left(1 + \frac{\mu}{\mu_k}\kappa_k\right)\phi_k(t) + \left(1 - \frac{\mu}{\mu_k}\right)\overline{\Phi_k(t)} = (1+\kappa)\varphi(t) + \widetilde{f}(t), \tag{5.6.8}$$

$$\left(\kappa - \frac{\mu}{\mu_k}\kappa_k\right)\overline{\phi_k(t)} + \left(\kappa + \frac{\mu}{\mu_k}\right)\Phi_k(t) = (1+\kappa)\Psi_k(t) + \widetilde{g}(t), \quad |t - a_k| = r_k, \tag{5.6.9}$$

where $\widetilde{f}(t) := f(t) + g(t)$, $\widetilde{g}(t) := -\kappa f(t) - g(t)$. The form of (5.6.8) and (5.6.9) is similar to an $\mathbb{R}$-linear condition, since φ and Ψ_k are analytic in $\mathbb{D}$, ϕ_k is analytic in $\mathbb{D}_k$, Φ_k is analytic in $\mathbb{D}_k$ except $z = a_k$, where its principal part has the form $\frac{r_k^2}{z-a_k}\phi_k'(a_k)$. Following our approach we deduce problem (5.6.8), (5.6.9) to a system of functional equations.

Using (5.6.8) we introduce the function

$$\Omega(z) := \begin{cases} \left(1 + \frac{\mu}{\mu_k}\kappa_k\right)\phi_k(z) - \sum_{m \neq k}\left(1 - \frac{\mu}{\mu_l}\right)\overline{\Phi_m(z_{(m)}^*)} - f^-(z), & |z - a_k| \leq r_k, \\ \quad k = 1, 2, \ldots, n, \\ (1+\kappa)\varphi(z) - \sum_{m=1}^{n}\left(1 - \frac{\mu}{\mu_m}\right)\overline{\Phi_m(z_{(m)}^*)} - f^+(z), & z \in \mathbb{D}, \end{cases}$$

where $\widetilde{f}(t) = f^+(t) - f^-(t)$, $t \in \partial\mathbb{D}$ is the representation of $\widetilde{f}(t)$ as the difference of functions $f^{\pm} \in \mathcal{H}_{\pm}(\partial\mathbb{D})$ by Sokhotsky–Plemelj formulae. The function $\Omega(z)$ is analytic in $\widehat{\mathbb{C}}$ except infinity, where its principal part has the form pz. The constant p is defined by

$$p := -\sum_{k=1}^{n}\left(1 - \frac{\mu}{\mu_k}\right)\overline{\phi_k'(a_k)}. \tag{5.6.10}$$

Hence, $\Omega(z) = p_0 + pz$, where p_0 is a constant. It follows from the definition of $\Omega(z)$ that

$$\left(1 + \frac{\mu}{\mu_k}\kappa_k\right)\phi_k(z) = \sum_{m \neq k}\left(1 - \frac{\mu}{\mu_m}\right)\overline{\Phi_m(z_{(m)}^*)} + p_0 + pz + f^-(z), \quad |z - a_k| \leq r_k, \tag{5.6.11}$$

$$(1 + \kappa)\varphi(z) = \sum_{m=1}^{n}\left(1 - \frac{\mu}{\mu_m}\right)\overline{\Phi_m(z_{(m)}^*)} + p_0 + pz + f^+(z), \quad z \in c l\mathbb{D}. \tag{5.6.12}$$

Differentiate (5.6.12) and substitute it into (5.6.7)

$$\Psi_k(z) = \Psi_1(z) + \frac{\lambda_{k1}(z)}{1+\kappa}\left[\sum_{m=1}^{n}\left(1 - \frac{\mu}{\mu_m}\right)\left[\overline{\Phi_m(z_{(m)}^*)}\right]' + p + f^{+\prime}(z)\right].$$

Then (5.6.9) can be written as the following

$$\left(\kappa - \frac{\mu}{\mu_k}\kappa_k\right)\overline{\phi_k(t)} + \left(\kappa + \frac{\mu}{\mu_k}\right)\Phi_k(t) = (1+\kappa)\Psi_1(t) \tag{5.6.13}$$

$$+\lambda_{k1}(t)\left[\sum_{m=1}^{n}\left(1 - \frac{\mu}{\mu_m}\right)\left[\overline{\Phi_m(z_{(m)}^*)}\right]' + p + f^{+\prime}(z)\right] + \tilde{g}(t), \quad |t - a_k| = r_k.$$

It is convenient to rewrite (5.6.13) in the form

$$(1+\kappa)\Psi_1(t) = \omega_k(t) + \left(\kappa - \frac{\mu}{\mu_k}\kappa_k\right)\overline{\phi_k(t)} \tag{5.6.14}$$

$$-\lambda_{k1}(t)\left(1 - \frac{\mu}{\mu_k}\right)\left[\overline{\Phi_k(t_{(k)}^*)}\right]' + F^+(t) - F^-(t) - p\lambda_{k1}(t),$$

where

$$\omega_k(z) := \left(\kappa + \frac{\mu}{\mu_k}\right)\Phi_k(z) - \lambda_{k1}(z)\sum_{m \neq k}\left(1 - \frac{\mu}{\mu_m}\right)\left[\overline{\Phi_m(z_{(m)}^*)}\right]', \quad z \in cl\mathbb{D},$$

and $-\lambda_{k1}(t)f^{+\prime}(t) + \tilde{g}(t) = F^+(t) - F^-(t)$ is the representation by Sokhotsky–Plemelj

5.6. AN ELASTIC PROBLEM FOR COMPOSITE MATERIALS.

formulae. Using (5.6.14) we introduce the function

$$\omega(z) := \begin{cases} \omega_k(z) - \sum_{m \neq k} \left\{ \left(\kappa - \frac{\mu}{\mu_m} \kappa_m \right) \overline{\phi_m(z^*_{(m)})} - \lambda_{m1}(t) \left(1 - \frac{\mu}{\mu_m}\right) \overline{[\Phi_m(z^*_{(m)})]} \right\} \\ \quad -F^-(z), \; |z - a_k| \leq r_k, \; k = 1, 2, ..., n, \\ \\ (1+\kappa)\Psi_1(z) - \sum_{m=1}^{n} \left\{ \left(\kappa - \frac{\mu}{\mu_m}\kappa_m \right) \overline{\phi_m(z^*_{(m)})} \right. \\ \quad \left. -\lambda_{m1}(t)\left(1 - \frac{\mu}{\mu_m}\right) \overline{[\Phi_m(z^*_{(m)})]} \right\} - F^+(z) + p\lambda_{k1}(z), \; z \in \mathbb{D}, \end{cases}$$

analytic in $\widehat{\mathbb{C}}$ except the points $z = a_k$. Analyzing the principal part of $\omega(z)$ and using (5.6.11) we conclude that

$$\omega(z) = \sum_{k=1}^{n} \frac{r_k^2 q_k}{z - a_k} + q_0, \tag{5.6.15}$$

where q_0 is a constant, $q_k = p + f'_k(a_k) + \phi'_k(a_k)\left(-1 + \kappa - \frac{\mu}{\mu_k}\kappa_k + \frac{\mu}{\mu_k}\right)$, $k = 1, 2, \ldots, n$, p has the form (5.6.10). It follows from the definition of $\omega(z)$ that

$$\left(\kappa + \frac{\mu}{\mu_k}\right)\Phi_k(z) = \sum_{m \neq k}\left\{ \left(\kappa - \frac{\mu}{\mu_m}\kappa_m\right)\overline{\phi_m(z^*_{(m)})} \right. \tag{5.6.16}$$

$$\left. + \lambda_{km}(z)\left(1 - \frac{\mu}{\mu_m}\right)\overline{[\Phi_m(z^*_{(m)})]'} \right\} + F^-(z) + \omega(z), \; |z - a_k| \leq r_k, \; k = 1, 2, ..., n.$$

$2n$ relations (5.6.11), (5.6.16) constitute a system of functional equations with respect to ϕ_k and Φ_k ($k = 1, 2, ..., n$), where $\phi_k(z)$, $\Phi_k(z) - \frac{r_k^2 \phi'_k(a_k)}{z - a_k} \in \mathcal{C}_A(\mathbb{D}_k)$; the known functions f^- and F^- also belong to $\mathcal{C}_A(\mathbb{D}_k)$.

In Chapter 4 we solved such systems by the method of successive approximations. Here we only note that the same method is applied at least if $1 - \frac{\mu}{\mu_m}$ and $\kappa - \frac{\mu}{\mu_m}\kappa_m$ are sufficiently small. This case corresponds to weakly inhomogeneous materials, when $\mu \approx \mu_m$ and $\kappa \approx \kappa_m$. Now we would like to present another method, the method of undetermined coefficients (see Addition Theorems in Subsection 4.9.2). Put

$$\phi_k(z) = \sum_{m=0}^{\infty} \alpha_{mk}(z - a_k)^m, \; \Phi_k(z) = \frac{r_k^2 \alpha_{1k}}{z - a_k} + \sum_{m=0}^{\infty} \beta_{mk}(z - a_k)^m, \; k = 1, 2, \ldots, n, \tag{5.6.17}$$

where α_{mk}, β_{mk} are unknown constants. Then

$$\overline{\phi_k(z^*_{(k)})} = \sum_{m=0}^{\infty} \overline{\alpha_{mk}} r_k^{2m} (z-a_k)^{-m}, \quad \overline{\Phi_k(z^*_{(k)})} = \overline{\alpha_{1k}} (z-a_k) + \sum_{m=0}^{\infty} \overline{\beta_{mk}} r_k^{2m} (z-a_k)^{-m},$$

$$\left[\overline{\Phi_k(z^*_{(k)})}\right]' = \overline{\alpha_{1k}} + \sum_{m=1}^{\infty} \overline{\beta_{mk}} r_k^{2m} (z-a_k)^{-(m+1)}, \quad k=1,2,\ldots,n.$$
(5.6.18)

Let us also represent the known functions in the form of expansions:

$$\lambda_{lk}(z) = \frac{r_l^2}{z-a_l} + \overline{a_l} - \overline{a_k} - \frac{r_k^2}{a_l - a_k} \sum_{m=0}^{\infty} \left(\frac{z-a_l}{a_k - a_l}\right)^m, \quad |z-a_l| \le r_l, \; l,k=1,2,\ldots,n; \; l \ne k.$$

For instance,

$$h_k(z) = \sum_{m=0}^{\infty} h_{mk} (z-a_k)^m,$$

where h_{km} are given constants. Reexpand the functions (5.6.18) on $(z-a_l)^m$ with $l \ne k$:

$$\overline{\Phi_k(z^*_{(k)})} = \overline{\alpha_{1k}}(z-a_l) + \overline{\alpha_{1k}}(a_l - a_k) + \sum_{m=0}^{\infty} \overline{\beta_{mk}} \frac{r_k^{2m}}{(a_l - a_k)^m} \left[\sum_{j=0}^{\infty} \left(\frac{z-a_l}{a_k - a_l}\right)^j\right]^m,$$

$$\left[\overline{\Phi_k(z^*_{(k)})}\right]' = \overline{\alpha_{1k}} + \sum_{m=0}^{\infty} \overline{\beta_{mk}} \frac{-mr_k^{2m}}{(a_l - a_k)^{m+1}} \left[\sum_{j=0}^{\infty} \left(\frac{z-a_l}{a_k - a_l}\right)^j\right]^{m+1},$$

$$\overline{\phi_k(z^*_{(k)})} = \sum_{m=0}^{\infty} \overline{\alpha_{mk}} \frac{r_k^{2m}}{(a_l - a_k)^m} \left[\sum_{j=0}^{\infty} \left(\frac{z-a_l}{a_k - a_l}\right)^j\right]^m.$$

Substitute these expansions into (5.6.11) and (5.6.16), and select and equate the coefficients with the same powers of $(z-a_l)$. As a result we obtain a linear algebraic system with respect to α_{mk} and β_{mk}.

This system can be solved by the method of reduction [114] which consists in changing the infinite sum $\sum_{m=0}^{\infty}$ in (5.6.17), (5.6.18) by the finite one $\sum_{m=0}^{N}$. The number N is determined by the desired accuracy. We do not declare that the method of reduction gives an exact solution. Here we only assert that this method allows us to obtain approximate analytical formulae which are useful in applications, because they contain parameters in symbolic form. We now proceed to deduce such simple

5.6. AN ELASTIC PROBLEM FOR COMPOSITE MATERIALS.

formulae. For the sake of clarity we consider the case, when the accuracy is defined by $N = 1$. Even in this simplest case we get new formulae not known earlier. For $N = 1$ (5.6.17) becomes

$$\phi_k(z) \approx \alpha_{0k} + \alpha_{1k}(z - a_k), \quad \Phi_k(z) \approx \frac{r_k^2 \alpha_{1k}}{z - a_k} + \beta_{0k} + \beta_{1k}(z - a_k). \qquad (5.6.19)$$

Substituting (5.6.19) in (5.6.11) and (5.6.16) and selecting the coefficients on $(z - a_k)^s$, for $s = 0, 1$, we obtain a system of $\mathbb{R}$-linear algebraic equations with respect to unknowns $p_0, q_0, \alpha_{0k}, \alpha_{1k}, \beta_{0k}, \beta_{1k}$. It is easy to check that the coefficient on $(z - a_k)^{-1}$ in (5.6.16) gives an identity. The remaining equations are

$$\left(1 + \frac{\mu}{\mu_k} \kappa_k\right) \alpha_{1k} = -\sum_{m \neq k} \left(1 - \frac{\mu}{\mu_m}\right) \frac{r_m^2 \overline{\beta_{1m}}}{(a_k - a_m)^2} - \left(1 - \frac{\mu}{\mu_k}\right) \overline{\alpha_{1k}} + p + f^{-\prime}(a_k), \qquad (5.6.20)$$

$$\left(1 + \frac{\mu}{\mu_k} \kappa_k\right) \beta_{1k} = -\sum_{m \neq k} \left(\kappa - \frac{\mu}{\mu_m} \kappa_m\right) \frac{r_m^2 \overline{\alpha_{1m}}}{(a_k - a_m)^2}$$

$$+ \sum_{m \neq k} \left(1 - \frac{\mu}{\mu_m}\right) \frac{r_m^2}{(a_k - a_m)^2} \left(\overline{\alpha_{1m}} + \overline{\beta_{1m}} \frac{-2|a_k - a_m|^2 + 3r_m^2}{(a_k - a_m)^4}\right) + F^{-\prime}(a_k) \qquad (5.6.21)$$

$$- \sum_{m \neq k} \frac{r_m^2}{(a_k - a_m)^2} \left[\left(\kappa + \frac{\mu}{\mu_m} - 1 - \frac{\mu}{\mu_m}\kappa_m\right)\alpha_{1m} - \sum_{s=1}^{n}\left(1 - \frac{\mu}{\mu_s}\right)\overline{\alpha_{1s}} + f^{+\prime}(a_m)\right],$$

$$\left(1 + \frac{\mu}{\mu_k} \kappa_k\right) \alpha_{0k} = \sum_{m \neq k} \left(1 - \frac{\mu}{\mu_m}\right) \overline{\beta_{0m}} + p_0 + \sum_{m \neq k} \left(-a_m \overline{\alpha_{1m}} + \frac{r_m^2 \overline{\beta_{1m}}}{a_k - a_m}\right) \qquad (5.6.22)$$

$$- a_k \left(1 - \frac{\mu}{\mu_k}\right) \overline{\alpha_{1k}} + f^-(a_k),$$

$$\left(\kappa + \frac{\mu}{\mu_k}\right) \beta_{0k} = \sum_{m \neq k} \left(\kappa - \frac{\mu}{\mu_m} \kappa_m\right) \overline{\alpha_{0m}} + q_0 + \sum_{m \neq k} \left\{\left(\kappa - \frac{\mu}{\mu_m}\kappa_m\right) \frac{r_m^2 \overline{\alpha_{1m}}}{a_k - a_m}\right.$$

$$+ \left(1 - \frac{\mu}{\mu_m}\right) \left(\overline{a_k} - \overline{a_m} - \frac{r_m^2}{a_k - a_m}\right) \left(\overline{\alpha_{1m}} - \frac{r_m^2 \overline{\beta_{1m}}}{(a_k - a_m)^2}\right)\right\} + F^-(a_k) + q_0 \qquad (5.6.23)$$

$$+ \sum_{m \neq k} \frac{r_m^2}{a_k - a_m} \left[-\sum_{s=1}^{n} \left(1 - \frac{\mu}{\mu_s}\right) \overline{\alpha_{1s}} + f'_m(a_m) + \overline{\alpha_{1m}} \left(-1 + \kappa - \frac{\mu}{\mu_m}\kappa_m + \frac{\mu}{\mu_m}\right)\right],$$

$$k = 1, 2, ..., n.$$

First we have to solve the system (5.6.20), (5.6.21) with respect to α_{1k}, β_{1k}. Then we solve the system (5.6.22), (5.6.23) with respect to α_{0k}, β_{0k}, p_0, q_0. The values α_{0k}, β_{0k}, p_0, q_0 do not impact on the stress distribution. Some of them stay undetermined. Thus we concentrate our attention on the system (5.6.20), (5.6.21). Frequently sets of equations are hard to solve even numerically. For instance, it is interesting to consider the case when the inclusions are far away from each other, i.e., the values $r_m |a_k - a_m|^{-1}$ are sufficiently small. Then we can apply the method of successive approximations with zero-th approximation

$$\left(1 + \frac{\mu}{\mu_k}\kappa_k\right)\alpha_{k1}^{(0)} = -\sum_{m\neq k}^{n}\left(1 - \frac{\mu}{\mu_m}\right)\overline{\alpha_{1m}^{(0)}} + f^{-\prime}(a_k),$$

$$\left(1 + \frac{\mu}{\mu_k}\kappa_k\right)\beta_{1k}^{(0)} = F^{-\prime}(a_k) + f^{+\prime}(a_m).$$

Solving this trivial system we substitute the result in the right-hand part of (5.6.20), (5.6.21) and obtain the first order approximation. We do not write the final results. Note only that even the first order approximation gives new analytical formulae for the stress distribution for the composite material under consideration.

5.7 Plane Stokes flow of the piece-wise homogeneous Newtonian liquid.

We consider a two-dimensional hydrodynamic problem derived by Stokes' equations (cf., e.g., [107])

$$\frac{\partial \sigma_{xx}}{\partial x} + \frac{\partial \sigma_{xy}}{\partial y} = 0, \quad \frac{\partial \sigma_{xy}}{\partial x} + \frac{\partial \sigma_{yy}}{\partial y} - \rho g = 0, \quad \frac{\partial v_x}{\partial x} + \frac{\partial v_y}{\partial y} = 0, \quad (5.7.1)$$

$$\sigma_{xx} = 2\mu\frac{\partial v_x}{\partial x} - p, \quad \sigma_{xy} = \mu\left(\frac{\partial v_x}{\partial x} + \frac{\partial v_y}{\partial y}\right), \quad \sigma_{yy} = 2\mu\frac{\partial v_x}{\partial x} - p,$$

where $\begin{pmatrix} \sigma_{xx} & \sigma_{xy} \\ \sigma_{xy} & \sigma_{yy} \end{pmatrix}$ is the stress tensor, (v_x, v_y) is the velocity vector, μ is the viscosity of the fluid, p is the pressure. We assume that liquid is in the gravitational field, and the gravitational forces are parallel to the y-axis. Equations (5.7.1) involve the external body force ρg corresponding to this gravitational field. Here, ρ is the density of liquid, g is the acceleration of gravity.

5.7. PLANE STOKES FLOW.

Let us consider the case, when ρ and μ are piece-wise constant functions

$$\rho(z) := \begin{cases} \rho_k, & z \in \mathbb{D}_k, \ k = 1, 2, \ldots, n \\ \rho_0 & z \in \mathbb{D}, \end{cases}, \quad \mu(z) := \begin{cases} \mu_k, & z \in \mathbb{D}_k, \ k = 1, 2, \ldots, n \\ \mu_0 & z \in \mathbb{D}, \end{cases}$$

and circular domains $\mathbb{D}_k$, $\mathbb{D}$ are displayed in Figure 4.3. We introduce the new unknown functions

$$p^0 := p - \rho g y, \quad \sigma_{xx}^0 := \sigma_{xx} - \rho g y, \quad \sigma_{yy}^0 := \sigma_{yy} - \rho g y.$$

Then equations (5.7.1) become

$$\frac{\partial \sigma_{xx}^0}{\partial x} + \frac{\partial \sigma_{xy}}{\partial y} = 0, \quad \frac{\partial \sigma_{xy}}{\partial x} + \frac{\partial \sigma_{yy}^0}{\partial y} = 0, \quad \frac{\partial v_x}{\partial x} + \frac{\partial v_y}{\partial y} = 0, \tag{5.7.2}$$

$$\sigma_{xx} = 2\mu \frac{\partial v_x}{\partial x} - p^0, \quad \sigma_{xy} = \mu \left(\frac{\partial v_x}{\partial x} + \frac{\partial v_y}{\partial y} \right), \quad \sigma_{yy}^0 = 2\mu \frac{\partial v_x}{\partial x} - p,$$

in each component $\mathbb{D}$ and $\mathbb{D}_k$. Equations (5.7.2) have such a form that the Kolosov-Muskhelishvili complex potentials can be introduced

$$v_x + i v_y = \frac{1}{\mu_0} \left[\varphi(z) - \overline{z \varphi'(z)} - \overline{\omega(z)} \right], \quad p^0 = -2 \operatorname{Re} \varphi'(z),$$

$$i \mathbf{N} = \varphi(z) + \overline{z \varphi'(z)} + \overline{\omega(z)}, \quad z = x + i y \in \mathbb{D}.$$

The vector $\mathbf{N}(z_1) - \mathbf{N}(z_2)$ denote the vector of forces applied to the arc $\widehat{z_1 z_2}$. Analogous potentials $\phi_k(z)$, $\psi_k(z)$ are introduced in $\mathbb{D}_k$.

We assume that the perfect contact holds between the components of liquid on each circle $\mathbb{T}_k$ dividing $\mathbb{D}$ and $\mathbb{D}_k$. This can be expressed by the zero jump of the velocity

$$[(v_x, v_y)]^{\pm} = 0, \tag{5.7.3}$$

and by the zero jump of the normal stresses

$$[\sigma_{xn}]^{\pm} = [\sigma_{yn}]^{\pm} = 0 \text{ on each } \mathbb{T}_k. \tag{5.7.4}$$

Here $[f]^{\pm}$ denote the jump $f(t) := \lim_{z \to t \ z \in \mathbb{D}_k} f(z) - \lim_{z \to t \ z \in \mathbb{D}} f(z)$. Using the complex potentials we write (5.7.3) as follows

$$\frac{1}{\mu_k} \left[\phi_k(t) - t \overline{\phi_k'(t)} - \overline{\psi_k(t)} \right] = \frac{1}{\mu} \left[\phi(t) - t \overline{\phi'(t)} - \overline{\psi(t)} \right], \quad |t - a_k| = r_k, \ k = 1, 2, \ldots, n. \tag{5.7.5}$$

It follows from (5.7.4) that

$$[\sigma_{xx}n_x + \sigma_{xy}n_y]^{\pm} = 0 \iff [\sigma^0_{xx}n_x + \sigma_{xy}n_y]^{\pm} = -gyn_x [\rho]^{\pm}, \qquad (5.7.6)$$

$$[\sigma_{xy}n_x + \sigma_{yy}n_y]^{\pm} = 0 \iff [\sigma_{xy}n_x + \sigma^0_{yy}n_y]^{\pm} = -gyn_y [\rho]^{\pm},$$

where (n_x, n_y) is the normal unit vector to L_k. Using the complex potentials it is possible to show that

$$\frac{d}{ds}\left[\phi_k(t) + t\overline{\phi'_k(t)} + \overline{\psi_k(t)} - \phi(t) + t\overline{\phi'(t)} + \overline{\psi(t)}\right] \qquad (5.7.7)$$

$$= \frac{iq_k}{2}(t - \bar{t}) \quad |t - a_k| = r_k, \ k = 1, 2, \ldots, n,$$

where $q_k := (\rho_k - \rho_0)g$, $\frac{d}{ds} = n_y \frac{\partial}{\partial x} - n_x \frac{\partial}{\partial y}$ is the derivative on the natural parameter s of the circle $\mathbb{T}_k$.

In order to find functions $\phi_k(z), \psi_k(z) \in \mathcal{C}^1_A(\mathbb{D}_k)$, $k = 1, 2, \ldots, n$, we represent functions $\phi(z), \psi(z)$ in the form

$$\phi(z) = \sum_{j=1}^{n} A_j \log(z - a_j) + \phi_0(z), \ \psi(z) = \sum_{j=1}^{n} (-\overline{A_j}) \log(z - a_j) + \psi_0(z), \qquad (5.7.8)$$

where $\phi_0(z), \psi_0(z) \in \mathcal{C}^1_A(\mathbb{D})$, and $\phi_0(\infty) = \psi_0(\infty) = 0$ with the conjugation conditions (5.7.5), (5.7.7). Branches of the logarithmic functions are defined as in Sections 4.1 and 4.4.

Let us integrate the first condition (5.7.8) on the natural parameter s. We have

$$t = r_k \exp\left(\frac{is}{r_k}\right) + a_k, \ |t - a_k| = r_k.$$

Hence,

$$\int_0^s (t - z_0)ds = \int_0^s (r_k \exp\left(\frac{is}{r_k}\right) + a_k)ds = ir_k^2 \exp\left(\frac{is}{r_k}\right) + ir_k^2 + a_k s.$$

Therefore,

$$f_k(t) := \int_0^s \frac{i}{2}(t - \bar{t})ds = \frac{1}{2}r_k^2 \exp\left(\frac{is}{r_k}\right) - r_k^2 + (a_k - \overline{a_k})\frac{is}{2} + \exp\left(-\frac{is}{r_k}\right).$$

5.7. PLANE STOKES FLOW.

Using the relation $\frac{is}{r_k} = \log \frac{t-a_k}{r_k}$ we obtain

$$f_k(t) = \frac{1}{2}r_k \left[t - a_k + \frac{r_k^2}{t-a_k} - 2r_k + (a_k - \overline{a_k})\log\frac{t-a_k}{r_k} \right], \quad k = 1,2,...,n.$$

Let us note that $f_k(t)$ is a real-valued function.

The conditions (5.7.5), (5.7.7) become

$$\phi_k(t) + t\overline{\phi_k'(t)} + \overline{\psi_k(t)} = \phi(t) + t\overline{\phi'(t)} + \overline{\psi(t)} + f_k(t), \tag{5.7.9}$$

$$\frac{1}{\mu_k}\left[\phi_k(t) - t\overline{\phi_k'(t)} - \overline{\psi_k(t)}\right] = \frac{1}{\mu_0}\left[\phi(t) - t\overline{\phi'(t)} - \overline{\psi(t)}\right], \quad |t - a_k| = r_k, \; k = 1,2,\ldots,n.$$

Let us introduce the new unknown functions

$$\Phi_k(z) = \left(\frac{r_k^2}{z-a_k} + \overline{a_k}\right)\phi_k'(z) + \psi_k(z), \quad |z-a_k| \leq r_k,$$

$$\Psi_k(z) = \left(\frac{r_k^2}{z-a_k} + \overline{a_k}\right)\phi'(z) + \psi(z), \quad z \in c\mathbb{D}, \; k = 1,2,\ldots,n.$$

The functions $\Phi_k(z)$ has a pole of first order at $z = a_k$. The function $\Psi_k(z)$ is represented as follows

$$\Psi_k(z) = \Psi_k^0(z) + \sum_{j=1}^{n}(-\overline{A_j})\log(z-a_j), \quad z \in c\mathbb{D},$$

where $\Psi_k^0 \in \mathcal{C}_A(\mathbb{D})$ vanishes at infinity. The functions $\Psi_k(z)$ are related by the identities

$$\Psi_1(z) - \left(\frac{r_1^2}{z-a_1} + \overline{a_1}\right)\phi'(z) = \Psi_k(z) - \left(\frac{r_k^2}{z-a_k} + \overline{a_k}\right)\phi'(z), \quad z \in c\mathbb{D}. \tag{5.7.10}$$

Let us rewrite (5.7.9) in the form

$$\left(1 + \frac{\mu_0}{\mu_k}\right)\phi_k(t) + \left(1 - \frac{\mu_0}{\mu_k}\right)\overline{\Phi_k(t)} = 2\phi(t) + f_k(t), \tag{5.7.11}$$

$$\left(1 - \frac{\mu_0}{\mu_k}\right)\overline{\phi_k(t)} + \left(1 + \frac{\mu_0}{\mu_k}\right)\Phi_k(t) = 2\Psi_k(t) + f_k(t), \quad |t-a_k| = r_k, \; k = 1,2,\ldots,n. \tag{5.7.12}$$

Following our approach we reduce problem (5.7.11), (5.7.12) to a system of functional equations. Let us represent the function $f_k(t)$ in the form $f_k(t) = f_k^-(t) - f_k^+(t)$, $|t - a_k| = r_k$, where

$$f_k^-(z) := -\frac{1}{2}r_k\left[z - a_k - 2r_k\right], \ |z - a_k| \leq r_k,$$

$$f_k^+(z) := \frac{1}{2}r_k\left[\frac{r_k^2}{z - a_k} + (a_k - \overline{a_k})\log\frac{z - a_k}{r_k}\right], \ |z - a_k| \geq r_k, \ k = 1, 2, \ldots, n.$$

One can note that it is not the Sokhotsky–Plemelj formulae, since the function $f_k^+(z)$ has a logarithmic discontinuous curve; a part of the cut connected the points $z = a_k$ and infinity located in $\mathbb{D}$.

Using (5.7.11) we introduce the function

$$\Omega(z) := \begin{cases} \left(1 + \frac{\mu_0}{\mu_k}\right)\phi_k(z) - \sum_{m\neq k}\left[\left(1 - \frac{\mu_0}{\mu_m}\right)\overline{\Phi_m(z_{(m)}^*)} - f_m^+(z)\right] + f_k^-(z), \\ |z - a_k| \leq r_k, \ k = 1, 2, \ldots, n, \\ 2\phi(z) - \sum_{m=1}^n\left[\left(1 - \frac{\mu_0}{\mu_m}\right)\overline{\Phi_m(z_{(m)}^*)} - f_m^+(z)\right], \ z \in \mathbb{D}, \end{cases}$$

which is analytic in $\widehat{\mathbb{C}}$ except of some points of $\mathbb{D}$, where it has logarithmic discontinuity with the principal part

$$-z\sum_{k=1}^n\left(1 - \frac{\mu_0}{\mu_k}\right)\overline{\phi_k'(a_k)}.$$

The function $\Omega'(z)$ is analytic in $\mathbb{C}$ and $\Omega'(\infty) = -\sum_{k=1}^n\left(1 - \frac{\mu_0}{\mu_k}\right)\overline{\phi_k'(a_k)}$. Then Liouville's theorem implies that

$$\Omega'(z) = -\sum_{k=1}^n\left(1 - \frac{\mu_0}{\mu_k}\right)\overline{\phi_k'(a_k)}, \ z \in \widehat{\mathbb{C}}.$$

Integrating this relation we obtain

$$\Omega(z) = -z\sum_{k=1}^n\left(1 - \frac{\mu_0}{\mu_k}\right)\overline{\phi_k'(a_k)} + p_0, \ z \in \widehat{\mathbb{C}}.$$

The constant p_0 is determined by the definition of $\Omega(z)$:

$$p_0 = \sum_{k=1}^n\left(1 - \frac{\mu_0}{\mu_k}\right)\overline{\psi_k(a_k)}.$$

5.7. PLANE STOKES FLOW.

This means that $\Omega(z)$ does not have logarithmic discontinuity. They disappear in the expression

$$\Omega(z) = 2\varphi(z) - \sum_{m=1}^{n}\left[\left(1 - \frac{\mu_0}{\mu_m}\right)\overline{\Phi_m(z^*_{(m)})} - f_m^+(z)\right], \quad z \in \mathbb{D}.$$

It can be true whenever the function

$$2\phi_0(z) = 2\phi(z) + \frac{1}{2}\sum_{k=1}^{n} r_k(a_k - \overline{a_k})\log\frac{z - a_k}{r_k}$$

is analytic in $\mathbb{D}$. Comparing with (5.7.8) we obtain

$$A_k = \frac{r_k}{4}(a_k - \overline{a_k}).$$

The constant $-4\pi A_k = -2ir_k \operatorname{Im} a_k$ expresses the external force applied to the particle occupying $\mathbb{D}_k$. It is not surprising that this force is purely imaginary, i.e., it is parallel to the y-axis that corresponds to the gravitational field. Let us write formulae to calculate $\varphi(z)$

$$2\phi(z) = 2\phi_0(z) - \frac{1}{2}\sum_{k=1}^{n} r_k(a_k - \overline{a_k})\log\frac{z - a_k}{r_k}, \qquad (5.7.13)$$

$$2\phi_0(z) = \Omega(z) + \sum_{m=1}^{n}\left[\left(1 - \frac{\mu_0}{\mu_m}\right)\overline{\Phi_m(z^*_{(m)})} - \frac{r_m^3}{2(z - a_m)}\right], \quad z \in \mathbb{D}.$$

The logarithmic terms of $\psi(z)$ are extracted by (5.7.8)

$$2\psi(z) = 2\psi_0(z) - \frac{1}{2}\sum_{k=1}^{n} r_k(a_k - \overline{a_k})\log\frac{z - a_k}{r_k}, \quad z \in \mathbb{D}. \qquad (5.7.14)$$

We introduce the new unknown functions

$$\varphi_k(z) := \left(1 - \frac{\mu_0}{\mu_k}\right)\phi_k(z), \quad \Theta_k(z) := \left(1 - \frac{\mu_0}{\mu_k}\right)\Phi_k(z) \in \mathcal{C}_A(\mathbb{D}). \qquad (5.7.15)$$

It follows from the definition of $\Omega(z)$ that

$$\varphi_k(z) = \sum_{m \ne k}\rho_m\overline{\Theta_m(z^*_{(m)})} + p_k - z\sum_{k=1}^{n}\rho_m\overline{\varphi'_k(a_k)} + h_{1k}(z), \quad |z - a_k| \le r_k, \ k = 1, 2, \ldots, n,$$

$$(5.7.16)$$

where
$$\rho_m := (\mu_m - \mu_0)/(\mu_m + \mu_0),$$

$$h_{1k}(z) := \left[\sum_{m \neq k} f_m^+(z) + f_k^-(z)\right] \bigg/ \left(1 - \frac{\mu_0}{\mu_k}\right), \quad p_k := p_0 \bigg/ \left(1 - \frac{\mu_0}{\mu_k}\right).$$

We have deduced the first n functional equations (5.7.16). In order to deduce the next ones we represent the function $f_k(t) - \bar{t} f'_k(t)$ in the following form

$$f_k(t) - \bar{t} f'_k(t) = g_k^+(t) - g_k^-(t), \quad |t - a_k| = r_k \text{ for each } k, \qquad (5.7.17)$$

where
$$g_k^-(z) := -\frac{1}{2} r_k [z - a_k - 2r_k], \quad |z - a_k| \leq r_k,$$

$$g_k^+(z) := \frac{1}{2} r_k \left[(a_k - \overline{a_k}) \log \frac{z - a_k}{r_k} + \frac{r_k^4 + r_k^2(a_k - \overline{a_k})}{(z - a_k)^3} + \frac{r_k^2 \overline{a_k}}{(z - a_k)^2} - \frac{a_k(a_k - \overline{a_k})}{z - a_k}\right],$$

$$|z - a_k| \geq r_k, \quad k = 1, 2, \ldots, n.$$

Similar to $\Omega(z)$ we introduce the function

$$\omega(z) := \begin{cases} \left(1 + \frac{\mu_0}{\mu_k}\right) \left[\Phi_k(z) - \overline{z^*_{(k)}} \phi'_k(z)\right] \\ \quad - \sum_{m \neq k} \left\{\left(1 - \frac{\mu_0}{\mu_m}\right) \left(\overline{\phi_m(z^*_{(m)})} - \overline{z^*_{(m)}} \overline{[\Phi_m(z^*_{(m)})]'}\right) - g_m^+(z)\right\} \\ \quad - g_k^-(z), \; |z - a_k| \leq r_k, \; k = 1, 2, \ldots, n, \\ \\ 2\left[\Psi_1(z) - \overline{z^*_1} \phi'(z)\right] - \sum_{m=1}^n \left\{\left(1 - \frac{\mu_0}{\mu_m}\right) \left(\overline{\phi_m(z^*_{(m)})} - \overline{z^*_{(m)}} \overline{[\Phi_m(z^*_{(m)})]'}\right) \\ \quad - g_m^+(z)\right\}, \; z \in cl\mathbb{D}. \end{cases}$$

The jump $\Delta_k := \lim_{z \to t \; z \in \mathbb{D}} \omega(z) - \lim_{z \to t \; z \in \mathbb{D}_k} \omega(z)$ is calculated as follows

$$\Delta_k = 2\left[\Psi_1(t) - \overline{t_1^*} \phi'(t)\right] - \left(1 - \frac{\mu_0}{\mu_k}\right) \left(\overline{\phi_k(z^*_{(k)})} - \overline{z^*_{(k)}} \overline{[\Phi_k(z^*_{(k)})]'}\right)$$

$$- \left(1 + \frac{\mu_0}{\mu_k}\right) \left[\Phi_k(z) - \overline{z^*_{(k)}} \phi'_k(z)\right] + g_k^+(t) - g_k^-(t), \; |t - a_k| = r_k \text{ for each } k.$$

5.7. PLANE STOKES FLOW.

It follows from (5.7.11) differentiated on t, (5.7.12), and (5.7.17) that $\Delta_k = 0$. By virtue of (5.7.10) we can write

$$\omega(z) = 2\psi(z) - \sum_{m=1}^{n} \left(1 - \frac{\mu_0}{\mu_m}\right) \left(\overline{\phi_m(z^*_{(m)})} - \overline{z^*_{(m)}} \left[\overline{\Phi_m(z^*_{(m)})}\right]'\right) + \sum_{m=1}^{n} g_m^+(z), \quad z \in c\mathbb{D}.$$

We extract the logarithmic terms in $\omega(z)$:

$$\omega(z) = 2\psi(z) + \sum_{m=1}^{n} \frac{r_m}{2}(a_m - \overline{a_m}) \log \frac{z - a_m}{r_m}.$$

Comparing with (5.7.14) we conclude that this is the function $2\psi_0(z)$ analytic in $\mathbb{D}$. Hence, by Liouville's theorem $\omega(z) \equiv q$ is a constant. The definition of $\omega(z)$ yields the next functional equations

$$\Theta_k(z) - \overline{z^*_{(k)}}\varphi'_k(z) = \sum_{m \neq k} \rho_m \left(\overline{\varphi_m(z^*_{(m)})} - \overline{z^*_{(m)}} \left[\overline{\Theta_m(z^*_{(m)})}\right]'\right) + h_{2k}(z) + q, \quad (5.7.18)$$

$$|z - a_k| \leq r_k, \quad k = 1, 2, \ldots, n,$$

where $h_{2k}(z) := \left[\sum_{m \neq k} g_m^+(z) + g_k^-(z)\right] / \left(1 - \frac{\mu_0}{\mu_k}\right)$. Thus $2n$ relations (5.7.16), (5.7.18) constitute a system of functional equations with respect to $2n$ functions $\varphi_k(z)$, $\Theta_k(z)$. If we know a solution of this system, then taking into account (5.7.15) we obtain

$$2\phi(z) = -z \sum_{m=1}^{n} \left(1 - \frac{\mu_0}{\mu_m}\right) \overline{\phi_m(z^*_{(m)})} + p_0 + \sum_{m=1}^{n} \left(1 - \frac{\mu_0}{\mu_m}\right) \overline{\Phi_m(z^*_{(m)})}$$

$$- \frac{1}{2} \sum_{m=1}^{n} \left[\frac{r_m^3}{z - a_m} - r_m \log \frac{z - a_m}{r_m}\right],$$

$$2\psi(z) = q + \sum_{m=1}^{n} \left\{\left(1 - \frac{\mu_0}{\mu_m}\right)\left(\overline{\phi_m(z^*_{(m)})} - \overline{z^*_{(m)}}\left[\overline{\Phi_m(z^*_{(m)})}\right]'\right) - f_m^+(z) + B_m(z)\right\},$$

where

$$B_m(z) := \frac{1}{2}r_m \left[\frac{r_m^2 + \overline{a_m}(a_m - \overline{a_m})}{z - a_m} + \frac{r_m^2(a_m - 2\overline{a_m})}{(z - a_m)^2} - \frac{r_m^4}{(z - a_m)^3}\right].$$

We now deduce some simple approximate formulae for φ_m and Θ_m. For the sake of simplicity we calculate only the derivatives φ'_m and Θ'_m. Anyway these derivatives completely determine the stress tensor and non-constant component of the vector velocity. We apply the method of addition theorems and represent the known and unknown functions in the form of their Taylor expansions

$$\phi_k(z) = \sum_{s=0}^{\infty} \alpha_s^k (z - a_k)^s, \quad \Theta_k(z) = \frac{r_k^2 \alpha_1^k}{z - a_k} + \sum_{s=0}^{\infty} \beta_s^k (z - a_k)^s,$$

$$h_{1k}(z) = \sum_{s=0}^{\infty} h_{1s}^k (z - a_k)^s, \quad h_{2k}(z) = \sum_{s=0}^{\infty} h_{2s}^k (z - a_k)^s, \quad |z - a_k| \leq r_k,$$
$$k = 1, 2, \ldots, n,$$

where α_s^k and β_s^k are undetermined coefficients. Substituting these expansions into (5.7.16), (5.7.18) and collecting the terms on $(z - a_k)^s$ we obtain an infinite set of $\mathbb{R}$-linear algebraic equations with respect to α_s^k and β_s^k, $s = 0, 1, 2, \ldots$; $k = 1, 2, \ldots, n$. If we restrict ourselves only on the first coefficient ($s = 1$), then we arrive at the following truncated set

$$\alpha_1^k = -\sum_{m \neq k} \rho_m \frac{r_m^2 \overline{\beta_1^k}}{(a_k - a_m)^2} + h_{11}^k - \rho_k \overline{\alpha_1^k}, \tag{5.7.19}$$

$$\beta_1^k = -\sum_{m \neq k} \rho_m \left[\frac{r_m^2}{(a_k - a_m)^2} \left(\overline{\alpha_1^k} - \frac{r_m^2 \overline{\beta_1^k}}{(a_k - a_m)^2} \right) + \left(\frac{r_m^2}{a_k - a_m} + \overline{a_m} \right) \frac{2 r_m^2 \overline{\beta_1^k}}{(a_k - a_m)^3} \right] + h_{21}^k,$$

where

$$h_{11}^k = \frac{\mu_k}{\mu_0 + \mu_k} \left[\frac{1}{2} r_k + \sum_{m \neq k} \frac{1}{2} r_m \left(\frac{r_m^2}{(a_k - a_m)^2} - \frac{a_k - \overline{a_k}}{a_k - a_m} \right) \right],$$

$$h_{21}^k = h_{11}^k - \frac{\mu_k}{\mu_0 + \mu_k} \sum_{m \neq k} \frac{1}{2} r_m \left(\frac{\overline{a_m}(a_m - \overline{a_m}) + r_m^2}{(a_k - a_m)^2} + 2 r_m^2 \frac{a_m - 2\overline{a_m}}{(a_k - a_m)^3} + \frac{3 r_m^2}{(a_k - a_m)^4} \right).$$

The functions φ_k and Θ_k can be approximately calculated by the formulae

$$\varphi_k(z) \approx \alpha_0^k + \alpha_1^k (z - a_k), \quad \Theta_k(z) \approx \frac{r_k^2 \alpha_1^k}{z - a_k} + \beta_0^k + \beta_1^k (z - a_k),$$

in which the constants α_0^k and β_0^k are not calculated, and α_0^k and β_0^k is a solution of (5.7.18). The functions $\phi(z)$ and $\psi(z)$ are calculated by the following formulae

$$\phi(z) \approx \sum_{k=1}^{n} \left(1 - \frac{\mu_0}{\mu_k}\right) \left(-a_k \overline{\alpha_1^k} + \frac{r_k^2 \overline{\beta_1^k}}{z - a_k} \right) - \frac{1}{2} \sum_{k=1}^{n} \left(\frac{r_{km}^3}{z - a_k} - r_k (a_k - \overline{a_k}) \log \frac{z - a_k}{r_k} \right),$$

5.7. PLANE STOKES FLOW.

$$\psi(z) \approx -\sum_{k=1}^{n}\left(1-\frac{\mu_0}{\mu_k}\right)\overline{z^*_{(k)}}\left(\overline{\alpha^k_1}-\frac{r_k^2\overline{\beta_1^k}}{(z-a_k)^2}\right)$$

$$-\frac{1}{2}\sum_{k=1}^{n}\left[\frac{r_{km}^3}{z-a_k}-r_k(a_k-\overline{a_k})\log\frac{z-a_k}{r_k}\right.$$

$$\left.+r_k\left(\frac{\overline{a_k}(a_k-\overline{a_k})+r_k^2}{z-a_k}+r_k^2\frac{a_k-2\overline{a_k}}{(z-a_k)^2}-\frac{r_k^4}{(z-a_k)^3}\right)\right].$$

Then the non-constant component of the velocity vector in $\mathbb{D}$ is calculated approximately by the formula

$$\mu_0 V(z) \approx -\sum_{k=1}^{n}\left(1-\frac{\mu_0}{\mu_k}\right)\left(-a_k\overline{\alpha^k_1}+\frac{r_k^2\overline{\beta_1^k}}{z-a_k}\right)-\frac{1}{2}\sum_{k=1}^{n}\frac{r_k^2}{(z-a_k)^3}+$$

$$+z\sum_{k=1}^{n}\left(1-\frac{\mu_0}{\mu_k}\right)\frac{r_k^2\beta_1^k}{(z-a_k)^2}-\frac{z}{2}\sum_{k=1}^{n}\left(\frac{r_k^3}{(z-a_k)^2}-r_k\frac{a_k-\overline{a_k}}{z-a_k}\right)$$

$$+\sum_{k=1}^{n}\left(1-\frac{\mu_0}{\mu_k}\right)\overline{z^*_{(k)}}\left(\overline{\alpha^k_1}-\frac{r_k^2\overline{\beta_1^k}}{(z-a_k)^2}\right)$$

$$+\frac{1}{2}\sum_{k=1}^{n}\left[\frac{r_k^3}{z-a_k}-r_k\left(\frac{a_k(\overline{a_k}-a_k)+r_k^2}{z-a_k}+r_k^2\frac{\overline{a_k}-2a_k}{(z-a_k)^2}-\frac{r_k^4}{(z-a_k)^3}\right)\right]$$

$$-i\sum_{k=1}^{n}r_k(a_k-\overline{a_k})\log\left|\frac{z-a_k}{r_k}\right|, \quad z\in\mathbb{D}.$$

Remark 5.4 *One can see that the approximate expression for $V(z)$ does not depend on the logarithmic branches. This is also true in the general case, because $2\mu_0 V(z) = \phi(z) - z\overline{\phi'(z)} - \overline{\psi(z)}$, and $\phi'(z)$ does not contain them and $\phi(z) - \overline{\psi(z)}$ contains only the real parts of the logarithms.*

We now proceed to justify a general approach to the system of functional equations (5.7.16), (5.7.18). As in the above example we find only derivatives of the unknown functions. Using (5.7.15) and changing $\Phi_k(z)$ by $\psi_k(z) - \overline{z^*_{(k)}}\phi_k(z)$ we introduce the function $\vartheta_k(z) := \left(1+\frac{\mu_0}{\mu_k}\right)\psi_k(z)$. Then (5.7.16), (5.7.18) becomes

$$\varphi_k(z) = \sum_{m\neq k}\rho_m\left(\overline{\vartheta_m(z^*_{(m)})}+z\overline{\varphi'_m(z^*_{(m)})}\right)+p_k-z\sum_{m=1}^{n}\rho_{mk}\overline{\varphi'_m(a_m)}+h_{1k}(z), \quad (5.7.20)$$

$$\vartheta_k(z) = \sum_{m \neq k} \rho_m \left[\overline{\varphi_m(z^*_{(m)})} - \overline{z^*_{(m)}} \left(\overline{\vartheta_m(z^*_{(m)})} - \overline{z\varphi'_m(z^*_{(m)})} \right)' \right] + h_{2k}(z) + q_k,$$
$$|z - a_k| \leq r_k, \ k = 1, 2, \ldots, n.$$

Differentiating (5.7.20) and using the relation $\left(\overline{z^*_{(m)}} \right)' = -r_m^2/(z-a_m)^2$ we obtain

$$\zeta_k(z) = \sum_{m \neq k} \rho_m \left(-\frac{r_m^2}{(z-a_m)^2} \overline{\eta_m(z^*_{(m)})} - \frac{r_m^2 z}{(z-a_m)^2} \overline{\zeta'_m(z^*_{(m)})} + \overline{\zeta_m(z^*_{(m)})} \right)$$
$$- \sum_{m=1}^{n} \rho_{mk} \overline{\zeta_m(a_m)} + h'_{1k}(z), \tag{5.7.21}$$

$$\eta_k(z) = \sum_{m \neq k} \rho_m \frac{r_m^2}{(z-a_m)^2} [-\overline{\zeta_m(z^*_{(m)})} - \overline{\eta_m(z^*_{(m)})}] + \frac{r_m^2}{(z-a_m)^2} \overline{\eta'_m(z^*_{(m)})}$$
$$+ \left(\frac{3z}{z-a_m} - 2 \right) \overline{\zeta'_m(z^*_{(m)})} + z \frac{r_m^2}{(z-a_m)^2} \overline{\zeta''_m(z^*_{(m)})} + h'_{2k}(z), \ |z-a_k| \leq r_k,$$
$$k = 1, 2, \ldots, n.$$

Here $\zeta_k(z) := \varphi'_k(z)$, $\eta_k(z) := \vartheta'_k(z)$. For the sake of briefness let us write (5.7.21) as an equation in $\mathcal{C}^2_{\mathcal{A}} := \mathcal{C}_{\mathcal{A}}(\cup_{k=1}^n \mathbb{D}_k) \times \mathcal{C}_{\mathcal{A}}(\cup_{k=1}^n \mathbb{D}_k)$

$$\Theta = \mathbf{A}\Theta + H, \tag{5.7.22}$$

where $\Theta(z) := (\zeta_k(z), \eta_k(z))^T$, $H(z) := (h'_{1k}(z), h'_{2k}(z))^T$ in each $\mathbb{D}_k$, and the operator $\mathbf{A}$ is defined by the right-hand part of (5.7.21). Following Section 4.3 we can show that $\mathbf{A}$ is a compact operator in $\mathcal{C}^2_{\mathcal{A}}$, because it is compounded by compact operators. For instance one of them is $\overline{\eta_m(z^*_{(m)})}$ and can be represented as a compact integral operator:

$$\overline{\eta'_m(z^*_{(m)})} = \frac{1}{2\pi i} \int_{\mathbb{T}_k^-} \frac{\eta_m(\tau) d\tau}{\tau - z^*_{(m)}}, \ z \in cl\mathbb{D}_k, \ m \neq k.$$

Lemma 5.2 *The set of linear algebraic equations*

$$x_k = \sum_{m \neq k} \rho_m \overline{y_m} + p_k, \ y_k = \sum_{m \neq k} \rho_m \overline{x_m} + q_k, \ k = 1, 2, \ldots, n, \tag{5.7.23}$$

with $0 \leq \rho_m < 1$, has a unique solution.

5.7. PLANE STOKES FLOW.

Proof. We demonstrate that the homogeneous set

$$x_k = \sum_{m \neq k} \rho_m \overline{y_m}, \quad y_k = \sum_{m \neq k} \rho_m \overline{x_m}, \quad k = 1, 2, \ldots, n, \qquad (5.7.24)$$

has only a trivial solution. Let us assume that x_k, y_k ($k = 1, 2, \ldots, n$) is a solution of (5.7.24). We introduce the quantities

$$x := \sum_{m=1}^{n} \rho_m \overline{y_m}, \quad y := \sum_{m=1}^{n} \rho_m \overline{x_m}.$$

It follows from (5.7.24) that

$$x = x_k + \rho_k \overline{y_k}, \quad y = y_k + \rho_k \overline{x_k}, \quad k = 1, 2, \ldots, n.$$

Hence,

$$x_k = \frac{x - \rho_k \overline{y}}{1 - \rho_k^2}, \quad y_k = \frac{y - \rho_k \overline{x}}{1 - \rho_k^2}, \quad k = 1, 2, \ldots, n. \qquad (5.7.25)$$

Substituting x_k and y_k from (5.7.25) into (5.7.24) we obtain n sets consisting of 2 equations on x and y

$$x \left(\frac{1}{1 - \rho_k^2} + \sum_{m \neq k} \frac{\rho_m}{1 - \rho_m^2} \right) - \overline{y} \left(\frac{\rho_k}{1 - \rho_k^2} + \sum_{m \neq k} \frac{\rho_m}{1 - \rho_m^2} \right) = 0,$$

$$x \left(\frac{\rho_k}{1 - \rho_k^2} + \sum_{m \neq k} \frac{\rho_m}{1 - \rho_m^2} \right) - \overline{y} \left(\frac{1}{1 - \rho_k^2} + \sum_{m \neq k} \frac{\rho_m}{1 - \rho_m^2} \right) = 0, \quad k = 1, 2, \ldots, n.$$

The determinants of these sets are equal

$$\left(\frac{1}{1 + \rho_k^2} + 2 \sum_{m \neq k} \frac{\rho_m}{1 - \rho_m^2} \right) \left(-\frac{1}{1 + \rho_k} \right) < 0, \quad k = 1, 2, \ldots, n.$$

Therefore, $x = y = 0$; (5.7.25) yields $x_k = y_k = 0$.
This proves the lemma. ∎

Theorem 5.6 *Equation (5.7.22) has a unique solution in C. This solution can be found by the method of successive approximations.*

Proof. According to the general line it is sufficient to prove that equation

$$\Theta = \lambda \mathbf{A}\Theta \qquad (5.7.26)$$

for $|\lambda| \leq 1$ has only a zero solution. One can assume that $0 < \lambda \leq 1$, since the general case $\lambda = |\lambda|\exp 2i\theta$ is reduced to the previous one by the change $z = \tilde{z}\exp i\theta$, $a_k = \tilde{a}_k \exp i\theta$, $\zeta_k(z) = \tilde{\zeta}_k(z)\exp i\theta$, $\eta_k(z) = \tilde{\eta}_k(z)\exp i\theta$. It turns out the case $0 < \lambda \leq 1$ is reduced to the case $\lambda = 1$. To show this we consider the function

$$\mu(x) := \mu_k \frac{1 - \rho_k x}{1 + \rho_k x}, \quad 0 \leq x \leq 1.$$

One can see that $\mu(0) = \mu_k$, $\mu(1) = \mu_0$, $\mu'(x) = \frac{2\rho_k \mu_k}{(1+\rho_k x)^2} \neq 0$, i.e., the function $\mu(x)$ monotonically changes from μ_k to μ_0, when x changes from 0 to 1. Let us change the viscosity μ_0 in (5.7.26) by $\mu(\lambda)$. Then we arrive to system (5.7.26) with $\lambda = 1$. We need to show now that the system $\Theta = \mathbf{A}\Theta$ has only a zero solution. It is sufficient to prove that the integrated system written in the form

$$\varphi_k(z) = \sum_{m \neq k} \rho_m \left(\overline{\vartheta_m(z^*_{(m)})} + \overline{z\varphi'_m(z^*_{(m)})} \right) + p_k - z \sum_{m=1}^{n} \rho_{mk}\overline{\varphi'_m(a_m)}, \qquad (5.7.27)$$

$$\vartheta_k(z) = \sum_{m \neq k} \rho_m \left[\overline{\varphi_m(z^*_{(m)})} - \overline{z^*_{(m)}} \left(\overline{\vartheta_m(z^*_{(m)})} - \overline{z\varphi'_m(z^*_{(m)})} \right)' \right] + q_k, \quad |z - a_k| \leq r_k,$$

$$k = 1, 2, \ldots, n,$$

has only constant solutions $\varphi_k(z) = x_k$, $\vartheta_k(z) = y_k$, where the quantities x_k, y_k satisfy set (5.7.23) having a unique solution.

Let $\varphi_k(z)$, $\vartheta_k(z)$ satisfy (5.7.27). Introduce the functions

$$\psi_k(z) := \vartheta_k(z)\left(1 + \frac{\mu_0}{\mu_k}\right)^{-1}, \quad \phi_k(z) := \varphi_k(z)\left(1 + \frac{\mu_0}{\mu_k}\right)^{-1}, \quad |z - a_k| \leq r_k,$$

$$\phi(z) = \frac{1}{2}\sum_{k=1}^{n}\left(1 - \frac{\mu_0}{\mu_k}\right)\overline{\vartheta_k(z^*_{(k)})},$$

$$\psi(z) = \frac{1}{2}\sum_{k=1}^{n}\left(1 - \frac{\mu_0}{\mu_k}\right)\left[\overline{\varphi_k(z^*_{(k)})} - \overline{z^*_{(k)}}\left(\overline{\vartheta_k(z^*_{(k)})} - \overline{z\varphi'_k(z^*_{(k)})}\right)'\right], \quad z \in cl\mathbb{D}.$$

5.8. NOTES AND COMMENTS.

These functions satisfy the conjugation problem

$$\phi_k(t) + \overline{t\phi_k(t)} + \overline{\psi_k(t)} = \phi(t) + \overline{t\phi'(t)} + \overline{\psi(t)} + R_k, \qquad (5.7.28)$$

$$\frac{1}{\mu_k}\left[\phi_k(t) - \overline{t\phi_k(t)} - \overline{\psi_k(t)}\right] = \frac{1}{\mu_0}\left[\phi(t) - \overline{t\phi'(t)} - \overline{\psi(t)}\right] + T_k, \ |t - a_k| = r_k,$$

where $k = 1, 2, \ldots, n$, and

$$R_k := \frac{1}{2}\sum_{m=1}^{n}\left(1 - \frac{\mu_0}{\mu_m}\right)(\overline{y_m} + x_m) - (\overline{y_k} + x_k),$$

$$T_k := \frac{1}{2\mu_0}\sum_{m=1}^{n}\left(1 - \frac{\mu_0}{\mu_m}\right)(-\overline{y_m} + x_m) - \frac{1}{\mu_k}(-\overline{y_k} + x_k), \ k = 1, 2, \ldots, n,$$

x_k, y_k are determined by (5.7.23). The functions $\phi(z)$ and $\psi(z)$ are bounded at infinity. Then the theorem of uniqueness [157], [206] implies that (5.7.28) has a unique solution up to an additive constant. But we know one of the solutions:

$$\phi_k(z) = x_k, \ \psi_k(z) = y_k, \ \phi(z) = \frac{1}{2}\sum_{m=1}^{n}\left(1 - \frac{\mu_0}{\mu_m}\right)\overline{y_m}, \ \psi(z) = \frac{1}{2}\sum_{m=1}^{n}\left(1 - \frac{\mu_0}{\mu_m}\right)\overline{x_m}.$$

Therefore, all solutions of (5.7.28) and, hence, (5.7.27) are constants. Then (5.7.26) has only a zero solution.

The theorem is proved. ∎

5.8 Notes and Comments.

5.1^0 The problems represented in Section 5.1 are closely related to the problems of Section 4.1. The structure of problem (5.1.6) is more general than the corresponding problems of Section 4.1. We only take constant coefficients here, when the mechanical sense of the problem holds. According to Section 4.1 using the method of factorization one could consider more general problems with Hölder-continuous coefficients.

5.2^0 Here we follow [186] in Subsection 5.2.1, and [291] in Subsections 5.2.2 and 5.2.3.

5.3^0 Here we follow [181]. The method of symmetry for algebraic and analytic curves was applied by many mathematicians. See, for instance, [136], [54] and works cited therein.

5.4^0 Here we follow [175, 183].

5.5^0 General problems of the steady heat conductions on the surfaces are studied sufficiently and completely in general works devoted to differential equations on manifolds. Here attention is paid to constructive applications of the complex analysis. These problems have applications also in porous media [6], [107], [227]. Subsection 5.5.2 follows [177].

5.6^0 Here we follow [176, 187].

5.7^0 Here we follow [182].

The present chapter and in general this book is devoted to problems for domains bounded by a finite numbers of components. One can find many problems solved in closed form in [168]-[170], [173, 178, 179] and works cited therein. The problems of mechanics of composite materials are based on boundary value problems with an infinite number of components. Periodic problems can be treated as similar types of problems as well as problems on tori. In particular the method of functional equations can be applied to such problems. We do not include them in the book and refer the reader to papers [190], [192]-[195], [203] devoted to the extension of the method of functional equations and to the wide literature devoted to other methods cited therein.

Bibliography

[1] M.J. Ablowitz & A.S. Fokas: *Complex Variables.* Cambridge University Press, Cambridge, 1997.

[2] M. Abramovitz & I.A. Stegun: *Handbook of Mathematical Functions.* Dover Publ., New York, 1972.

[3] N.Kh. Abudarova & Yu.V. Obnosov: *On a nonlinear mixed Riemann-Hilbert boundary value problem.* In: Kraevye zadachi i ikh pril. Cheboksary (1986), 4-11 (in Russian).

[4] V.M. Adamyan, D.Z. Arov, and M.G. Krein: *On infinite Hankel matrices and generalized problems by Carathéorory-Fejér and by F.Riesz.* Funkc. Analiz i Pril. 2 (1968), No.1, 1-19 (in Russian).

[5] V.M. Adamyan, D.Z. Arov, and M.G. Krein: *Infinite Hankel matrices and generalized problems of Carathéorory-Fejér and I. Schur.* Funkc. Analiz i Pril. 2 (1968), No.4, 1-17 (in Russian).

[6] P.M. Adler: *Porous Media: Geometry and Transport.* Butterworth & Heinmann, London, 1992.

[7] T. Akaza: *Singular sets of some Kleinian groups.* Nagoya Math. J. 26 (1966), 127-143.

[8] T. Akaza & K. Inoue: *Limit sets of geometrically finite free Kleinian groups.* Tohoku Math. J. 36 (1984), 1-16.

[9] N.I. Akhiezer: *Elements of the Theory of Elliptic Functions.* Nauka, Moscow, 1970 (in Russian); English transl.: AMS, Providence, Rhode Island, 1990.

[10] L.A. Aksent'ev: *Construction of the Schwarz operator by the method of symmetry.* In: Trudy Seminara po Kraevym Zadacham, Kazan, 1967, vyp. 4, 3-10 (in Russian).

[11] L.A. Aksent'ev, N.B. Il'inskii, M.T. Nuzhin et al.: *Theory of inverse problems for analytic functions and its applications.* Itogi nauki i tekhniki VINITI. Math. Analysis. 18 (1980), 67-124 (in Russian).

[12] E.P. Aksent'eva: *To the investigation of the nonlinear boundary value problem of Hilbert type, I and II.* Izv.Vuz. Mathematika 5(96)(1970), 14-23 and 6(97)(1970), 16-21(in Russian).

[13] E.P. Aksent'eva: *On solvability of a nonlinear boundary value problem of Hilbert type.* In: Trudy Seminara po Kraevym Zadacham, Kazan, 1971, vyp. 8, 22-30 (in Russian).

[14] E.P. Aksent'eva: *On a case of a nonlinear boundary value problem of Hilbert type.* In: Trudy Seminara po Kraevym Zadacham, Kazan, 1972, vyp. 9, 23-30 (in Russian).

[15] E.P. Aksent'eva: *To the investigation of the nonlinear boundary value problem of Riemann type.* Izv.Vuz. Mathematika, 1973, n. 6, 3-7 (in Russian).

[16] E.P. Aksent'eva & A.S. Kalugina: *To the solution of nonlinear Hilbert problem with constant coefficients.* In: Trudy Seminara po Kraevym Zadacham, Kazan, 1975, vyp. 12, 9-18 (in Russian).

[17] E.P. Aksent'eva: *About nonlinear Carleman's problem with constant coefficients.* In: Trudy Seminara po Kraevym Zadacham, Kazan, 1978, vyp. 15, 16-23 (in Russian).

[18] A.G. Alekhno: *On a nonlinear boundary value problem on a simple closed piecewise smooth contour with a single wedge-point.* In: I.V. Gaishun et al. (Eds.), Boundary Value Problems, Special Functions and Fractional Calculus, Proceedings of International Conference Dedicated to 90th Birthday of Academician F.D. Gakhov, Minsk, 1996, 9-19 (in Russian).

[19] A.B. Aleksandrov: *Norm of Hilbert transform in the spaces of Hölder functions.* Funkc. Analiz i Pril. 9 (1975), n. 2, 1-4 (in Russian).

[20] I.A. Aleksandrov & A.S. Sorokin: *The problem of Schwarz for multiply connected domains.* Sib.Math.Zh. 13 (1972), n. 5, 971-1001 (in Russian).

[21] S.N. Antonzev, V.N. Monakhov and A.V. Kazhikhov: *Boundary Value Problems of the Mechanics of Nonhomogeneous Fluids.* Nauka, Moscow 1983 (in Russian); English transl.: North-Holland, Amsterdam, 1990.

[22] B.D. Annin & G.P. Cherepanov: *Elastic-plastic Problem.* Nauka, Moscow, 1983 (in Russian).

[23] G.V. Arzhanov: *On nonlinear boundary value problem of Riemann type.* Doklady AN SSSR 132 (1960), n. 6, 1227-1230 (in Russian).

[24] G.V. Arzhanov: *On nonlinear boundary value problem of Riemann type.* Sib. Math. Zhurn. 2 (1961), n. 4, 481-504 (in Russian).

[25] G.V. Arzhanov: *On solvability of homogeneous nonlinear boundary value problem of power type.* Izv.Vuz. Mathematika (1978), n. 8, 8-18 (in Russian).

[26] S. Axler, P. Bourdon, and W. Ramey: *Harmonic Functions Theory.* Springer-Verlag, New York, 1992.

[27] N.S. Bakhvalov & G.P. Panasenko: *Averaging of processes in periodic media.* Nauka, Moscow, 1984.

[28] N.K. Bari & S.B. Stechkin: *Best approximations and different properties of two conjugate functions.* Trudy Moscow Math. Obstch. 1975, n. 5, 485-522 (in Russian).

[29] A.V. Batyrev: *Nonlinear Hilbert problem.* In: Abstracts of 7th All-Union Conference on the Theory of Functions of Complex Variable. Rostov-on-Don (1964), 19 (in Russian).

[30] H. Begehr: *Boundary value problems for analytic and generalized analytic functions.* In: Methods, Trends and Applications of Complex Analysis (E. Lanckau & W. Tutschke, Eds.), Akademie-Verlag, Berlin, 1983.

[31] H. Begehr & G.N. Hile: *Nonlinear Riemann boundary value problems for a nonlinear elliptic system in the plane.* Math.Zeitschr. 179 (1982), 241-261.

[32] H. Begehr & A. Dzhuraev: *An Introduction to Several Complex Variables and Partial Differential Equations.* Addison-Wesley Longman, Harlow, 1997.

[33] H. Begehr & R.P. Gilbert: *Das Randwert-Normproblem für ein fastlineares elliptisches System und eine Anwendungen.* Ann. Acad. Sci. Fenn. Ser. A.I. Math. 3 (1977), 179-184.

[34] H. Begehr & G.C. Hsiao: *Nonlinear boundary value problems of Riemann-Hilbert type.* Cont. Math. 11 (1982), 139-153.

[35] H. Begehr & G.C. Wen: *Nonlinear Elliptic Boundary Value Problems and their Applications.* Addison-Wesley Longman, Harlow, 1996.

[36] E.N. Bereslavskij: *On integrating in closed form of a class of functions and its applications.* Dif. uravn. 25 (1989), n. 6, 1048-1051 (in Russian).

[37] S. Bergman: *The Kernel Function and Conformal Mapping.* AMS, Providence, Rhode Island, 1970.

[38] M.Z. Berkolaiko: *On cojugation operator in the spaces of Hölder type.* Soob. AN GruzSSR 99 (1980), n. 2, 281-284 (in Russian).

[39] M.Z. Berkolaiko & Ja.B. Rutitski: *On operators in the spaces $\mathcal{H}_{\phi,E}$.* Sib. math zhurn. 24 (1983), n. 3, 18-33 (in Russian).

[40] K.J. Binns & P.J. Lawrensen: *Analysis and Computation of Electric and Magnetic Field Problems.* Clarendon Press, Oxford, 1963.

[41] R.P. Boas: *Entire Functions.* Academic Press, New York, 1954.

[42] B. Bojarski: *On a boundary value problem of the theory of analytic functions.* Dokl. AN SSSR 119 (1958), 199-202 (in Russian).

[43] B. Bojarski: *On generalized Hilbert boundary value problem,* Soobsch. AN GruzSSR 25 (1960), n. 4, 385-390 (in Russian).

[44] B. Bojarski: *Connection between complex and global analysis: some analytical and geometrical aspects of Riemann-Hilbert transmission problem.* In: Methods, Trends and Applications of Complex Analysis (E. Lanckau & W. Tutschke, Eds.), Akademie-Verlag, Berlin, 1983, 97-110.

[45] M.Kh. Brenerman & B.A. Kac: *An estimate of the norm of singular integral and its application in some boundary value problems.* Izv. vuz.Mathematika, 1985, n. 1, 8-17 (in Russian).

[46] R.B. Burckel: *Iterating analytic self-maps of discs.* Amer. Math. Month., 88 (1981), 396-407.

[47] G.G. Chaevskii: *Nonlinear conjugation problem on Riemann surfaces in the case of the contour homological to zero.* Doklady AN BSSR 29 (1960), n. 2, 123-126 (in Russian).

[48] V.G. Cherednichenko: *Inverse Logarithmic Potential Problem.* VSP, Zeist, 1996.

[49] G.P. Cherepanov: *On a nonlinear boundary value problem for analytic functions appeared in some visco-elastic problems.* Doklady AN SSSR 147 (1962), n. 3, 566-568 (in Russian).

[50] G.P. Cherepanov: *Visco-elastic problem in the case of antiflat deformation.* Prikl. Math. i Mech. 26 (1962), vyp. 4, 17-23 (in Russian).

[51] G.P. Cherepanov: *Boundary value problems with analytic coefficients.* Doklady AN SSSR, 161 (1965), n. 2, 312-314 (in Russian).

[52] G.P. Cherepanov: *Mechanics of Brittle Distruction.* Nauka, Moscow, 1974 (in Russian).

[53] L.I. Chibrikova: *Basic Boundary Value Problems for Analytic Functions.* KSU, Kazan, 1977 (in Russian).

[54] L.I. Chibrikova & L.G. Salekhov: *To solution of a general linear conjugation problem of the analytic function theory in the case of algebraic curves.* Trudy sem. po kraev. zadach., 1968, n. 5, 224-229 (in Russian).

[55] A.R. Cickishvili: *On construction of analytic functions which map conformally half-plane onto circular polygon.* Dif. uravn. 21 (1985), n. 4, 646-656 (in Russian).

[56] J.W. Cohen & O.J. Boxma: *Boundary Value Problems in Queueing System Analysis.* North-Holland, Amsterdam, 1983; Russian transl.: Mir, Moscow, 1987.

[57] E.F. Collingwood & A.J. Lohwater: *The Theory of Cluster Sets.* Cambridge UP, Cambridge,1966; Russ. transl.: Mir, Moscow, 1971.

[58] C.C. Cowen: *Iteration and the solution of functional equation for analytic functions in the unit disk.* Trans. Amer. Math. Soc. 265 (1981), n. 1, 69-95.

[59] J. Crank: *Free and Moving Boundary Problems.* Clarendon Press, Oxford, 1984.

[60] R.V. Craster: *Conformal mappings involving curvilinear quadrangles.* IMA J. of Appl. Math. 57 (1996), 181-191.

[61] R.V. Craster: *The solution of a class of free boundary value problems.* Proc. R. Soc. Lond. A 453 (1997), 607-630.

[62] R.V. Craster & Viêt Hà Hoàng: *Applications of Fuchsian differential equations to free boundary problems.* Proc. R. Soc. Lond. A 454 (1998), 1241-1252.

[63] I.I. Daniljuk: *Irregular Boundary Value Problems on the Plane.* Nauka, Moscow, 1975 (in Russian).

[64] L.E. Dunduchenko: *On the Schwarz formula for an n-connected domain.* Dopovedi AN URSR (1966), n. 5, 1386-1389 (in Ukranian).

[65] P. Duren: *Theory of H^p-spaces.* Academic Press, New York, 1970.

[66] A. Dzhuraev: *Systems of Equations of Composite Type.* Addison-Wesley Longman, Harlow, 1989.

[67] A. Dzhuraev: *Methods of Singular Integral Equations.* Addison-Wesley Longman, Harlow, 1992.

[68] A. Dzhuraev: *On singular integral equations approach to generalized analytic functions.* In: Generalized Analytic Functions, H. Florian et al. (Eds.), Kluwer AP, Dordrecht, 1998, 17-25.

[69] M.A. Efendiev: *Nonlinear Hilbert problem in an annulus.* Doklady AN Azerb.SSR 32 (1976), n. 11, 3-5 (in Russian).

[70] M.A. Efendiev: *Nonlinear Hilbert problem for meromorphic functions in a disc.* Izv. AN Azerb.SSR, Ser. Fiz.-Tekh. i Mat. Nauk (1976), n. 5, 53-58 (in Russian).

[71] M.A. Efendiev: *A Fredholm degree for a quasi-linearlike mapping of quasi-cylindric domains and the nonlinear Hilbert problem in annalus.* Izv. AN Azerb.SSR, Ser. Fiz.-Tekh. i Mat. Nauk (1979), n. 5, 18-23 (in Russian).

[72] M.A. Efendiev: *"Topologically nontrivial" and meromorphic solutions of non-linear Hilbert problem for multiply connected domain.* Izv. AN Azerb.SSR, Ser. Fiz.-Tekh. i Mat. Nauk (1982), n. 5, 18-21 (in Russian).

BIBLIOGRAPHY

[73] M.A. Efendiev: *A nonlinear Hilbert problem with boundary conditions smoothly immersed in* $\mathbf{R}^2$. Izv. AN Azerb.SSR, Ser. Fiz.-Tekh. i Mat. Nauk (1983), n. 3, 7-9 (in Russian).

[74] M.A. Efendiev: *Topological and constructive description of the solutions of a nonlinear Hilbert problem.* Preprint No. 2, Institut Phys. Akad. Nauk Azerb. SSR, Baku 1988 (in Russian).

[75] M.A. Efendiev & W.L. Wendland: *Nonlinear Riemann–Hilbert problems for multiply connected domains.* Nonlinear Analysis. Theory, Methods and Applications 27 (1996), 37-58.

[76] M.A. Efendiev & W.L. Wendland: *Nonlinear Riemann–Hilbert problems without transversality.* Math. Nachr. 183 (1997), 73-89.

[77] C.M. Flliot & J.R. Ockendon: *Weak and Variational Methods for Moving Boundary Problems.* Pitman Publ., Boston, 1982.

[78] Yu.P. Emetz: *Boundary Value Problems of Anisotropic Electrodynamics.* Naukova Dumka, Kiev, 1987 (in Russian).

[79] Yu.P. Emetz & Yu.V. Obnosov: *Compact analogous of heterogeneous system with the structure of check-board field.* J. Techn. Phys., 60 (1990), n. 8, 59-66 (in Russian).

[80] A. Erdelyi, F. Oberhettinger & F.G. Tricomi: *Higher Transcendental Functions.* Vol. 1, AMS, New York, 2nd ed., 1951; Russian transl.: Nauka, Moscow, 1965.

[81] V.T. Erofeenko: *Addition Theorems.* Nauka i Tekhnika, Minsk, 1988 (in Russian).

[82] M.A. Evgrafov: *Analytic Functions.* Nauka, Moscow, 1968 (in Russian).

[83] G. Fayolle, R. Iasnogorodski, and V. Malyshev: *Random Walks in the Quarter Plane. Algebraic Methods, Boundary Value Problems and Applications.* Springer-Verlag, Berlin, 1999.

[84] L. Ford: *Automorphic functions.* New York, 1929; Russian transl.: ONTI, Moscow, 1936.

[85] A.Sh. Gabib-Zade: *Investigation on a nonlinear Hilbert problem.* Doklady AN AzSSR **14** (1958), n. 4, 275-278 (in Russian).

[86] D. Gaier: *Konstruktive Methoden der konformen Abbildung.* Springer-Verlag, Berlin, 1964.

[87] D. Gaier: *Lectures on Complex Approximation.* Birkhäuser, Boston-Basel-Berlin, 1987.

[88] F.D. Gakhov & L.I. Chibrikova: *Solution of the problems of mechanics of continuous media by the reduction to the boundary value problems for automorphic functions.* In: Int. Symposium on Applications of the Theory of Functions to the Mechnics of Continuous Media (1964, Tbilisi), Nauka, Moscow, 1965, vol. II, p.208-218 (in Russian).

[89] F.D. Gakhov: *On a nonlinear boundary value problem generalizing the Riemann boundary value problem.* Doklady AN SSSR, 181 (1968), n. 2, 271-274 (in Russian).

[90] F.D. Gakhov: *On a nonlinear boundary value problem with admissible zeroes on the contour.* Doklady AN SSSR, 210 (1973), n. 6, 21-24 (in Russian).

[91] F.D. Gakhov: *Boundary Value Problems.* Nauka, Moscow, 1977 (3rd edition) (in Russian); Engl. transl. of 1st ed.: Pergamon Press, Oxford, 1966.

[92] F.D. Gakhov & Ju.I. Cherskii: *Equations of Convolution Type.* Nauka, Moscow, 1978 (in Russian).

[93] L.A. Galin: *Elastic-plastic Problems.* Nauka, Moscow, 1984.

[94] I.V. Gaishun, A.A. Kilbas, et al. (Eds.): *Boundary Value Problems, Special Functions and Fractional Calculus.* Proceeding of International Conference Dedicated to 90th Birthday of Academician F.D. Gakhov. Belarus, Minsk, Feb 16-20, 1996. Belgosunversitat, Minsk, 1996.

[95] J. Garnett: *Bounded Analytic Functions.* Academic Press, New York, 1980; Russian transl.: Mir, Moscow, 1984.

[96] V.N. Gavdzinskii & I.M. Spitkovskii: *On a method of effective construction of factorization.* Ukr. matem. Zhurn. 34 (1982), n. 1, 15-19 (in Russian).

[97] R.P. Gilbert & J.L. Buchanan: *First Order Elliptic Systems.* Academic Press, New York, 1983.

BIBLIOGRAPHY

[98] I. Gohberg & N. Krupnik: *Introduction to the Theory of One-Dimensional Singular Integral Operators,* vol. I, II, Birkhäuser-Verlag, Basel, 1991.

[99] G.M. Goluzin: *Solution of basic plane problems of mathematical physics for the case of Laplace equation and multiply connected domains bounded by circles (method of functional equations).* Math. zb. 41: 2 (1934), 246-276.

[100] G.M. Goluzin: *Solution of spatial Dirichlet problem for Laplace equation and for domains embounded by finite number of spheres.* Math. zb. 41: 2 (1934), 277-283.

[101] G.M. Goluzin: *Solution of plane heat conduction problem for multiply connected domains enclosed by circles in the case of isolated layer.* Math. zb. 42: 2 (1935), 191-198.

[102] G.M. Goluzin: *Geometric Theory of Functions of Complex Variable.* Nauka, Moscow 1966 (2nd ed.) (in Russian); Engl. transl. by AMS, Providence, RI 1969.

[103] N.V. Govorov & N.K. Kuznetsov: *On nonlinear conjugation boundary value problem for a disconnected contour.* In: Theory of Function, Functional Analysis and Applications, Kharkov, 1974, vyp. 20, 49-63 (in Russian).

[104] N.V. Govorov: *Riemann's Boundary Problem with Infinite Index.* Nauka, Moscow, 1986; English transl.: Birkhäuser-Verlag, Basel, 1994.

[105] A.I. Guseinov: *On conformal mapping of a circle on a domain close to circle.* Trudy math. sectora AN AzSSR, 1946, n. 2, 18-22 (in Russian).

[106] A.I. Guseinov & Kh.Sh. Mukhtarov: *Introduction to the Theory of Nonlinear Singular Integral Equations.* Nauka, Moscow, 1980 (in Russian).

[107] J. Happel & H. Brenner: *Low Reynolds Numbers Hydrodynamics.* Prentice-Hall, Englewood Cliffs, New Jersey, 1965.

[108] S.Ya. Havinson: *Extremal problems for certain classes of analytic functions in finitely connected regions.* AMS Trans. Ser. 2, 5 (1957), 1-34.

[109] K. Hoffman: *Banach Spaces of Analytic Functions.* Prentice-Hall, Englewood Cliffs, N.J., 1962; Russian transl.: Izdat. Inostr. Lit., Moscow, 1963.

[110] I.I. Ibragimov: *Selected Questions of Analytic Function Theory.* Elm, Baku, 1984 (in Russian).

[111] I. Imai: *Applied Hyperfunction Theory.* Kluwer AP, Dordrecht, 1992.

[112] B.B. Iskenderov: *Nonlinear boundary value problem with a shift.* Uch. zap. Azerb. inst. nefti i khim., ser. 9, 1977, n. 3, 46-50 (in Russian).

[113] B.B. Iskenderov: *On a nonlinear boundary value problem for piecewise analytic functions.* Nauch. trudy Azerb. univ., ser. phiz.-math.nauk, 1979, n. 5, 30-39 (in Russian).

[114] L.V. Kantorovich & G.P. Akilov: *Functional Analysis.* Nauka, Moscow, 1984 (3rd edition) (in Russian).

[115] V.V. Kashevskij: *A nonlinear boundary value problem in the case of composite contour.* Izv. AN BSSR, ser. phys.-math.nauk, 1982, n. 4, 44-48 (in Russian).

[116] V.V. Kashevskij: *Mixed boundary value problem on a finite Riemann surface.* In: Problems of hydrodynamics of a big speed and of boundary value problems, Krasnodar, 1982, 49 (in Russian).

[117] M.V. Keldysh: *Conformal mapping of multiply connected domains on the canonical domains.* Uspekhi Math.nauk 6 (1939), 90-119.

[118] D. Khavinson: *Factorization theorems for different classes of analytic functions in multiply connected domains.* Pacif. J. Math., 108 (1983), n. 2, 295-318.

[119] D. Khavinson: *Symmetry and uniform approximation by analytic functions.* Proc. AMS 101 (1987), n. 3, 475-483.

[120] B.V. Khvedelidze: *Linear jump boundary value problems of the theory of functions, singular integral equations and certain of their applications.* Trudy Tbil. Mthem. Inst. AN GruzSSR, XXIII (1956), 3-158.

[121] S.N. Kiyasov: *A study of a nonlinear Riemann boundary value problem in the case of analytically extendable coefficients.* In: Kraevye zadachi i ikh prilozh., Cheboksary (1986), 54-59.

[122] S.N. Kiyasov: *On a quastion of solvability of linear fractional Riemann boundary value problem.* In: Trudy seminara po kraevym zadacham, Kazan' (1987), vyp. 23, 116-129.

[123] S.N. Kiyasov: *Linear fractional Riemann boundary value problem and its application to factorization of certain classes of Hölder matrix functions of the second order.* Russian Mathematics (Izv. VUZ) 39 (1995), n. 9, p. 23-29.

[124] I.I. Komjak: *Nonlinear boundary value problem of Riemann type with positive exponents.* Izv. AN BSSR, ser. phys.-math.nauk, 1970, n. 6, 83-87 (in Russian).

[125] I.I. Komjak: *On two-dimensional singular integral equations with analytic coefficients.* Dif. uravn. 16 (1980), n. 5, 908-916 (in Russian).

[126] I.I. Komjak & V.V. Mityushev: *On the solution of the general boundary value problem for annulus.* Vesti AN BSSR, ser. phys.-math.nauk, 1983, n. 3, 25-33 (in Russian).

[127] P. Koosis: *Introduction to H_p Spaces.* Cambridge Univ. Press, Cambridge, 1980; Russian transl.: Mir, Moscow, 1984.

[128] M.A. Krasnosel'skii et al.: *Approximate Methods for Solution of Operator Equations.* Nauka, Moscow, 1969 (in Russian); Engl. transl.: Wolters-Noordhoff, Groningen, 1972.

[129] M.A. Krasnosel'skii et al.: *Integral Equations. A Reference Text.* Nauka, Moscow, 1968 (in Russian); Engl. transl.: Noordhoff Internatinal, Leyden, 1975.

[130] V.G. Kravchenko & A.M. Nikolaichuk: *On equivalence of certain type of Riemann boundary value problem for the system of n pairs of functions to a complete singular equation with Cauchy kernel.* Dif. uravn. 9 (1973), n. 2, 343-376 (in Russian).

[131] N.V. Krylov: *Lectures on Elliptic and Parabolic Equations in Hölder Spaces.* AMS, Providence, Rhode Island, 1996.

[132] M. Kuczma, B. Chosewski, and R. Ger: *Iterative functional equations.* Encyclopedia Math. Appl., 32, Cambridge University Press, Cambridge, 1990.

[133] N.K. Kuznetsov: *On a nonlinear problem of a type $(\Phi^+)^\alpha = G(\Phi^-)^\beta$ for an open contour.* Nauch. Trudy Kubanskogo univ., 1980, vyp. 180, 65-71 (in Russian).

[134] N.K. Kuznetsov: *On a nonlinear problem of a type $(\Phi^+)^\alpha = G(\Phi^-)^\beta$ for an open contour.* Nauch. Trudy Kubanskogo univ., 1980, vyp. 180, 72-78 (in Russian).

[135] N.K. Kuznetsov: *On a nonlinear conjugation problem on the plane with two cuts.* Izv.Vuz. Mathematika (1977), n. 11, 34-45 (in Russian).

[136] N.V. Lambin: *Method of symmetry and its application to boundary value problems.* BGU, Minsk, 1960 (in Russian).

[137] M.A. Lavrentiev, B.V. Shabat: *Methods of the Theory of Functions of Complex Variable.* Nauka, Moscow, 1973 (in Russian).

[138] J. Lehner: *Discontinuous Groups and Automorphic Functions.* AMS, Providence, Rhode Island, 5th print., 1990.

[139] B.Ya. Levin: *Distribution of Zeros of Entire Functions*, Gostekhizdat, Moscow, 1956; Engl. transl. by AMS, Providence, Rhode Island., 1980.

[140] A.A. Gol'dberg, B.Ya. Levin, and I.V. Ostrovskii: *Entire and meromorphic functions.* In: A.A. Gonchar, V.P. Khavin, and N.K. Nikol'skii (Eds.): *Complex Analysis I*, Itogi nauki i tekhniki VINITI, Moscow, 1991, 5-186; Engl. transl. by Springer-Verlag, Berlin, 1997.

[141] I.K. Lifanov: *Singular Integral Equations and Discrete Vortices.* VSP, Utrecht-Tokyo, 1996.

[142] G.S. Litvinchuk: *On stability of a boundary value problem of analytic function theory.* Doklady AN SSSR 174 (1967), n. 6, 1268-1270 (in Russian).

[143] G.S. Litvinchuk: *Two theorems on stability of partial indices for Riemann boundary value problem and their applications.* Izv. Vuz. Mathematika 12 (1967), n. 6, 47-57 (in Russian).

[144] G.S. Litvinchuk: *Boundary Value Problems and Singular Integral Equations with Shift.* Nauka, Moscow, 1977 (in Russian).

[145] G.S. Litvinchuk & I.M. Spitkovskii: *Sharp estimates of defect numbers of generalized Riemann boundary value problem, factorization of Hermitian matrix-functions, and some problems of approximation by meromorphic functions.* Mathem. zb. 117 (1982), n. 2, 196-215 (in Russian).

[146] G.S. Litvinchuk & I.M. Spitkovskii: *Factorization of Measurable Matrix Functions.* Akademie-Verlag, Berlin and Birkhäuser-Verlag, Basel, 1987.

[147] G.S. Litvinchuk & E.I. Zverovich: *Boundary value problems with shift for analytic functions and singular functional equations.* Uspekhi Math. nauk 23 (1968), n. 3, 67-121 (in Russian).

[148] J.-K. Lu: *Boundary Value Problems for Analytic Functions.* World Scientific, Singapore, 1993.

[149] J.-K. Lu & H.T. Cai: *Mathematical Theory in Periodic Plane Elasticity.* Gordon & Breach, Tokyo (published).

[150] R. Magnanini: *A fully nonlinear boundary value problem for Laplace equation in dimension two.* Applicable Analysis 39 (1990), 185-192.

[151] A.I. Markushevich: *On a boundary value problem of analytic function theory.* Uch. zapiski MGU 1 (1946), vyp. 100, 20-30 (in Russian).

[152] A.I. Markushevich: *Theory of Analytic Functions.* Nauka, Moscow, Vol. 2, 1967.

[153] M. Maliszewski: *On certain boundary value problem with weight.* Demonstr. Math. VII (1974), n. 3, 297-311.

[154] A.M. Meirmanov: *The Stefan Problem.* Nauka, Moscow, 1986 (in Russian); English transl.: DeGruyter, Berlin, 1992.

[155] E. Meister: *Randwertaufgaben der Funktionentheorie.* B.G. Teubner, Stuttgart, 1983.

[156] L.G. Mikhailov: *New Class of Singular Integral Equations and its Applications to Differential Equations with Singular Coefficients.* AN TadzhSSR, Dushanbe, 1963 (in Russian); English transl.: Akademie-Verlag, Berlin, 1970.

[157] S.G. Mikhlin: *Integral Equations*, Pergamon Press, New York, 1964.

[158] S.G. Mikhlin, N.F. Morozov & M.V. Paukshto: *The Integral Equations of the Theory of Elasticity.* Teubner, Stuttgart-Leipzig, 1995.

[159] S.G. Mikhlin & S. Prössdorf: *Singuläre Integraloperatoren.* Akademie-Verlag, Berlin 1980; English transl.: Akademie-Verlag, Berlin, 1986.

[160] L.M. Milne-Thomson: *Theoretical Hydrodynamics* (5th ed.). Macmillan & Co Ltd, London, 1968.

[161] V.V. Mityushev: *Solution of linear functional equation with shift into domain in class of analytic functions.* Vesti Akad.navuk BSSR.Ser.Phys.-Math. 1983, n. 5, 117. Preprint N 2278-83, VINITI (in Russian).

[162] V.V. Mityushev: *On solution of the general boundary value problem for some concentric circumferences.* Vesti Akad.navuk BSSR.Ser.Phys.-Math. 1983, n. 6, 104. Preprint N 2279-83, VINITI (in Russian).

[163] V.V. Mityushev: *On solution to Markushevich's boundary value problem for a circular domain.* Vesti AN BSSR. Ser.Phys.-Math. (1985), n. 1, 119 (in Russian).

[164] V.V. Mityushev: *On application of the incomplete factorization method.* Doklady AN BSSR 29 (1985), n. 8, 688-690 (in Russian).

[165] V.V. Mityushev: *Solution of the boundary real value problem (Markushevich's problem) for an annulus in the special case.* Izv. Vuz. Mathematika (1986), n. 3, 67-69 (in Russian).

[166] V.V. Mityushev: *On solution of the $\mathbb{R}$-linear boundary value problem for a complex circular contour.* Vesti AN BSSR. Ser.Phys.-Math. (1986), n. 5, 118 (in Russian).

[167] V.V. Mityushev: *On solvability of the equation $f(sz) = Ff(z)$ in a class of analytic functions.* Vesti AN BSSR. Ser.Phys.-Math. (1987), n. 1, p.117. Preprint N 8829-B86, VINITI (in Russian).

[168] V.V. Mityushev: *On certain boundary value problems of the analytic functions theory to be reduced on functional equations.* Vesti AN BSSR. Ser.Phys.-Math. (1987), n. 5, p.113. Preprint N 8000-B86, VINITI (in Russian).

[169] V.V. Mityushev: *Steady heat conduction of the fibrous material with imperfect contact between materials.* Ingen.Phys.J., **53** (1987), n. 1, p.154-155. Preprint N 1049-B87, VINITI (in Russian).

[170] V.V. Mityushev: *On the solutions of the $\mathbb{R}$-linear boundary value problem (Markushevich's problem) on the torus and cylinder.* Izv. Vuz. Mathematika. Preprint N 7546-B87, VINITI (in Russian).

[171] V.V. Mityushev: *Heat and wave equations on the graphs.* Diff. uravn. **24** (1988), n. 9, 1053-1060 (in Russian).

BIBLIOGRAPHY

[172] V.V. Mityushev: *Functional equations for analytic and harmonic functions.* Diff. uravn. **25** (1989), n. 10, p.1816. Preprint N 2660-B89, VINITI (in Russian).

[173] V.V. Mityushev: *Solution of a problem of crack theory by functional equations method.* Theor. Appl. Mech. (1989), n. 16, p.89-93 (in Russian).

[174] V.V. Mityushev: *Functional equation for analytic functions with singularity in fixed point.* Vesti AN BSSR. Ser.Phys.-Math. Preprint N 1501-B89, VINITI (in Russian).

[175] V.V. Mityushev: *On a general boundary $\mathbb{R}$ -linear problem and its applications in the mechanics of reactive media.* Vesti AN BSSR. Ser.Phys.Math. (1990), n. 3, p.114 (in Russian).

[176] V.V. Mityushev: *Some mathematical problems of the heat conduction theory.* 1. Izv. Vuz. Energetika. (1987) Preprint N 7709-B87, VINITI; 2. (1989) Preprint N 5348-B89, VINITI. 3. (1990) N 3995-B90, VINITI (in Russian).

[177] V.V. Mityushev: *Steady heat conduction two-dimensional plates with covers and branches.* Engen. Phys. J. (1990). Preprint N 1638-B90, VINITI (in Russian).

[178] V.V. Mityushev, M.E. Tolochko *On solution of the boundary $\mathbb{R}$-linear value problem on the real axis and two parallel lines.* Vesti AN BSSR. Ser.Phys.-Math. (1990) Preprint N 1638B90, VINITI (in Russian).

[179] V.V. Mityushev: *Functional equations for harmonic functions in multiply connected domain.* Diff. uravn. 26 (1990), n. 12, p.2176-2178 (in Russian).

[180] V.V. Mityushev: *On solution of the boundary value problem arisen in the queueing system analysis.* Vesti AN BSSR (1991), n. 2, p.117-118 (in Russian).

[181] V.V. Mityushev: *On solution of the boundary $\mathbb{R}$- linear problem for certain doubly connected domain.* Vestnik BGU (1991) Preprint N 4019-B91, VINITI (in Russian).

[182] V.V. Mityushev: *Plane Stokes's flow of the piece-homogeneous Newton fluid.* Vestnik BGU (1991) Preprint N 4059-B91, VINITI (in Russian).

[183] V.V. Mityushev & N. Rylko: *Plane Stokes's flow of viscous fluid with reactive inclusions.* In: Proc. Int. Conf. Reo. Heat Phys. Symp. Minsk, 1991.

[184] V.V. Mityushev: *Eigenvalues of the $\mathbb{R}$-linear problems.* Izv. Vuz. Mathematika (1992), n. 11, p.35-38 (in Russian).

[185] V.V. Mityushev: *Solution of a non-local problem for the Lavrentev-Bitsadze's equation.* Diff. uravn. 28 (1992), n. 8, p.1461-1464 (in Russian).

[186] V.V. Mityushev: *Plane problem for the steady heat conduction of material with circular inclusions.* Arch. Mech. 45 (1993), n. 2, p.211-215.

[187] V.V. Mityushev: *A method of functional equations for boundary value problems of continuous media.* Reports Math. Phys. 33 (1993), n. 1-2, p.137-147.

[188] V.V. Mityushev: *Solution of the Hilbert boundary value problem for a multiply connected domain.* Slupskie Prace Mat.-Przyr. 9a (1994), 37-69.

[189] V.V. Mityushev: *Generalized method of Schwarz and addition theorems in mechanics of materials containing cavities.* Arch. Mech. 47 (1995), n. 6, p.1169-1181.

[190] V.V. Mityushev: *Rayleigh's integral and the square array of cylinders.* Arch. Mech. 47 (1995), n. 1, p.27-37.

[191] V.V. Mityushev: *The first order approximation of the effective conductivity of a family of nonlinear composites.* J. Tech. Phys. 36 (1995), 429-432.

[192] V.V. Mityushev: *Two-dimensional steady heat conduction of composite materials.* In: Proc of IX Int. Symp. Heat and Mass Transfer, Augustow (1995), Part 2, 93-100.

[193] V.V. Mityushev: *Application of functional equations to effective heat conduction of composite materials.* Slupsk, WSP Publisher, 1996, 155 pp. (in Polish).

[194] V.V. Mityushev: *Transport properties of regular array of cylinders.* ZAMM 77 (1997), 2, 115-120.

[195] V.V. Mityushev: *Functional equations and its applications in mechanics of composites.* Demonstr. Math. 30 (1997), n. 1, 64-70.

[196] V.V. Mityushev: *Riemann problem on the double of a multiply connected circular region.* Ann. Pol. Math. 67 (1997), n. 1, 1-14.

[197] V.V. Mityushev: *Transport properties of finite and infinite composite materials and Rayleigh's sum.* Arch. Mech. 49 (1997), n. 2, 345-358.

[198] V.V. Mityushev & N. Rylko: *Boundary value problems, the Poincaré series, the method of Schwarz and composite materials.* (1997) In: Proc of Int. Congres IMACS 97, Berlin, v. 1, 165-170.

[199] V.V. Mityushev: *On construction of Abelian differentials on the closed Riemannian surface.* Slupskie Prace Mat.-Przyr. (published).

[200] V.V. Mityushev: *Convergence of the Poincaré series for classical Schottky groups.* Proceedings Amer. Math. Soc. 126 (1998), n. 8, 2399-2406.

[201] V.V. Mityushev: *Hilbert boundary value problem for multiply connected domains.* Complex Variables 35 (1998), 283-295.

[202] V.V. Mityushev: *Conformal mapping of the domain bounded by a circular polygon with zero angles.* Annals. Pol. Math. (published).

[203] V.V. Mityushev: *Steady heat conduction of a material with an array of cylindrical holes in the nonlinear case.* IMA J. of Appl. Math. 61 (1998), 91-102.

[204] V.V. Mityushev: *A linear functional equation with a singularity at a fixed point.* Aequationes Math. **57** (1999), 37-44.

[205] N.I. Muskhelishvili: *Singular Integral Equations.* Nauka, Moscow, 1968 (3rd edition) (in Russian); English translation of the 1st ed.: Wolters-Noordhoff, Groningen,1946.

[206] N.I. Muskhelishvili: *Some Basic Problems of Mathematical Elasticity Theory.* Nauka, Moscow, 1966 (5th edition) (in Russian); English translation of the 1st ed.: Wolters-Noordhoff, Groningen, 1953.

[207] V.K. Natalevich: *Nonlinear singular integral equations and nonlinear boundary value problems of the theory of analytic functions.* Uch. zap. Kazanskogo Univ. 112 (1952), kn. 10, 155-190 (in Russian).

[208] V.K. Natalevich: *On nonlinear singular integral equations and nonlinear boundary value problems of the theory of analytic functions.* Doklady AN SSSR 83 (1952), n. 1, 19-22 (in Russian).

[209] V.K. Natalevich: *On a nonlinear boundary value problem for analytic functions.* Nauch. trudy Novocherkasskogo Polytechn. Inst. 26 (1955), 455-459 (in Russian).

[210] Yu.V. Obnosov: *On nonlinear boundary value problem of Hilbert type.* Izv. Vuz. Mathematika (1973), n. 10, 42-49 (in Russian).

[211] Yu.V. Obnosov: *On solution of nonlinear boundary value problem of Hilbert type.* Izv. Vuz. Mathematika (1975), n. 3, 82-91; n. 4, 42-51 (in Russian).

[212] Yu.V. Obnosov: *Solution of a nonlinear mixed boundary value problem of the theory of analytic functions.* Izv. Vuz. Mathematika (1975), n. 5, 96-102 (in Russian).

[213] Yu.V. Obnosov: *Solution of homogeneous power Hilbert problem with constant exponent.* In: Trudy Seminara po Kraevym Zadacham, Kazan, 1978, vyp. 15, 99-107 (in Russian).

[214] Yu.V. Obnosov: *To the solution of linear Hilbert boundary value problem in a special case.* Izv. Vuz. Mathematika (1979), n. 9, 29-40 (in Russian).

[215] Yu.V. Obnosov: *Solution of a nonlinear mixed boundary value problem of the theory of analytic functions.* In: Complex Analysis and Applications'81, Sept 20-27, 1981, Varna, Bulgaria, Academy of Science Pbl., Sofia, (1984), 549-563 (in Russian).

[216] Yu.V. Obnosov: *Solution of a nonlinear mixed Cherepanov's boundary value problem in the class of holomorphic functions in the case $n \leq 2$.* In: Theory of Functions of Complex Variable and Boundary Value Problems. Cheboksary, (1982), 70-76 (in Russian).

[217] Yu.V. Obnosov: *Some nonlinear boundary value problems of the theory of analytic functions folvable in closed form.* In: Scientific Articles of Jubileen Seminar on Boundary Value Problems Dedicated to 75th birthday of Academician F.D. Gakhov, Universitetskoe Pbl., Minsk, 1985, 86-95 (in Russian).

[218] Yu.V. Obnosov: *Exact solution of a problem of $\mathbb{R}$-linear conjugation for the right triangular check-board field.* Izv. Vuz. Mathematika (1994), n.8, 55-66 (in Russian).

[219] Yu.V. Obnosov: *Exact solution of a problem of $\mathbb{R}$-linear conjugation for rectangular check-board field.* Proc. R.Soc. Lond. A 452 (1996), 2423-2442.

[220] D. Oestreich: *Singular integral equations applied to free surface seepage from non-linear channels.* Mathematical Methods in the Applied Sciences 12 (1990), 209-219.

[221] V.V. Panasyuk, M.P. Savruk, and Z.T. Nazarchuk: *Methods of Singular Integral Equations in Two-dimensional Diffraction Problems.* Naukova Dumka, Kiev, 1984 (in Russian).

[222] D. Partyka: *The Generalized Neumann-Poincaré Operator and its Spectrum.* Dissertationes Mathematicae, v. 366, Warsaw, 1997.

[223] A.S. Peters: *The solution of a certain nonlinear Riemann-Hilbert problem with an application.* Comm. and Appl. Math. 26 (1973), n. 2, 87-104.

[224] S.K. Pichorides: *On the best values of the constants in the theorems M. Riesz, Zygmund and Kolmogorov.* Studia Meth. 44 (1972), n. 2, 165-179.

[225] V.P. Platonov, O.I. Marichev, et al. (Eds.): *Scientific Articles of Jubileen Seminar on Boundary Value Problems Dedicated to 75th Birthday of Academician F.D. Gakhov.* Minsk, Feb 18-20, 1981. Universitetskoe Pbl., Minsk, 1985.

[226] B.E. Pobedrya: *Mechanics of Composites.* MGU, Moscow, 1984 (in Russian).

[227] P.Ya. Polubarinova-Kochina: *Theory of the Ground Water Flow.* Nauka, Moscow, 1977 (in Russian).

[228] H. Poincaré: *Oeuvres.* Gauthier-Villart, Paris, v. 2, 1916; v. 4, 1950; v. 9, 1954.

[229] P.Ya. Polubarinova-Kochina: *On additional parameters on the examples of circular 4-polygon.* Prikl.Math. i Mech., 55 (1991), n. 2, 222-227 (in Russian).

[230] C. Pommerenke: *Univalent Functions.* Vandenhoeck and Ruprecht, Göttingen, 1975.

[231] I.I. Privalov: *Boundary Values of Analytic Functions.* Gostekhizdat, Moscow, 1950 (in Russian); German transl.: DVW, Berlin, 1956.

[232] S. Prössdorf: *Einige Klassen Singulärer Gleichungen.* Akademie-Verlag, Berlin, 1974 (in German); English transl.: North-Holland, Amsterdam, 1978; Russian transl.: Mir, Moscow, 1979.

[233] V.A. Prokhorov: *On a theorem of Adamjan-Arov-Krejn.* Math. zbornik, 184 (1993), n. 1, 89-104 (in Russian).

[234] V.A. Prokhorov: *Rational approximation of analytic functions.* Math. zbornik, 184 (1993), n. 2, 3-32 (in Russian).

[235] L.P. Primachuk: *Nonlinear conjugation boundary value problem for two holomorphic functions on an open arc.* Vestnik BGU, (1981), n. 1, 44-46 (in Russian).

[236] V.M. Radygin & O.V. Golubeva: *Application of complex functions to physics and tekhnics.* Vysshaja Shkola, Moscow, 1983 (in Russian).

[237] M. Reissig & E. Wegert: *Nonlinear boundary value problems for elliptic systems in the plane.* Complex Variables, 27 (1995), 193-210.

[238] S. Richardson: *Hele-Shaw flow with a free boundary produced by the injections of a fluid into a narrow channel.* J. Fluid Mech. 56 (1972), 609-618.

[239] S.V. Rogosin: *On solvability of nonlinear conjugation boundary value problem of power type.* Doklady NAN Belarusi 43 (1999), n. 3, 36-40 (in Russian).

[240] W. Rudin: *Real and Complex Analysis.* McGraw-Hill, London-New York, 1970.

[241] N.A. Rysjuk: *Nonlinear boundary value problem of Riemann Type with real exponents in the class of automorphic functions.* Vesti AN BSSR, ser. phys.-math. nauk (1973), n. 4, 51-56 (in Russian).

[242] N.A. Rysjuk: *On Nonlinear boundary value problem of Riemann type.* Doklady AN BSSR,21 (1977), n. 4, 299-301 (in Russian).

[243] N.A. Rysjuk: *To the question of nonlinear generalization of Rieman boundary value problem.* In: Boundary Value Problems, Special Functions and Fractional Calculus, Proceedings of International Conference Dedicated to 90th Birthday of Academician F.D. Gakhov, Minsk, 1996, 332-334 (in Russian).

[244] S.G. Samko, A.A. Kilbas, and O.I. Marichev: *Fractional Integrals and Derivatives. Theory and Applications.* Nauka i Tekhnika, Minsk, 1989 (in Russian); English transl.: Gordon and Breach, New York, 1993.

[245] M.V. Samokhin: *To the question of existence of analytic functions with a given modulus of boundary values.* Mat. Sb. 187 (1996), n.1, 113-120 (in Russian).

[246] G.P. Sendecky: *Multiple circular inclusion problem in longitudinal shear deformation.* J. Elasticity 1 (1971), 83-86.

[247] B.V. Shabat: *Introduction to Complex Analysis.* Nauka, Moscow, vol. I, II, 1975 (in Russian).

BIBLIOGRAPHY

[248] J.H. Shapiro: *Composition Operators.* Springer-Verlag, New York-Berlin, 1993.

[249] A.I. Schnirel'man: *The degree of quasi-linearlike mapping and the nonlinear Hilbert problem.* Mat. Sb. 89 (1972) n. 3, 366-389 (in Russian).

[250] Yu.V. Sidorov, M.V. Fedorjuk and M.I. Shabunin: *Lectures on the Theory of Functions of Complex Variable.* Nauka, Moscow, 1989 (3rd ed.) (in Russian).

[251] A. Signorini: *Sopra un problema al contoro nella teoria delle funzioni di variabile complessa.* Ann. Mat. Pura Appl. Ser. 3, 25 (1916), 253-273.

[252] B. Smith, P. Björstad, and W. Gropp: *Domain decomposition. Parallel multilevel methods for elliptic partial differential equations.* Cambridge University Press, Cambridge, 1996.

[253] P.V. Solov'ev: *On a boundary value problem in the theory of analytic functions.* Doklady AN SSSR 33 (1941), n. 3, 190-192 (in Russian).

[254] J. Springer: *Introduction to Riemann surfaces.* AMS, Providence, Rhode Island, 1981, 2nd ed.; Russian transl.: Izd. Inostr. Lit., Moscow, 1960.

[255] S. Stoilov: *Teoria Functiilor de o Variabilā Complexā.* Ed. Acad. RP Române, vol. I, II, 1954 (in Romanian); Russian transl.: Izd. Inostr. Lit., Moscow, 1962.

[256] A. Susea: *On Generalized homogeneous Riemann boundary value problem.* Studii si cercetari mat. Acad. RSR 19 (1967), n.1, 155-161 (in Romanian).

[257] R. Thurman: *Maximal capacity, Robin capacity, and minimum energy.* Indiana Univ. Math. J. 46 (1997), n. 2, 621-636.

[258] E.K. Timofeev: *On a nonlinear boundary value problem.* Izv. Vuz. Mathematika (1969), n. 4, 92-102 (in Russian).

[259] M.E. Tolochko: *On solvability of nonlinear boundary value problem of Riemann type for multiply connected domain.* Lithv. Math. Zb. 17 (1977), n. 3, 136-137 (in Russian).

[260] M.E. Tolochko: *On homogeneous nonlinear boundary value problem of power type in the exceptional case.* In: Int. Conference on Boundary Value Problems, Special Functions and Fractional Calculus (dedicated to the 90th birthday of Academician F.D. Gakhov), Minsk, February 16-20, 1996, Abstracts, Minsk, BSU, 1996, p. 100 (in Russian).

[261] M.B. Tursunkulov: *Investigation of a nonlinear Hilbert problem with shift.* In: The Questions of Integration of Differential Equations, Tashkent, 1966, 232-248 (in Russian).

[262] E.V. Tyurikov: *On a class of nonlinear problems of cojugacy with a shift for analytic functions.* Doklady AN SSSR 290 (1986), 796-800 (in Russian).

[263] Kh.Kh. Valiev: *On a nonlinear singular integral equation.* Uchen. zap. Bash.univ., Ufa, 1965, vyp. 20, n. 1, 111-113 (in Russian).

[264] Kh.Kh. Valiev: *On solution of a type of nonlinear singular integral equation.* Trudy Ufimsk. avia. inst., Ufa, 1971, vyp. 26, 101-104 (in Russian).

[265] G. Valiron: *Lectures on the General Theory of Integral Functions.* Privat, Toulouse, 1923.

[266] I.N. Vekua & A.K. Rukhadze: *The problem of the torsion of circular cylinder reinforced by transversal circular beam.* Izv. AN SSSR, 1933, n. 3, 373-386.

[267] I.N. Vekua: *On nonlinear Riemann boundary value problem.* Trudy Tbilissk. Math. Inst.11 (1942), 109-139 (in Russian).

[268] I.N. Vekua: *Generalized Analytic Functions.* Nauka, Moscow, 1988 (2nd edition) (in Russian).

[269] N.P. Vekua: *Systems of Singular Integral Equations.* Nauka, Moscow, 1970 (2nd edition) (in Russian) ; English transl. of the 1st ed.: P.Noordhoff N.V., Groningen,1967.

[270] A. Visintin: *Models of Phase Transitions.* Birkhä user, Boston-Basel-Berlin, 1997.

[271] V.I. Vlasov & D.B. Volkov: *The Dirichlet problem for a circle with cut.* Vychisl. zentr AN SSSR, 1986, 40 pp (in Russian).

[272] V.I. Vlasov & S.L. Skorokhodov: *Analytical solution of the Dirichlet problem for the Poisson equation for a class of polygon domain.* Vychisl. zentr AN SSSR, 1988, 33 pp (in Russian).

[273] V.I. Vlasov & D.B. Volkov: *Solution of the Dirichlet problem for the Poisson equation in some domains with a compound boundary.* Vychisl. zentr AN SSSR, 1989, 44 pp (in Russian).

[274] E. Wegert: *Topological methods for strongly nonlinear Riemann-Hilbert problems for holomorphic functions.* Math. Nachr. 134 (1987), 201-230.

[275] E. Wegert: *Nonlinear Riemann-Hilbert problems and their relationship to extremal problems for holomorphic functions.* Math. Nachr. 137 (1988), 141-157.

[276] E. Wegert: *Boundary Value Problems for Holomorphic Functions and Singular Integral Equations.* Akademie-Verlag, Berlin, 1992.

[277] E. Wegert, G. Khimchiachvili, and I. Spitkovsky: *Nonlinear transmission problems.* Mem. Differential Equations Math.Phys. 12 (1997), 223-230.

[278] E. Wegert, G. Khimchiachvili, and I. Spitkovsky: *On totally real non-compact manifolds globally foliated by analytic discs.* Indiana Univ. Math. J. (published).

[279] E. Wegert & L.v. Wolfersdorf: *On a class of quasi-linear Riemann-Hilbert problems.* Zeitschrift für Analysis und ihre Anwendungen, 6 (1987), 235-240.

[280] G.C. Wen: *Conformal Mappings and Boundary Value Problems.* AMS, Providence, Rhode Island, 1992.

[281] G.C. Wen: *Approximate Methods and Numerical Analysis for Elliptic Complex Equations.* Gordon & Breach, Tokyo (to appear).

[282] G.C. Wen & H. Begehr: *Boundary Value Problems for Elliptic Equations and Systems.* Longman Scientific and Technical, London, 1990.

[283] W. Wendland: *Elliptic Systems in the Plane.* Pitman, London, 1979.

[284] R. Wilson: *Introduction to the Graph Theory.* Edinburgh, 1972; Russian transl.: Mir, Moscow, 1977.

[285] L.v. Wolfersdorf: *On a boundary value problem for a special class of first order semilinear elliptic systems in the plane.* Zeitschrift für Analysis und ihre Anwendungen, 2 (1) (1983), 37-40.

[286] L.v. Wolfersdorf: *On strongly nonlinear Poincaré boundary value problems for harmonic functions.* Zeitschrift für Analysis und ihre Anwendungen, 3 (5) (1984), 385-399.

[287] L.v. Wolfersdorf: *On the theory of the nonlinear Riemann-Hilbert problem for holomorphic functions.* Complex Variables, 3 (1984), 323-346.

[288] L.v. Wolfersdorf: *On the theory of the nonlinear Hilbert problem for holomorphic functions.* In: Partial Differential Equations with Complex Analysis (H. Begehr & A. Jeffrey, Eds.), Notes in Math. Sci. 262 (1992), 134-149.

[289] N.N. Yukhanonov: *On a quasilinear general boundary cojugacy problem for analytic functions on the unit circle.* Dokl. AN TadzhSSR 26 (1983), 687-691.

[290] S.I. Yurchenko, I.M. Mel'nik and L.I. Spinko: *Nonlinear Riemann boundary value problem with discontinuous coefficients on a Riemann manifold.* Diff. Uravnenija, Rjazan' (1979), n. 14, 160-170 (in Russian).

[291] T.N. Zhorovina & V.V. Mityushev: *A mixed boundary value problem for multiply connected domains.* Vestnik BGU. Ser.1 (1996) Preprint VINITI N 2932-B94 (in Russian).

[292] K. Zhu: *Operator Theory in Function Spaces.* Marcel Dekker, New York, 1990.

[293] V.A. Zmorovich: *On a generalization of the Schwarz integral formula on n-connected domains.* Dopovedi URSR (1958), n. 5, 489-492 (in Ukranian).

[294] E.I. Zverovich: *Boundary value problems of analytic functions in Hölder classes on Riemann surfaces.* Uspekhi Mat. nauk 26 (1) (1971), 113-179 (in Russian).

[295] A. Zygmund: *Trigonometric Series.* **1**, Cambridge University Press, Cambridge, 1959; Russian transl.: Mir, Moscow, 1965.

Index

ℂ-linear conjugation problem for analytic functions, 37
ℝ-linear problem, 52
ℂ-linear conjugation problem, 2

a fixed point of the operator, 20
addition theorems, 187
almost bounded , 41
analytic functions, 14
analytic solution, viii

Blaschke product, 24
BMO, 17
BMOA, 18
Borel-Valiron inequality, 48
boundary, 10
boundary value problem of multiplication type, 5

canonical function, 38
Cauchy formula, 27
Cauchy index, 2
Cauchy integral, 26
Cauchy principal value of a singular integral, 28
Cauchy type integral, 28
central index, 47
circle, 9
closed arc, 11
closed curve, 11
closure, 10
complex conjugation, 10

complex Green function, 31, 33
complex plane, 9
complex potential, 22
conformal mapping, 50
conjugate boundary function, 25
conjugation problem of power type, 57
continuous arc, 11
continuous curve, 11
continuous functions, 13
contractive operator, 20

Decomposition Theorem, 23
disc, 9
distribution of zeros of entire function, 47

entire function, 45
entire function of finite order, 45
entire function with respect to points, 75

factorization, 38

general nonlinear conjugation problem of power type, 5
generalized modulus problem, 7
GMS, 181

Hölder-continuous functions, 14
half-plane, 9
Hardy space, 17
harmonic conjugate, 25

harmonic measure, 34
Hilbert problem, 2
Hilbert transform, 26

index of the problem, 37
infinetely differentiable functions, 14
initial point of approximation, 18

jump problem, 39

knot of a curve, 85

Lebesgue space, 16
level, 145
line, 9
linear-fractional boundary value problem, 5
Lipschitz-continuous functions, 14
log-bounded function, 75
Logarithmic Conjugation Theorem, 22
loop, 10
Lyapunov curve, 13

Markushevich problem, 4
maximal term, 47
measure space, 16
method of successive approximation, 18
mixed Cherepanov's problem, 7
mixed problems, 7
modified Dirichlet problem, 37
modulus problem, 6
multiply connected domain, 11

non-tangential limit, 24
nonlinear boundary value problem of power type, 4
nonlinear Riemann-Hilbert problem of power type, 6
nontangential boundary function, 24

open arc, 11

order of entire function, 46
oriented curve, 11

path, 10
Poisson integral, 23
Poisson kernel, 23
polynomial Riemann–Hilbert nonlinear problem, 6
Pumping Principle, 21

regularizing factor, 43
Riemann map, 44
Riemann mapping theorem, 50
Riemann problem, 2
Riemann sphere, 9
Riemann–Hilbert problem, 2, 42

Schauder spaces, 14
Schwarz kernel, 23
Schwarz operator, 3, 30
Schwarz problem, 2
Schwarz's Reflection Principle , 10
simple curve, 11
simple functional equation, 131
simply connected domain, 11
singular integral, 28
small Hölder spaces, 14
smooth curve, 13
Sokhotsky-Plemelj formulae, 29
solution in closed form, vii, viii
standard complementary projectors, 30
symmetry, 10

tangency of an arc at its end-point, 75
the spectral radius, 19
type of entire function, 46

vector-matrix ℂ-linear problem, 6
Villat's formula, 36
VMO, 18

VMOA, 18

weighted Hölder spaces, 14
winding number, 2, 27